# Origametry

Origami, the art of paper folding, has a rich mathematical theory. Early investigations go back to at least the 1930s, but the twenty-first century has seen a remarkable blossoming of the mathematics of folding. Besides its use in describing origami and designing new models, it is also finding real-world applications from building nano-scale robots to deploying large solar arrays in space.

Written by a world expert on the subject, *Origametry* is the first complete reference on the mathematics of origami. It brings together historical results, modern developments, and future directions into a cohesive whole. Over 180 figures illustrate the material while numerous "diversions" provide jumping-off points for readers to deepen their understanding. This book is an essential reference for researchers of origami mathematics and its applications in physics, engineering, and design. Educators, students, and enthusiasts will also find much to enjoy in this fascinating account of the mathematics of folding.

# Origametry

## Mathematical Methods in Paper Folding

THOMAS C. HULL

Western New England University

CAMBRIDGE
UNIVERSITY PRESS

## CAMBRIDGE
### UNIVERSITY PRESS

Shaftesbury Road, Cambridge CB2 8EA, United Kingdom

One Liberty Plaza, 20th Floor, New York, NY 10006, USA

477 Williamstown Road, Port Melbourne, VIC 3207, Australia

314–321, 3rd Floor, Plot 3, Splendor Forum, Jasola District Centre, New Delhi – 110025, India

103 Penang Road, #05–06/07, Visioncrest Commercial, Singapore 238467

Cambridge University Press is part of Cambridge University Press & Assessment,
a department of the University of Cambridge.

We share the University's mission to contribute to society through the pursuit of
education, learning and research at the highest international levels of excellence.

www.cambridge.org
Information on this title: www.cambridge.org/9781108478724

DOI: 10.1017/9781108778633

First published 2021

*A catalogue record for this publication is available from the British Library*

*Library of Congress Cataloging-in-Publication data*
Names: Hull, Thomas, 1969– author.
Title: Origametry : mathematical methods in paper folding / Thomas Hull.
Description: New York : Cambridge University Press, 2020. | Includes index.
Identifiers: LCCN 2020022971 (print) | LCCN 2020022972 (ebook) |
ISBN 9781108478724 (hardback) | ISBN 9781108746113 (ebook)
Subjects: LCSH: Origami–Design. | Origami–Mathematics.
Classification: LCC TT870 .H85 2020 (print) | LCC TT870 (ebook) | DDC 736/.982–dc23
LC record available at https://lccn.loc.gov/2020022971
LC ebook record available at https://lccn.loc.gov/2020022972

ISBN    978-1-108-47872-4    Hardback
ISBN    978-1-108-74611-3    Paperback

This book is dedicated to Paul Chapman, who introduced me to origami and gave me my first origami book.

# Contents

**Introduction** *page* 1

**Part I Geometric Constructions** 9

**1      Examples and Basic Folds** 11
1.1     Constructing an Equilateral Triangle 11
1.2     Dividing a Segment into 1/*n*ths 13
1.3     Trisecting an Angle 16
1.4     Folding a Regular Heptagon 19
1.5     The Basic Origami Operations 23
1.6     Historical Remarks 29

**2      Solving Equations via Folding** 31
2.1     Folding Conics 31
2.2     Solving Second Degree Equations 36
2.3     Solving Third Degree Equations 38
2.4     Cubic Curves and Beloch's Fold 45
2.5     Historical Remarks 46

**3      Origami Algebra** 48
3.1     Definitions and Origami Arithmetic 48
3.2     The Field of Origami Numbers 50
3.3     Other Characterizations of Origami Numbers 55
3.4     Other Work: Origami Rings 57

**4      Beyond Classic Origami** 58
4.1     Multifolds 58
4.2     Lang's Angle Quintisection 60
4.3     Describing Multifold Origami Operations 63
4.4     Solving Equations with Multifolds 66
4.5     Constructions with Curved Creases 69
4.6     Open Problems 71

**Part II  The Combinatorial Geometry of Flat Origami**                               73

**5       Flat Vertex Folds: Local Properties**                                        75
         5.1   Definitions and Flat-Foldability                                        75
         5.2   Mountain-Valley Parity                                                  81
         5.3   Necessary and Sufficient Conditions                                     85
         5.4   Cone Folds                                                              88
         5.5   Counting Valid Mountain-Valley Assignments                              90
         5.6   The Configuration Space of Flat Vertex Folds                            96
         5.7   Matrix Model for Flat Vertex Folds                                     104
         5.8   Open Problems                                                          105
         5.9   Historical Remarks                                                     105

**6       Multiple-Vertex Flat Folds: Global Properties**                             107
         6.1   Impossible Crease Patterns                                             107
         6.2   Generalized Kawasaki: Necessary Conditions and the Folding Map         110
         6.3   Generalized Maekawa                                                    112
         6.4   Justin's Theorem and Paper with Holes                                  114
         6.5   Global Flat-Foldability                                                119
         6.6   Flat-Foldability Is NP-Hard                                            127
         6.7   Open Problems                                                          135

**7       Counting Flat Folds**                                                       137
         7.1   Two-Colorable Crease Patterns                                          137
         7.2   Phantom Folds of the Miura-ori                                         139
         7.3   The Stamp-Folding Problem                                              145
         7.4   Tethered Membrane Lattice Folding                                      152
         7.5   Open Problems                                                          157

**8       Other Flat-Folding Problems**                                               159
         8.1   How Many Times Can We Fold a Sheet of Paper?                           159
         8.2   Can Any Shape Be Folded and Unfolded?                                  161
         8.3   Origami Design                                                         165
         8.4   The Rumpled Ruble, or Margoulis Napkin Problem                         172
         8.5   The Fold-and-Cut Problem                                               174

**Part III  Algebra, Topology, and Analysis in Origami**                             179

**9       Origami Homomorphisms**                                                     181
         9.1   Symmetry Groups of Flat Origami                                        181
         9.2   Examples of Origami Homomorphisms                                      184
         9.3   Applications to Origami Tessellations                                  186
         9.4   Open Problems                                                          189

| | | |
|---|---|---|
| **10** | **Folding Manifolds** | 191 |
| | 10.1 Isometric Foldings | 191 |
| | 10.2 The Local Structure of the Singular Set | 193 |
| | 10.3 Robertson's Theorem | 199 |
| | 10.4 The Angle Sum and Recovery Theorems | 203 |
| | 10.5 Maekawa and Kawasaki for Isometric Foldings | 205 |
| | 10.6 Open Problems | 215 |
| | 10.7 Historical Remarks | 215 |
| | | |
| **11** | **An Analytic Approach to Isometric Foldings** | 217 |
| | 11.1 Lipschitz Continuous and Rigid Maps | 217 |
| | 11.2 Applications to Dirichlet Problems | 223 |
| | | |
| **Part IV Non-flat Folding** | | 229 |
| | | |
| **12** | **Rigid Origami** | 231 |
| | 12.1 Matrix Model and Necessary Conditions | 232 |
| | 12.2 The Gauss Map | 239 |
| | 12.3 The Gauss Map and Rigid Origami | 246 |
| | 12.4 Another Generalization of Maekawa | 251 |
| | 12.5 Open Problems | 254 |
| | 12.6 Historical Remarks | 255 |
| | | |
| **13** | **Rigid Foldings** | 256 |
| | 13.1 Infinitesimal Rigid Foldability | 256 |
| | 13.2 Angle Relationships for Single-Vertex Rigid Foldability | 271 |
| | 13.3 An Intrinsic Condition for Rigid Vertices | 280 |
| | 13.4 Open Problems | 288 |
| | 13.5 Historical Remarks | 289 |
| | | |
| **14** | **Rigid Origami Theory** | 290 |
| | 14.1 Complexity of Rigid Foldability with Optional Creases | 290 |
| | 14.2 Configuration Spaces of Rigid Foldings | 296 |
| | 14.3 Self-Foldability | 304 |
| | 14.4 Open Problems | 317 |
| | | |
| | *References* | 319 |
| | *Index* | 330 |

# Introduction

## A Growing Interest

Origami is the art of paper folding, as is likely known by anyone picking up this book. Less known is how diverse origami is as an art form. Most people who have practiced origami have folded paper cranes, fishes, and frogs, or perhaps some of the many "playground" origami models that children teach to each other, like paper airplanes, fortune-tellers, and ninja throwing stars. Those who catch the paper-folding bug, however, learn how to fold dragons, insects, and octopi, each from an uncut square. Those who go further might see pictures like those in Figure 1, where two origami models, one rather simple and the other very complex, are shown with their respective **crease patterns** (the pattern of creases you would see if you were to unfold the model). Others might explore **origami tessellations**, which are origami models where the crease pattern forms a regular tiling of the plane. Others might become addicted to **modular origami**, where multiple, sometimes hundreds, of pieces of paper are all folded in the same way and then locked together to form beautiful polyhedral objects.

After staring at crease patterns and folding intricate geometric and representational origami models, one may start to suspect that there are mathematical rules at play in origami. Is there any inherent geometry to these crease patterns? Is there a way to predict into what shape they fold? Is there a limit to the complexity of shapes that can be folded? How would we make the previous question into a precise conjecture that could be proven?

On a somewhat different note, anyone interested in paper folding might have noticed a pronounced increase in the number of scientific news reports, viral web

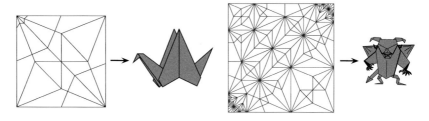

**Figure 1** The classic flapping bird (crane) with its crease pattern, and Maekawa's Devil (Kasahara and Maekawa, 1983) with its crease pattern.

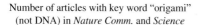

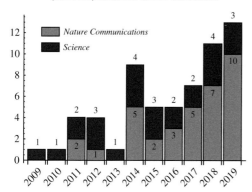

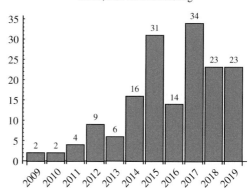

**Figure 2** The rise of papers featuring origami applications in *Nature Communnications*, in *Science*, and at the American Physics Society's March Meeting.

videos, and research articles on the use of origami in science and engineering since the year 2013. The National Science Foundation (of the United States of America) offered about 12 large grants in each of the years 2012 and 2013 in their Emerging Frontiers in Research and Innovation program on "Origami Design for Integration of Self-assembling Systems for Engineering Innovation (ODISSEI)." Each of these four-year, approximately $2 million grants spurred a bevy of research in physics and engineering university departments across the country. Evidence of this growth in interest can be seen in the number of papers relating to origami applications in high-profile journals like *Science* and *Nature Communications*. (See Figure 2, where the numbers indicate papers with keyword "origami" excluding papers on DNA-origami, which has nothing to do with the folding of 2-dimensional sheets.)

At the same time, more and more sessions and papers at physics and engineering conferences have been devoted to origami applications since 2013 – far more than the NSF ODISSEI grants would generate on their own. See Figure 2 again for the number of papers with "origami" in their title from the American Physics Society's annual March Meeting over a period of ten years. Similar increases in origami applications in physics can be seen at engineering conferences. For example, at the ASME (American Society of Mechanical Engineers) 2019 Conference on Smart Materials, Adaptive Structures, and Intelligent Systems, there was one short course on origami mechanics, one keynote speaker with expertise in origami mechanics, one symposium of talks on "bioinspired adaptive origami systems," a best paper award given to Jakob Faber for the paper "Bioinspired spring origami," and 23 papers presented on origami research in mechanical engineering.

Why are so many scientists, including the National Science Foundation, excited about origami? Origami enthusiasts are not surprised; one very well-known origami model is called the **Miura-ori**, also known as the **Miura map fold**. It was invented by a Japanese astrophysicist named Koryo Miura who was searching for a way to send large solar panels into outer space. He came up with an ingenious origami fold (see Section 7.2 and Figure 7.3 on page 139) that collapses a large sheet into a small area, and it does so in a very nice way. The Miura-ori can open and close in a simple, smooth motion. In fact, Miura proved that his model will open and close **rigidly**, meaning that the regions of material between the creases (the faces of the crease pattern) can remain perfectly flat, as if made of stiff metal, as the model opens and closes with the creases acting like hinges. This mean that the Miura-ori could be used for solar panel deployment: the solar cells can be the faces of the crease pattern, joined by hinges, and then folded into a compact package to be incorporated into a space satellite design, opening once the satellite is in orbit.

Koryo Miura invented his Miura-ori in the 1970s, and since then it has been the subject of much research. But such scientific interest accelerated in the 2010s. It proved to be an inspirational model for mechanical engineers impressed by the smooth rigid folding of the model. Physicists were excited by the Miura-ori's property of exhibiting **negative Poisson ratio**. That is, when a normal object, say a block of cheese, is squeezed in one direction, it will expand in the other directions, thus exhibiting what physicists call positive Poisson ratio. But if you compress a material, say, sitting in the $xy$-plane, in direction of the $x$-axis and this also makes it **compress** in the direction of the $y$-axis, then the material is said to have negative Poisson ratio, a very unusual property for a material to have. In fact, physical materials that have such unusual properties are called **metamaterials**, and fabricating them has become a hot area of research in physics. The Miura-ori exhibits negative Poisson ratio, in that when the paper is mostly unfolded and we squeeze it in one direction, it will automatically compress in the other direction. Furthermore, the Miura-ori is very simple to fabricate. It is just made from a planar sheet of material folded with a regular crease pattern! Might origami lead to other, simple-to-make metamaterials?

There are many other origami models besides the Miura-ori that open and close nicely, or that exhibit interesting mechanical behavior when folded. In the 2010s engineers sought out such origami models as inspiration for making devices with novel mechanical behavior. Furthermore, anything created from origami could be easily manufactured from a flat material, lending itself to mass production and inexpensive fabrication costs. In addition, scientists learned that origami mechanisms possessed the uncanny ability to function **independent of scale**. For example, the smooth opening-and-closing of the Miura-ori works for large structures, like solar panel arrays, and it also works for tiny structures, like a heart stent or a nano-scale piece of graphene.

Once this feature of origami became well-known, robotics engineers became very excited. Normal robotics mechanisms, like a ball-and-joint robot arm, operate only for certain sizes of scale. If a robot can be designed using origami folding motion principles, then the resulting robot could, in theory, work at any scale. Thus, robotics engineers began searching for origami models that exhibited interesting motions, like twisting, expanding, or gripping. Astrophysics engineers searched for more origami-inspired ways to deploy structures in outer space. Architects began to see origami as a means to unfurl small structures, like tents, or even large structures, like the dome of a sports arena. Physicists sought new ways to make and understand metamaterials from origami. And all of these applications required a better understanding of the mathematics of paper folding in order to achieve their ends.

But none of that was the inspiration for me to write this book.

## Why I Wrote This Book

I began practicing origami when I was eight years old. I did it because it was fun, and trying to figure out origami instructions from books was a challenge that resulted in an interesting and pretty object as a reward. My interest in origami grew in parallel to my interest in mathematics. I did not realize it at the time, but this was not a coincidence. The same things that appealed to me about the mathematics that I was learning in school, namely the precise rules one had to follow and the simple elegance of the patterns that mathematics exhibited, were also present in origami. The simple rules of origami are no scissors or glue; only folding is allowed. (Although this is a modern restriction. Origami is hundreds of years old, dating to early Edo-era Japan at the very latest, and records indicate that back then making use of cuts was often permitted.) The folded models that result from origami have a simple beauty to them. A folded paper crane is, after all, a piecewise-planar approximation to an actual crane. Even complex origami models are usually an abstraction of the figure that they are supposed to represent. And geometric origami models, like origami tessellations and modular origami, directly represent patterns and shapes found in mathematics.

Therefore it is no surprise that during my college years my origami and mathematics interests, and worlds, began to converge. It is not an exaggeration to say that everything that I learned in college about polyhedral geometry was from my studies in modular origami. And in some of the more advanced origami books that I acquired, there were mentions of **origami theorems**! (Specifically, Kawasaki's and Maekawa's Theorems, see pages 81 and 85.) These glimpses at a connection between origami and mathematics made me search for other references. This was in the late 1980s, pre-World Wide Web, so just trying to Google was not an option. I had to pour through printed volumes of *Mathematics Reviews* and found only a few research papers that discussed the mathematics of origami. (One was by Toshikazu Kawasaki and Masaaki Yoshida (Kawasaki and Yoshida, 1988), and it forms the basis of Chapter 9 of this book.)

During graduate school I amassed a growing collection of science and mathematical articles on origami. I also started writing my own papers on origami mathematics. One impetus for writing papers was that as I investigated origami, I discovered proofs, conjectures, and examples that I couldn't find in the literature. Another motivation, however, was that the actual mathematics that I uncovered (from sometimes obscure references) and discovered (on my own) was very pretty! The simplicity of Maekawa's Theorem and the multiple ways to think about its proof was elegant mathematics. It deserved to be known to the greater mathematics community. Writing papers was one way to distribute what I was finding. So was giving talks, and I gave a lot of them.

As I continued to find mathematics papers that dealt with paper folding, I noticed two patterns emerging. One was that over and over again it seemed that researchers were reinventing the wheel, without knowledge of work in the area of mathematical paper folding that had come before. This is not too surprising. Before the days of the Internet, it was very hard (as I was learning while in graduate school!) to find each and every paper that had been published on a rather obscure topic, such as paper folding. But seeing this did make me think that someone needed to gather all these papers and unify the disparate work that had been done.

The second realization was cause for amazement. I was continually stunned by the number of different areas of mathematics – calculus, geometry, number theory, abstract algebra, differential topology, analysis – that had been applied to origami. Being a physical activity, it made sense that there would be many ways to model paper folding. But the sheer variety of different approaches one could take was surprising, exciting, and rather intimidating to a graduate student. As such, I found myself locating some papers on origami mathematics and thinking to myself, "I'm not ready to read this one, but perhaps in a few years I will be." For some papers, "a few years," turned out to be over a decade as the need to graduate, find a job, and learn how to be a professor took priority.

Yet the need was clear to me: a book needed to be written on the many mathematical methods that may be applied to the assorted aspects of origami. The diverse research that had been done needed to be unified, and the far-ranging approaches needed to be gathered in one place.

For me this became a labor of love. Discovering a paper like Stewart A. Robertson's "Isometric folding of Riemannian manifolds" (1977–1978) meant many joyous months of learning new results and connecting them with previously known (to me) work. Then, years later, stumbling upon a paper like Lawrence and Spingarn's "An intrinsic characterization of foldings of Euclidean space" (1989), which has so many similarities to Robertson's paper, meant more months of head-scratching as I tried to discern, reconcile, and then combine their two approaches.

## Outline of the Book

The book you now hold in your hands is the result of such efforts. It took a long time to write. Much of Part I, on Geometric Constructions in origami, was first written in 2006 while on sabbatical leave from Merrimack College. Of course, in the intervening 14 years new research was done on this topic, and I've tried to keep those chapters updated. Still, in 2006 it was possible for me to contain most of everything known about origami geometric constructions in less then 100 pages. Now, if I wanted to include full proofs of all the important results in just this one aspect of origami mathematics, it would probably have to be a book of its own. Therefore I have had to cut some corners and refer readers to the literature for some of the more current results (such as Nishimura's proof (2013) that 2-fold origami can solve arbitrary quintic equations).

Part II is where I try to solidify an area of origami mathematics that I like to call the Combinatorial Geometry of Flat Origami. This concerns only origami models that lie in a flat plane when all the creases are folded. The previously mentioned theorems of Maeakawa and Kawasaki exemplify this area. Maekawa's Theorem states that the difference between the mountain and valley creases that meet at a vertex in a flat origami crease pattern must always be two. This is a combinatorics result, but its proof must at some level rely on the geometry (or, if viewed the right way, the topology) of the folded paper. Kawasaki's Theorem is a necessary and sufficient condition that determines if a collection of creases meeting at a vertex can fold flat, and this condition relies solely on the angles between the creases. Thus, this is a purely geometric result. However, Maekawa's and Kawasaki's Theorems are actually intertwined. We first see this in Chapter 5 when expanding our flat-folding perspective from planar paper to folding cones, and we see it again in Chapter 6 when looking through the lens of Justin's Theorem, a result on flat-foldable crease patterns with multiple vertices. The interplay of combinatorics and geometry is present in many aspects of mathematics, but origami provides a lovely example with enough results and depth to make it a separate sub-genre in origami mathematics.

In Part III I chose to gather three aspects of origami mathematics that correspond to three mathematical fields that all students learn in graduate school: algebra, analysis, and topology. For algebra we find a delightful origami homomorphism that relates, using only the geometry of a flat-foldable crease pattern, the symmetry group of a crease pattern with the symmetry group of the flat-folded model. For topology we dive deeply into the work of Stewart Robertson (and Lawrence and Spingarn) on the folding of Riemannian manifolds, in which we find surprising and elegant generalizations of Kawasaki's and Maekawa's Theorems. For analysis we describe more recent work of Dacorognna, Marcelini, and Paolini that uses high-dimensional isometric foldings (the $n$-dimensional analog to flat origami) to solve certain Dirichlet partial differential equations.

Part IV turns to Non-flat Folding, in particular rigid origami. This is the natural setting for many applications of origami in physics and engineering. The mathematical models needed to prove that the Miura-ori fold can open and close with the crease pattern's faces remaining rigid are found here, as are many, many results on rigid origami folding angle relationships, complexity, and configuration spaces. Rigid origami is a very active research area, and this part of the book only represents my take on the subject, as a mathematician. Luckily, one of the world experts on rigid origami, professor Tomohiro Tachi of the University of Tokyo, is writing a full monograph on the subject (Tachi, 2020).

## Features of the Book

Throughout the book I have tried to provide a thorough set of useful references, although as applied origami has grown more and more popular, including a complete bibliography of all the current literature has become impossible. When applicable, I have included historical notes as a way to provide context to the many, and sometimes confusing, reference points in the literature.

**Diversions**    There are many places in this book where details of a certain aspect or problem of origami mathematics are omitted. This was done because if absolutely all details of every proof or idea were included, then this book would have been twice as long and rejected by the publisher. In addition, there are many delightful examples and excursions in the various approaches to mathematical origami. It really is a shame that I could not include them all. Instead, I sprinkled many **diversions** throughout the book. Some of these are interesting, fun exercises left for the reader to explore. Others are meant to be parts of proofs that are fairly straightforward and therefore can be left to the reader (often with references to the literature for full details).

This book is a monograph, not a textbook (although it could be used as a text for a graduate-level or reading course), and therefore it didn't seem appropriate to end each chapter with a list of exercises (nor would this be appropriate for many parts of the book). Spreading diversions through the chapters seemed to be a good compromise for including many of the juicy tidbits of origami mathematics without overburdening the exposition.

**Open Problems**    I have also provided, at the end of all appropriate chapters, a list of some of the open problems in the related fields of origami mathematics. This, of course, runs the risk of dating the book in a few years, as many of the topics covered in this text are currently areas of active research. But hopefully these problems will encourage readers to join the fun that is origami mathematics.

I also need to point out that not all aspects of origami mathematics are covered in this book. The field has simply grown too large for one monograph to cover. The largest omission is the topic of origami using curved creases. While this topic is touched on very briefly in Section 4.5, it is only done in the context of origami geometric constructions. Modeling folded structures made from curved creases is possible using differential geometry. This is a very interesting subject and very much an active area of research. In fact, new work on curved crease origami is being developed so quickly as to make planning a monograph-level exposition on the topic somewhat impossible, as it could become painfully dated before the book saw print. Indeed, work is being done at the time of this writing by several leading researchers on mathematical and computational origami toward writing a stand-alone text on curved creases.

## Acknowledgements

Finally, there are many people and organizations that need to be thanked who helped in the completion of this book. Large sections of this book were completed during sabbatical leaves, first from Merrimack College and then from Western New England University. The latter, in fact, was spent enjoying the hospitality of the University of Tokyo, where several graduate students, in particular Akito Adachi, test-read chapter drafts of this book. Parts of this book were written while under grant support from the National Science Foundation, namely grants EFRI ODISSEI-1240441 and

DMS-1906202. I am indebted to the following individuals for extensive conversations or correspondence on all matters of origami mathematics over the past 10+ years: Zach Abel, Hugo Akitaya, Roger Alperin, Hannah Alpert, Bryan Gin-ge Chen, David Cox, Erik Demaine, Martin Demaine, David Eppstein, Arthur Evans, Johnna Farnham, Tomoko Fuse, Jacques Justin, Toshikazu Kawasaki, Jason Ku, Robert J. Lang, Anna Lubiw, Jun Maekawa, Jeannine Mosely, Michelle Normand, Jessica Ginepro Notestine, Joseph O'Rourke, Emanuele Paolini, Aubrey Rumbolt, Christian Santangelo, and Tomohiro Tachi. Last, but not least, this book absolutely benefitted by my partnership with sarah-marie belcastro, who has been both endlessly supportive and helpfully critical.

# Part I

## Geometric Constructions

One of the more easily recognized areas of overlap between origami and mathematics is in the realm of Euclidean geometry. Specifically, all origami folds produce creases and points of intersection and thus are performing some kind of geometric construction. In the first part of this book, we will fully explore the kinds of constructions that are possible with origami.

The history of this topic in paper folding is interesting and longer than one might think. Because of the central role that straightedge and compass (SE&C) constructions played in classical Greek geometry, as well as in geometry education throughout the Western world, it is not surprising to learn that many people thought of replacing these traditional construction tools with a folded piece of paper. Fedrich Froebel was one of the earliest educators to capitalize on this idea, incorporating a number of folding construction activities as ways in which to teach elementary geometry. (See (Heerwart, 1920; Liebschner, 1992).) Inspired by this, T. Sundara Row of India wrote the classic text *Geometric Exercises in Paper Folding* in the late 1800s. Row's descriptions of how to fold hexagons, pentagons, and other geometric figures in paper proved so popular among students and educators that it remained in print for over sixty years (Row, 1901). In the 1930s Italian mathematician Margherita P. Beloch extended such work by proving that origami can solve general cubic equations (Beloch, 1936). SE&C constructions can only solve quadratics, and it seems that all references to origami geometric constructions prior to Beloch, and for a good fifty years afterward, hinted not at all at the possibility of origami being able to do more. So Beloch's work was truly groundbreaking. However, it remained in complete obscurity until the 1980s, by which time several people had discovered ways to trisect angles (Husimi, 1980; Justin, 1984; Martin, 1985) and double cubes (Martin, 1985; Messer, 1986). Since then origami geometric constructions have found their place next to other construction methods (see (Martin, 1998; Cox, 2004)) as well as having emerged as an area of research in their own right (see (Auckly and Cleveland, 1995; Alperin, 2000)).

In Chapters 1–3 we will assume that all our crease lines are straight and that every time we fold the paper we make only one crease. This means that a sequence of crease lines can only be made by making one fold at a time. The cases of curved creases and simultaneous folds (multifolds) are compelling, complicated, controversial, and covered in Chapter 4.

# 1    Examples and Basic Folds

SE&C constructions are much easier to analyze than origami constructions. The main reason for this is because it is easy to classify all the possible operations that can be performed with the tools of SE&C.

**1:** Given two points, we can use the straightedge to draw a line connecting them.

**2:** Given a point $p$ and a length $r$, we can use the compass to draw a circle with radius $r$ centered at the point $p$.

**3:** We can locate the points of intersection between combinations of circles and lines.

Studies of these basic SE&C operations can be found in a number of classic geometry texts; see (Courant and Robbins, 1941; Martin, 1998), for example. The idea is to show that, given a line segment of unit length, we can construct segments with the length of any rational number $a/b \in \mathbb{Q}$ as well as any expression involving rational numbers and the operation of taking square roots. Proving that these are the **only** kinds of lengths that can be constructed with SE&C requires carefully considering the kinds of equations we obtain when locating the intersection points of two circles, a circle and a line, or two lines, and then proving that repeated use of such intersections gives us the smallest field extension of the rationals that is closed under square roots. See (Cox, 2004, Section 10.1).

Performing such an analysis on straight-crease, single-fold origami is more perplexing because the tools we have are more flexible. There are many different ways in which we can determine a crease when folding, say, a square sheet of paper, especially if there are pre-creases already made in the paper. Classifying all the basic operations of origami and proving that there cannot be any more has been a controversial topic, let alone studying the algebra of such folds.

Before trying to formulate a list of basic origami operations, we will first familiarize ourselves with paper folding's variety by way of some construction examples. The majority of readers will not have seen these types of explicit geometric paper folding methods before, and exposure to such examples can be a big intuition-builder before undertaking more abstract analysis. Plus, they're fun. Readers are encouraged to try them.

## 1.1   Constructing an Equilateral Triangle

The following challenge appeared in *Mathematics Magazine* (Vol. 67, No. 2, April 1994, p. 123):

---

**Diversion 1.1**    Starting with a square sheet of paper, fold it to produce a square having three-fourths its area. Only five folds are allowed.

---

The puzzle is referenced as coming from a book called *Mathematical Brain Benders* by Stephen Barr (1982). This puzzle is especially fun for origami practicioners who immediately conjecture that they can do it in fewer than five folds.

This challenge is similar to the following: Starting with a square piece of paper, fold it into a perfect equilateral triangle. To accomplish this, one would need to construct a $60°$ angle, which could be done by folding the sides of a $30°$-$60°$-$90°$ triangle in the square. In other words, we would need to construct line segments of length 1, 2, and $\sqrt{3}$ in our paper. It is standard, however, to always assume that our starting square has side length 1, so it would be more feasible to create a $30°$-$60°$-$90°$ triangle with side lengths $1/2$, 1, and $\sqrt{3}/2$. Constructing $\sqrt{3}/2$ is exactly what we would need for the 3/4-area puzzle as well.

There are many ways to fold a $30°$-$60°$-$90°$ triangle in a square. In fact, it is not hard to find explicit methods for doing this in origami instruction books, especially books on modular origami (like (Fuse, 1990)), although such books usually do not mention that they are performing such a construction.

Figure 1.1 shows a standard method for producing such a $30°$-$60°$-$90°$ triangle. This can be shown synthetically, or we can merely note that if the square has side length 1, then $PB$ must also have length 1, since it is the image of a side of the square under the fold. By symmetry, $AP$ must also have length 1, and thus we have that the points $APB$ form an equilateral triangle. (This immediately gives us that the fold made in Figure 1.1 produced the desired angles.)

A variation on this challenge is to fold an equilateral triangle of maximum area within our square piece of paper. Utilizing analytic techniques to discover what triangle orientation gives the maximal area can be a good exercise for calculus students (see (Hull, 2012)), but developing a folding method is another matter. Figure 1.2 shows the standard method of doing this as presented by Emily Gingras (Merrimack College class of 2003). The first picture is her "proof without words" that the angle $\theta$ shown is $15°$, which proves that the other pictures give the proper equilateral triangle.

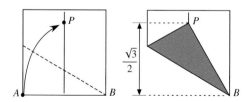

**Figure 1.1**  Producing a $30°$-$60°$-$90°$ triangle: First fold the square in half and unfold. Then fold the lower left corner up to the crease line, while making the crease go through the lower right corner.

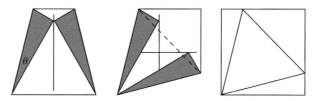

**Figure 1.2** A "proof without words" for constructing the maximal equilateral triangle.

Both of these methods involve constructing a line segment whose length is an expression involving square roots, $\sqrt{3}/2$ in the first case and $2/\sqrt{2+\sqrt{3}}$ in the second. What kind of folding operation produced these lengths? In both cases we had a point being folded to a line (point $A$ being folded to the half-way crease in Figure 1.1) where we also make sure that the crease passes through a second point (point $B$ in Figure 1.1). This operation will be explored further in Chapter 2.

As an extra challenge, readers can try to use the method from Figure 1.2 to discover the classic method that paper folders use to fold a square into a regular hexagon with maximal area.

## 1.2 Dividing a Segment into 1/*n*ths

The problem of dividing the side of a piece of paper into $n$ equal lengths is one which has been a favorite of origami geometry enthusiasts. References to it and various solutions for the cases where $n = 3, 5,$ or 7 can be found in some origami books (see (Kasahara and Takahama, 1987; Kasahara, 1988)), on origami email lists, and on a variety of webpages. The challenge, of course, are the cases when $n$ is odd, since folding lengths in half is simple and can generate all even numbers once the odds have been handled.

Such division methods have practical applications in origami as well. Many origami models start off by asking the folder to first divide the square into thirds or into a $5 \times 5$ grid. Interestingly, the most common method used by origamists to fold thirds is to use the method shown in Figure 1.3. The idea is to "eyeball" it by curving the paper into an S shape and easing the creases into their proper places. With practice this can

**Figure 1.3** Folding thirds, the multifold method.

(1)              (2)              (3)          1/3

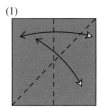 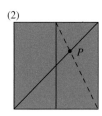 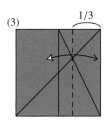

**Figure 1.4** Folding thirds exactly. (1) Crease a diagonal and the 1/2 vertical crease. (2) Make a crease that connects the midpoint of the top edge and the lower right corner. Let this crease intersect the diagonal at $P$. (3) Make a crease at $P$ perpendicular to the bottom and top sides. Then this last crease will be 1/3 from the right side.

be done very quickly and accurately, but it violates our rule of one fold at a time since it requires making two creases simultaneously (which is called a 2-fold or a multifold; these will be discussed in Chapter 4).

A mathematically precise, one-fold-at-a-time way to fold a square of paper into thirds is shown in Figure 1.4. While the origins of this method are unclear, Lang (1988) refers to it as the **crossing diagonals method**. The correctness of this method can be proven by similar triangles or by noticing that the point $P$ is at the intersection of the lines $y = x$ and $y = -2x + 2$, where we assume that the square has side length 1 and lower left corner is at the origin. Thus $P = (2/3, 2/3)$.

This method can be generalized for arbitrary odd values of $n = 2k + 1$. Instead of making a vertical crease at the line $x = 1/2$, make it at $x = (2k - 1)/(2k)$. This is feasible because any odd factor of $2k$ is a smaller odd number than $n$. Thus, by induction, we can assume that dividing the side of our square into $1/(2k)$ths can be done. Then, using the same method as the 1/3 case, our point $P$ would be at the intersection of the lines $y = x$ and $y = -2kx + 2k$, so $P = (2k/(2k + 1), 2k/(2k + 1))$, giving us a landmark for dividing the side into $1/n$ths.

The crossing diagonals method is sometimes used by origami designers; see John Montroll's Chess Board (Montroll, 1993), for example. However, since the method requires several crease lines to be made across the paper, it isn't viewed as ideal. A better method in this regard is based on **Haga's Theorem** (Kasahara and Takahama, 1987), which states that if we fold a corner of a square (or rectangular) sheet of paper to a point on a nonadjacent side, then several similar triangles can be found and the resulting crease can mark the sides of the paper at interesting lengths.

In particular, if we mark a point at $(1/(2k), 1)$ on the square and fold the lower left corner (the origin) to this point, as seen in Figure 1.5, then triangles $A$ and $B$ are similar. Also, triangle $A$ is a right triangle and one leg, $x$, and the hypothenuse make up a side of the square, so the hypothenuse is $1 - x$. The Pythagorean Theorem then gives us that $x = (2k + 1)(2k - 1)/(8k^2)$. Letting the short leg of triangle $B$ be $y$, the similarity relation gives us

$$\frac{y}{1/(2k)} = \frac{1 - 1/(2k)}{(2k + 1)(2k - 1)/(8k^2)},$$

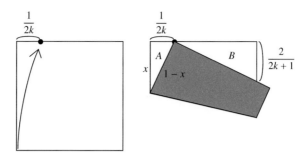

**Figure 1.5**  Haga's Theorem applied to the odd division problem.

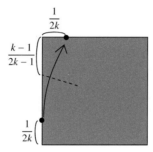

**Figure 1.6**  Noma's method.

which simplifies, amazingly enough, to $y = 2/(2k + 1)$. Thus if divisions of $1/n = 1/(2k + 1)$ are desired, constructing $1/(2k)$ and the one fold of Haga's Theorem will do the trick.

---

**Diversion 1.2** (Geretschläger, 2002, 2008)    Prove that the perimeter of triangle $B$ in Figure 1.5 is always half the perimeter of the original square.

---

Haga's Theorem contains many other geometric morsels. See (Husimi and Husimi, 1979; Haga, 2002; Geretschläger, 2002) for more information.

However, it is possible to make any $1/n$ divisions along the side of a square without folding any creases all the way across the paper. The idea is to perform folds that only require making pinch marks on the perimeter of the paper. This would clearly be attractive for origami designers, making it possible to create any $a/b$ mark on the perimeter without marring the paper's interior with extraneous creases.

Masamichi Noma (1992) developed such a method, and it is summarized in Figure 1.6. The idea is, if divisions of $1/n = 1/(2k − 1)$ are the goal, to make pinch marks at length $1/(2k)$ on the left side of the top edge of the square and at the bottom side of the left edge. This gives us two marked points on the paper's perimeter. If we fold these two points together, we can pinch the paper only on the left side, so as to avoid

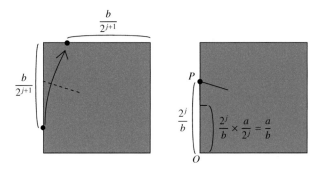

**Figure 1.7** Noma's method used to construct $a/b$.

making a crease all the way across the square. This crease will intersect the left edge $(k-1)/(2k-1)$ from the top corner.

---

**Diversion 1.3**   Prove that Noma's method works.

---

Robert J. Lang has synthesized Noma's method, among others, to generate algorithms for producing folding sequences of pinch marks to create any rational length divisions. In (Lang, 2003) he suggests the following to apply Noma's method to create an arbitrary rational length $a/b$ for integers $a < b$:

(1)  Let $2^j$ be the largest power of 2 smaller than $b$.
(2)  Construct lengths $b/2^{j+1}$ along the top and left sides of the square, as shown in Figure 1.7. (This is easy since the denominators are just powers of 2.)
(3)  Bring these two points together to make a crease pinch along the left side at point $P$.
(4)  Then length $OP$ ($O$ being the lower left corner) will be $2^j/b$. (The same work needed in Figure 1.6 shows this.)
(5)  Divide segment $OP$ into $1/2^j$ths (which is easy). Taking $a$ of these from $O$ gives a length $(2^j/b)(a/2^j) = a/b$.

In all three of these division methods, none of the basic folding operations used are very complex. Each case involved only the "moves" of folding a crease between two existing points or folding one point onto another point. This is hardly surprising because only rational lengths were being constructed. But the variety and ingenuity of these methods are nonetheless a marvel.

## 1.3     Trisecting an Angle

The hallmark of origami geometric constructions has been the fact that paper folding can, fairly easily, trisect angles. The first known method for doing this was created by

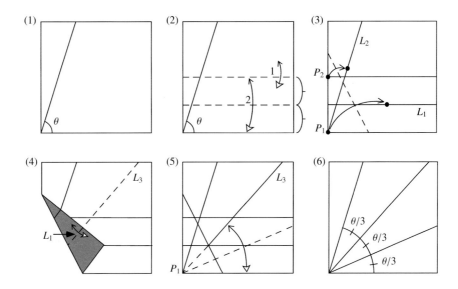

**Figure 1.8** Abe's angle trisection method.

Hisashi Abe (Husimi, 1980) sometime in the late 1970s. His method for trisecting an arbitrary acute angle $\theta$ is shown in Figure 1.8 and proceeds as follows:

(1) Position your angle $\theta$ in the lower left corner of the square, as shown.
(2) Make a horizontal crease, labeled 1 in the figure, parallel to the bottom edge, and then fold and unfold the bottom edge to this crease line (labeled 2). Line 1 can be made at any height, although if $\theta$ is less than 45° then crease 1 might need to be closer to the bottom edge for the next step to be possible.
(3) Then, using the labeling in the figure, fold the corner $P_1$ onto $L_1$ **while at the same time** making point $P_2$ land on line $L_2$. This will require curling the paper over, lining up these two points onto their lines, and then pressing the crease flat.
(4) Leaving this last crease folded, you'll see part of $L_1$ reflected on this flap of paper. Refold this crease, extending it through the rest of the paper to crease line $L_3$.
(5) Unfold step (3) and extend the left side of $L_3$ – it will hit the corner $P_1$. Then fold the bottom side of the square to $L_3$ to bisect the angle $L_3$ makes with the bottom.
(6) Voilà! The angle $\theta$ has been trisected.

One way to prove that Abe's method works is shown in Figure 1.9. First, we need to establish that when crease $L_3$ is extended in step (5) of Figure 1.8, it will intersect point $P_1$. If we let $F$ be the point where $L_3$ intersects the crease line from step (3), and draw the segment $P_1F$ on the unfolded paper, then the acute angles between (the unextended) $L_3$ and $L_1$ and between $P_1F$ and $L_1$ are equal, since the fold in step (3) superimposes them. (This is angle $\alpha$ in Figure 1.9(a).) Thus this angle acts as a vertical angle, and $P_1F$ and $L_3$ must form a straight line.

(a)                                    (b)

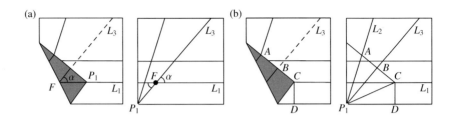

**Figure 1.9** Proof of Abe's trisection.

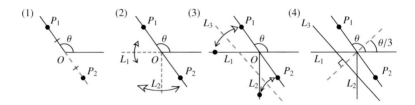

**Figure 1.10** Justin's angle trisection method.

Then we can label points $A, B, C,$ and $D$ as in Figure 1.9(b), where $C$, $A$, and $B$ are the images of $P_1$, $P_2$, and the point in between, respectively, under the step (3) trisection fold. (For $D$ we drop a perpendicular to the bottom of the square.) This gives us that the lengths $AB$, $BC$, and $CD$ are all congruent, and thus $\triangle ABP_1$, $\triangle BCP_1$, and $\triangle CDP_1$ are congruent right triangles, giving us the trisection.

While Abe's method as shown in Figure 1.8 only works for acute $\theta$, readers are encouraged to explore how it can be extended for obtuse angles.

By now readers will have probably noticed the unusual folding step in this method that seems to give us the trisection, namely step (3) in Figure 1.8. In this step we have $P_1$ being folded to a line, which by itself is similar to what we had to do when constructing equilateral triangles in Section 1.1. But one point going to one line is not enough to uniquely determine a crease, and for step (3) we choose to nail down where $P_1$ will go on $L_1$ by also requiring $P_2$ to fold onto line $L_2$. This "two points folding onto two lines" origami operation is a move that is rarely seen in origami instructions, but it turns out to be the key that gives paper folding more muscle than SE&C constructions. (We'll see exactly why in Chapter 2.) In fact, any straight-crease, single-fold origami construction that goes beyond the constructible range of SE&C will require a move such as this.

Independently of Abe, the French mathematician Jacques Justin (1984) also developed an angle trisection method at roughly the same time. (See also (Justin, 1986b).) Justin's method, shown in Figure 1.10, allows the starting angle $\theta$ to be obtuse and positioned in the interior of the square. The method is as follows (see the corresponding pictures in Figure 1.10):

(1) Let $\theta$ be an angle at a point $O$. Let $P_1$ be a point on one side of $\theta$, and extend the line $P_1O$ so that we can find a point $P_2$ on this line but on the other side of $O$ so that $P_1O \cong P_2O$.

(2) Extend the other side of the angle $\theta$ to become the line $L_1$. Fold line $L_2$ to be perpendicular to $L_1$ at the point $O$. (This is a move we haven't seen before; it involves folding $L_1$ onto itself making the crease go through $O$.)

(3) Now fold $P_1$ onto $L_1$ and $P_2$ onto $L_2$ simultaneously to create line $L_3$.

(4) Finally, fold a crease perpendicular to $L_3$ that goes through point $O$. This line will make an angle of $\theta/3$ with one side of angle $\theta$.

---

**Diversion 1.4**    Prove that Justin's trisection method works.

---

There is a lot of interesting mathematics to be explored in this "two points to two lines" origami move. Questions to ponder might include: Given any two points and two lines, can this operation always be performed? Does it always result in a unique crease? What kinds of numbers (segment lengths) is it constructing for us? We will address these questions in Chapters 2 and 3.

## 1.4    Folding a Regular Heptagon

The easiest regular polygons to fold from a square are, well, the square, regular octagon, 16-gon, and other $2^n$-gons. We saw earlier that equilateral triangles are not too hard to fold from a square, and this admits regular hexagons and dodecagons relatively easily.

---

**Diversion 1.5**    Devise a way to fold a regular pentagon from a square piece of paper. (See (Morassi, 1989) for an analysis of approximate and exact methods for this.)

---

However, the smallest regular $n$-gon that origami can produce that SE&C cannot is the heptagon.

The first published instructions for folding a regular heptagon appears to be those of Scimemi (1989) and, independently, Geretschläger (1997b) (although Justin (1986b) also provides the basic ingredients for such a construction). Both of these methods are nearly identical, however, which is not surprising because both follow a classic algebraic approach to the heptagon problem, as can also be seen in the non-folding heptagon construction given by Gleason (1988). Gleason's construction assumes the tools of straightedge, compass, and an angle trisector. Since we know origami can trisect angles, Gleason's method could be used, as is, to fold a regular heptagon. Alperin (2002) did just this, but such a strategy produces a lengthy and inelegant folding

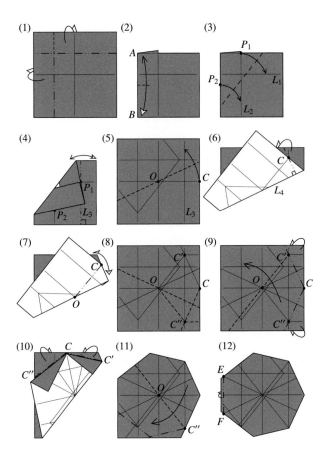

**Figure 1.11**  Folding a regular heptagon.

procedure. (We should note that mathematical purity demands that we find a folding procedure that is mathematically exact. However, when one physically makes a fold there will always be error present. So for practicality's sake it is always better to find folding sequences that help minimize error, either by being short or by encompassing folds that are easy to perform.)

Figure 1.11 shows a more cleaned-up way to fold a regular heptagon than those previously given. (Scimemi (1989) doesn't give an explicit folding sequence, and Geretschläger (1997b) has more folds than are necessary.) Our procedure is as follows:

(1) First, crease the paper in half from top to bottom and left to right. Then fold the top 1/4 behind and then the left 1/4 behind.
(2) Make a pinch crease on the left side by bringing points $A$ and $B$ together.
(3) Now we're ready for the fold that does the "magic." Fold point $P_1$ onto line $L_1$ and point $P_2$ onto line $L_2$ at the same time.

(4) Notice where $P_1$ went after step (3). Mountain fold a vertical crease, perpendicular to the bottom edge, on the underneath layer of paper, creating line $L_3$. Crease sharply and then unfold everything.

(5) Notice where $L_3$ is on the unfolded sheet. Fold $C$, the midpoint on the right side, to line $L_3$ so that the crease goes through the center $O$ of the paper.

(6) Step (5) created the folded edge $L_4$. Fold the right flap of paper behind, making the crease go through $C$ while being perpendicular to $L_4$.

(7) Now fold and unfold line $OC$. (This crease already exists, but you want it to be made through **all** layers of paper.) Then unfold everything.

(8) Line $CC'$ is one side of our heptagon. ($C'$ is the image of $C$ under the fold in step (5).) Repeat steps (5)–(7) on the bottom half of the paper, creating point $C''$.

(9) Fold $CC'$ and $CC''$ behind. Then valley fold $OC'$, extending it across the paper.

(10) Use the images of $CC'$ and $CC''$ to fold two more sides of our heptagon. Then unfold $OC'$.

(11) Repeat steps (9)–(10) on the bottom half.

(12) Fold the left side behind with crease $EF$ to complete the heptagon.

To see why this works, let us set up a coordinate system for the paper as follows: let $O$, the center of the paper, be the origin, and let the side of the square be of length 4. (We choose these coordinates to more easily illustrate the connection with Gleason's analysis in (Gleason, 1988).) Our goal is to show that the point $C'$ in Figure 1.11 has coordinates $(2\cos(2\pi/7), 2\sin(2\pi/7))$, and thus points $C$, $C'$, and $C''$ form three vertices of a heptagon of radius 2. Since these points are used to generate the other vertices in a logical way, this would prove the folded heptagon's validity.

The points in step (3) of Figure 1.11 are $P_1 = (0, 1)$ and $P_2 = (-1, -1/2)$, where $L_1$ is the $x$-axis and $L_2$ is the $y$-axis. Suppose that $P_1$ gets folded to the point $P_1' = (t, 0)$ on $L_1$ and $P_2$ gets folded to the point $P_2' = (0, s)$ on $L_2$.

The segment $P_1 P_1'$ has slope $-1/t$, and the crease line in step (3) must be the perpendicular bisector to this segment. So the slope of the crease line must be $t$ and pass through the midpoint of $P_1 P_1'$, which is $(t/2, 1/2)$. Thus one formula for the crease line in step (3) is

$$y = tx - \frac{t^2}{2} + \frac{1}{2}.$$

On the other hand, segment $P_2 P_2'$ has slope $(2s+1)/2$ and midpoint $(-1/2, (2s-1)/4)$. Thus, another formula for our crease line is

$$y = \frac{-2}{2s+1}x - \frac{1}{2s+1} + \frac{2s-1}{4}.$$

Our aim is to find the value of $t$ (the $x$-coordinate of $P_1'$), since this determines the location of line $L_3$ in step (5) and thus the location of $C'$. Equating the slopes of our two line equations gives $s = -(t+2)/(2t)$. This can then be substituted into the equation we get by equating the constant terms of our two line equations, resulting in a single equation in $t$. After simplifying, this becomes

$$t^3 + t^2 - 2t - 1 = 0. \tag{1.1}$$

Sure enough, $t = 2\cos(2\pi/7)$ satisfies this equation, proving that $L_3$ is in the proper place. (The other roots of Equation (1.1) are real and negative, and thus are not values of $t$ that would make the fold in step (3) of Figure 1.11 work.) For readers who do not immediately believe our claims as to the solutions of Equation (1.1), we present an argument from (Gleason, 1988).

Consider the vertices of a regular heptagon as the seventh roots of unity in the complex plane, that is, the complex solutions of $z^7 - 1 = 0$. Factoring out the obvious $z - 1$ term for the $z = 1$ corner, we get the equation for the remaining six corners: $z^6 + z^5 + z^4 + z^3 + z^2 + z + 1 = 0$. Let $A = \cos(2\pi/7) + i\sin(2\pi/7)$ (the principle seventh root of 1). Since the reciprocals of complex numbers on the unit circle are the same as the complex conjugates, we have that $1/A = A^6$, $1/A^2 = A^5$, and $1/A^3 = A^4$. Plugging these into our factored heptagon equation, we see that $A$ satisfies

$$A^3 + A^2 + A + 1 + \frac{1}{A} + \frac{1}{A^2} + \frac{1}{A^3} = 0. \tag{1.2}$$

But we also have that $A + 1/A = A + \overline{A} = 2\cos(2\pi/7)$. Furthermore, notice that

$$A^2 + \frac{1}{A^2} = \left(A + \frac{1}{A}\right)^2 - 2 \quad \text{and} \quad A^3 + \frac{1}{A^3} = \left(A + \frac{1}{A}\right)^3 - 3\left(A + \frac{1}{A}\right).$$

Substituting these into Equation (1.2), we get

$$\left(A + \frac{1}{A}\right)^3 + \left(A + \frac{1}{A}\right)^2 - 2\left(A + \frac{1}{A}\right) - 1 = 0.$$

Therefore, $A + 1/A = 2\cos(2\pi/7)$ is a solution to Equation (1.1). Similar machinations show that other two roots of Equation (1.1) are $2\cos(4\pi/7)$ and $2\cos(6\pi/7)$.

Figure 1.12 provides a geometric interpretation of what's going on. If $A$ is the principle root of $(z^7 - 1)/(z - 1)$, then $1/A$ is just $A$ reflected about the real axis, and $(A + 1/A)/2$ is the midpoint of the segment connecting these two points. Since this midpoint is on the real axis, it's just the real part of $A$, which is $\cos(2\pi/7)$. The same thing holds for the other roots $A^2$ and $A^3$. Therefore the roots of $(z^7 - 1)/(z - 1)$ after the substitution $z = A + 1/A$ will just be twice the real parts of the seventh roots of unity, excluding $z = 1$. There are only three such numbers, and the equation after substitution simply becomes Equation (1.1).

In fact, Equation (1.1) is the standard equation one tries to solve when confronting the regular heptagon construction problem. (See (Martin, 1998), for example.) In a very real sense, this equation is being solved in step (3) of the folding sequence. It is interesting to note that both the folding methods of Scimemi (1989) and of Geretschläger (1997b) incorporate basically the same fold seen in step (3) of Figure 1.11 to create a crease with the proper slope for the heptagon construction. That two independent researchers came upon the same fold to solve Equation (1.1) is not a coincidence – they were both trying to solve the same equation via folding. We will see in Chapter 2 how to go about solving general cubic equations with such folding operations.

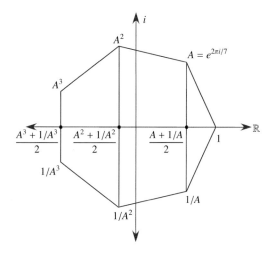

**Figure 1.12** Geometric interpretation of $A + 1/A$ and the like.

The Basic Origami Operations

Now that we have seen several examples of origami constructions, we are in a better position to consider classifying the **basic origami operations**, or **BOOs** for short. This turns out to be more problematic than one might expect.

As seen at the beginning of this chapter, it is easy to classify what operations are possible under SE&C because we know exactly what our tools can do. The examples we've seen show that origami admits many different types of operations. When trying to make a list of them, it is not clear if one is really a special case of another, or whether we have found them all. For example, in the 1980s Huzita and Scimemi (1989) developed the following list of operations for origami:

**O1:** Given two points $P_1$ and $P_2$, we can fold a crease line connecting them.
**O2:** Given two lines, we can locate their point of intersection, if it exists.
**O3:** Given two points $P_1$ and $P_2$, we can fold the point $P_1$ onto $P_2$ (perpendicular bisector).
**O4:** Given two lines $L_1$ and $L_2$, we can fold the line $L_1$ onto the line $L_2$ (angle bisector).
**O5:** Given a point $P$ and a line $L$, we can make a fold line perpendicular to $L$ passing through the point $P$ (perpendicular through a point).
**O6:** Given two points $P_1$ and $P_2$ and a line $L$, we can, whenever possible, make a fold that places $P_1$ onto the line $L$ and passes through the point $P_2$.
**O7:** Given two points $P_1$ and $P_2$ and two lines $L_1$ and $L_2$, we can, whenever possible, make a fold that places $P_1$ onto $L_1$ and also places $P_2$ onto $L_2$.

---

**Diversion 1.6**    Under what conditions will operations O6 and O7 be possible? (This will be addressed in the next chapter.)

---

Readers who enjoy paying attention to details can examine this list to check it for completeness and optimality. That is, are any BOOs missing? Can any of BOOs O1–O7 be performed via a combination of the other BOOs?

We will address the latter question later. For the former, we can try to distinguish origami operations by what combination of points and lines they use. For example, should the following be considered an additional BOO?

**O8:** Given a point $P$ and two nonparallel lines $L_1$ and $L_2$, we can make a fold perpendicular to $L_2$ that places $P$ onto $L_1$.

Jacques Justin included this operation in his list of BOOs, which may predate the Huzita–Scimemi list (see (Justin, 1986b)). Also, Hatori (2003) independently proposed this BOO as an addition much later.

Since BOO O8 takes two lines and a point as input, and none of the Huzita–Scimemi BOOs do this, we could consider this to be a separate BOO. Sticklers for optimality may disagree.

---

**Diversion 1.7**    Show that O8 can be performed via a sequence of BOOs O1–O6.

---

However, with the addition of BOO O8, we can prove that no more basic operations are possible.

**Theorem 1.1**    *If we only allow one fold at a time, and assuming all our creases are straight lines, then the only folding operations possible are O1–O8.*

Robert J. Lang first proved this in 2003 (Lang, 2003) using vector geometry methods. We present a more elementary proof from Hull (2005) that follows Lang's basic argument.

*Proof*    When we do origami, we only have two types of things to fold to each other: points and lines. Sometimes when we fold one of these to another, it uniquely determines the fold, like when folding one point to another point. But other times, like when folding a point to a line, there is still a **degree of freedom** left that must be removed before a specific fold is determined. The possible combinations of points and lines that can be folded to each other are as follows (note that we can ignore O2, since it involves no folding):

**Case 1:**  Fold a point to another point – no degree of freedom.
**Case 2:**  Fold a point to itself – one degree of freedom (the angle of the crease going through the point).

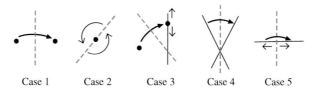

**Figure 1.13** The five cases of points/lines being folded to each other.

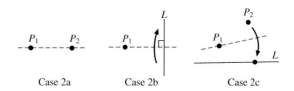

**Figure 1.14** The subcases of folding a point $P_1$ to itself.

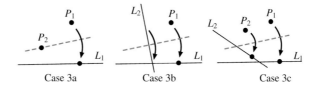

**Figure 1.15** The subcases of folding a point $P_1$ to a line $L_1$.

**Case 3:** Fold a point to a line – one degree of freedom (the point can be anywhere on the line).

**Case 4:** Fold a line to another line – no degree of freedom.

**Case 5:** Fold a line to itself – one degree of freedom (see Figure 1.13).

Cases 1 and 4 are operations O3 and O4, respectively. The other cases all have a degree of freedom that needs to be removed to determine what folds they can become.

**Case 2:** *Fold a point $P_1$ to itself.* The only creases that can fold a point $P_1$ to itself are creases going through the point $P_1$, and there are an infinite number that do this. To specify a unique crease we need to combine this with another operation that also possesses a degree of freedom. This gives us three subcases, as illustrated in Figure 1.14:

**Case 2a:** Also fold another point $P_2$ to itself. This is operation O1.

**Case 2b:** Also fold a line $L$ to itself. This is the perpendicular line operation O5.

**Case 2c:** Also fold another point $P_2$ to a line $L$. This is operation O6.

**Case 3:** *Fold a point $P_1$ to a line $L_1$.* Again, we have a degree of freedom because the point could be folded anywhere on the line. Thus we need to combine this with another degree of freedom (see Figure 1.15):

**Case 3a:** Also fold another point $P_2$ to itself. This again gives us operation O6.

**Case 3b:** Also fold another line $L_2$ to itself. This gives us operation O8.

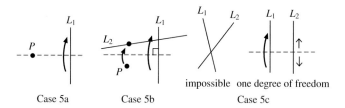

**Figure 1.16**  The subcases of folding a line $L_1$ to itself.

**Case 3c:** Also fold another point $P_2$ to another line $L_2$. This is operation O7.
We can now see that we've covered all seven of our BOOs. To make sure there are no more, however, we need to look at the last case.

**Case 5:** *Fold a line $L_1$ to itself.* Folding a line $L_1$ to itself is really just folding a crease perpendicular to $L_1$. But since this crease could intersect $L_1$ at any point, this gives us a degree of freedom. Combining with other operations gives the following (as shown in Figure 1.16):

**Case 5a:** Also fold a point $P$ to itself. This gives us operation O5 again.

**Case 5b:** Also fold a point $P$ to another line $L_2$. This gives us O8 again.

**Case 5c:** Also fold another line $L_2$ to itself. If these two lines intersect, then this is impossible, since the resulting crease would have to be perpendicular to both. Thus the two lines would have to be parallel, but this still results in one degree of freedom. Thus this combination is redundant—we would still need another operation to specify a unique crease, bringing us back to Case 5a or 5b.

This exhausts all the possibilities of folding points and lines to points and lines, completing the proof.                                                                    □

Operations O1–O8 encompass everything that straight-crease, single-fold origami can do. This list does contain redundancies, however, and to eliminate them we need to be more specific about what is given to us at the start of our constructions.

For example, Alperin (2000) assumes the paper to be the entire complex plane with the given constructed points 0 and 1. He then chooses operations O1–O4, O6, and O7 to be his list of construction operations. (Actually, Alperin and several other writers use the word "axiom" to refer to allowed folds. But since O6 and O7 are not always possible, "operations" seems a more appropriate term.) One could equivalently assume that the four points $(\pm 1, \pm 1)$ are given, which, when the lines $y = \pm 1$ and $x = \pm 1$ are folded using O1, simulate the boundary of a square piece of paper.

Certainly the given points we start with and the operations O1–O8 guarantee that every crease line that is constructed will have a constructed point on it somewhere, as well as every point having a constructed line passing through it. This observation has led several people, including Martin (1998) and Hatori (2003) to prove that all we really need to characterize origami constructions are operations O2 and O7.

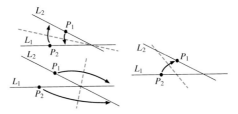

**Figure 1.17**  Generating O3 and O4 from O7.

**Theorem 1.2**  *Assuming that we are given at least two constructed points contained in nonparallel constructed lines (which may be identical), then any straight-crease, single-fold origami construction from this starting set can be completely described by combinations of operations O2 and O7.*

*Proof*  By Theorem 1.1, we know that operations O1–O8 are all we need to consider. We obviously need to keep O2, but the others can be shown to be special cases of O7 where either we have $L_1 = L_2$ or some of the points $P_1, P_2$ lie on the lines $L_1, L_2$. This needs to be done carefully.

For the operations O3 and O4, let $P_1$ be on $L_2$ and $P_2$ be on $L_1$. (We know that all constructed lines contain constructed points and vice versa, so we can assume this from the premises of O3 and O4.) Then there will be at most three ways in which $P_1$ and $P_2$ can be folded onto $L_1$ and $L_2$, respectively, as shown in Figure 1.17. The first two cases will amount to folding $L_1$ onto $L_2$, or bisecting one of the angles made at their intersection, producing O4. (If $L_1$ and $L_2$ are parallel, then there will be only one way to do this.) The other case folds $P_1$ onto $P_2$, giving us O3.

In O6, we are given points $P_1$, $P_2$ and line $L_1$. Let $L_2$ be any line through $P_2$. So long as O6 is possible, we can have O7 fold $P_2$ to itself on line $L_2$ and fold $P_1$ onto $L_1$. (Note that we are not concerned here with when O6 is possible – we will take that up in the next chapter.)

For O1 and O5, let $L_1$ and $L_2$ be lines containing $P_1$ and $P_2$, respectively. Then, assuming $L_1$ and $L_2$ are not parallel, there are at most three different folds we could make that will leave $P_1$ on $L_1$ and $P_2$ on $L_2$: (a) making the crease pass through $P_1$ and $P_2$, (b) making the crease pass through $P_1$ and be perpendicular to $L_2$, and (c) making the crease through $P_2$ and perpendicular to $L_1$. While some of these may be identical, case (a) is operation O1. Cases (b) and (c) would give us O5. If $L_1$ and $L_2$ do happen to be parallel, then $P_1$ and $P_2$ could not have been the original two points in the construction, so there exist other points and lines that we can use to construct a different line through one of $P_1$ or $P_2$.

For O8 we can take an arbitrary point $P_2$ on $L_2$ and use O7 to fold $P_1$ onto $L_1$ while folding $P_2$ to a (probably) different place on $L_2$ to ensure that the crease line will be perpendicular to $L_2$. This covers all the operations O1–O8.          □

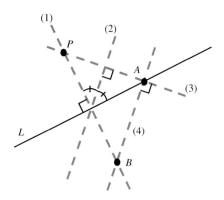

**Figure 1.18** Constructing the reflection of $P$ about $L$.

**Remark 1.3**  A common concern regarding the list of BOOs O1–O8 is whether or not we unfold the paper after each operation. Geometrically, all we care about is that each BOO creates a new line in the plane and that the intersection of lines creates new points. From this view, it does seem that every time we perform a BOO, we unfold the paper immediately to consider the new line formed in the paper (plane).

However, in practice origamists often leave the paper folded in order to perform an operation using the folded layers of paper as a guide. This was done several times in the heptagon construction in Section 1.4, for example. It seems conceivable that repeated applications of a BOO, especially O7, on a piece of paper without unfolding it might lead to a construction that could not be achieved by performing one BOO at a time, unfolding after each.

Yet this is not the case. The locations of any points or lines that are moved in the process of folding the paper flat with a BOO can be reconstructed if we immediately unfold the paper. (And therefore there is no need to keep the paper folded.) To prove this, suppose that $L$ is a crease line constructed by some BOO. All we need to do is show that for any point $P$ or other line $L'$, we can construct the reflections of $P$ and $L'$ about $L$.

One method for constructing the reflection of $P$ about $L$ is shown in Figure 1.18. First use O5 to crease a line (1) perpendicular to $L$ passing through $P$. Then use O4 to fold this crease line onto $L$, bisecting the angle between them to make the crease (2). Then use O5 again with $P$ and the crease (2), labeling $A$ as the point where this crease (3) intersects $L$. Finally, perform O5 again with the crease line (3) and the point $A$ to make the crease (4). Where this crease intersects the crease line (1), called $B$ in Figure 1.18, is the reflection of $P$ about $L$.

Constructing the reflection of a line $L'$ about $L$ can be handled similarly. All we need is a point or two constructed on $L'$; by reflecting these about $L$, we can then create the reflection of $L'$ by using O1.

This does, of course, use the convention that we can think of our sheet of paper as being as large as we wish and that the boundary lines of the paper are just constructed lines like any other. In any case, we conclude that while in practice it is often much more efficient to leave the paper folded after performing an origami operation, we can keep our list of BOOs simply to O1–O8 (or O2 and O7, if we want to be really efficient) by unfolding the paper after each step without losing any origami construction power.

Some of the subtleties in the work of this section, especially Theorem 1.1, seem to be ignored by much of the literature in this area. Several researchers (see (Alperin, 2000), for example) create definitions of what it means to be origami constructible (as we will in Chapter 3) referring to the basic origami operations mentioned here, but they do so with no argument as to whether more operations might be possible. Other papers, for example (Auckly and Cleveland, 1995) and parts of (Alperin, 2000), ask what can be constructed by a deliberately reduced set of origami operations. Investigators in the origami community were very concerned with the question of whether more operations existed (see (Hull, 1996)), whereby Lang's proof of Theorem 1.1 is viewed as a breakthrough.

In some sense it doesn't matter which of the moves O1–O8 we choose for an official list of basic origami operations. As we will see in the next chapter, it is O7 that separates origami from SE&C constructions, and being able to reference the other operations makes for very convenient notation.

## 1.6 Historical Remarks

The first person to seriously analyze origami geometric constructions seems to have been T. Sundara Row (1901). The first person to introduce operation O7 seems to have been Margherita Beloch (1936). (See the remarks in Section 2.5 for more information.) However, none of these early researchers made a formal list of possible origami construction operations. The first such list to see print seems to have been in Jacques Justin's 1986 paper (Justin, 1986b). Justin states that his list was inspired by an unpublished list created by Peter Messer (1984). Messer's list contains the operations O1–O7, but not O8. Justin's list contains all O1–O8. It appears that Scimemi independently developed a list of origami construction operations, which became those listed in (Huzita and Scimemi, 1989) and only included O1–O7.

Complicating things further, George Martin published a paper in 1985 (Martin, 1985) that defines origami constructions using only operations O2 and O7, and he seems to have developed this without any knowledge of Beloch's work. Martin cites Row and a publication of Yates, but he also cites (Dayoub and Lott, 1977), which describes how one can use a Mira, a geometric construction tool that allows one to reflect points about a line in the same way origami does, to trisect an arbitrary angle. In doing so Dayoub and Lott use the Mira to perform an operation very similar to O7 in origami. Martin then refines this method for the Mira, publishing his own paper on Mira constructions that contains exactly the form of operation O7 where the two given lines are perpendicular, whereupon Martin proves that the Mira can be used to construct cube roots (Martin, 1979). So it could be that Martin was inspired by the Mira to devise his version of operation O7 for origami.

Both Justin's paper (Justin, 1986b) and the Huzita–Scimemi paper (Huzita and Scimemi, 1989) were published in hard-to-find publications, or at least in periodicals that weren't indexed in the standard mathematical abstracts. This is likely why subsequent work, like (Auckly and Cleveland, 1995; Geretschläger, 1997a; Alperin, 2000; Hatori, 2003) make no mention of Messer, Justin, or Scimemi (or Beloch, for that matter).

# 2 Solving Equations via Folding

That origami is actually solving equations when we fold is the most powerful concept for developing folding procedures for a given construction. The fact that origami can solve cubic equations, making it more powerful than the straightedge and compass, was first discovered by the Italian mathematician Margherita Beloch in the 1930s (Beloch, 1936). In this chapter we will explore the multitude of ways in which the solving of equations, and what kind, can be done by straight-crease, single-fold origami.

## 2.1 Folding Conics

We begin by analyzing the basic origami operation O6:

**O6:** Given two points $P_1$ and $P_2$ and a line $L$, we can, whenever possible, make a fold that places $P_1$ onto the line $L$ and passes through the point $P_2$.

There is a very easy and delightful activity that demonstrates exactly what this folding operation does. In fact, this activity has been used as a classroom exercise for geometry, algebra, and even calculus classes for more than a century. (See (Row, 1901, p. 116) and (Lotka, 1907; Yates, 1943; Bruckheimer and Hershkowitz, 1977; Smith, 2003).)

The activity is illustrated in Figure 2.1. Take a piece of paper and mark a point $P_1$ somewhere below the center of the paper. Let $L$ be the bottom side of the paper. Then perform operation O6 over and over again, folding $P_1$ to many different places on $L$. (You can think of this as choosing $P_2$ to be at various spots on the left and right sides of the paper.) This will create numerous crease lines on the paper, and if enough of them are made, they begin to outline the form of a parabola.

To be more precise, we have the following theorem:

**Theorem 2.1** *When folding a point P to a line L, the crease line made will be tangent to the parabola with focus P and directrix L.*

There are many ways to prove this. We first provide a synthetic argument.

*Proof* Let $P'$ be the point on $L$ to which $P$ is folded and call the crease line $C$. Let $X$ be the point on $C$ so that $\overline{XP'} \perp L$. Since the act of folding reflects one side of the

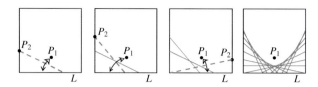

**Figure 2.1** Activity exploring operation O6.

paper across $C$, and $X \in C$, we have that $\overline{XP}$ reflects onto $\overline{XP'}$ when the fold is made. Thus $|XP| = |XP'|$, which means that $X$ is a point on the parabola with focus $P$ and directrix $L$.

Now let $Q$ be another point on $C$, $Q \neq X$. By the same argument as in the previous paragraph, since $Q \in C$ and $P$ reflects to $P'$ under the fold, we have $|QP| = |QP'|$. Since $Q$ is not $X$, $\overline{QP'}$ is not perpendicular to $L$. Thus $|QP|$ cannot equal the perpendicular distance from $Q$ to $L$. This proves that $X$ is the only point on $C$ that is also on the parabola with focus $P$ and directrix $L$, and therefore $C$ is tangent to this parabola. $\square$

Theorem 2.1 can also be proven by calculating the equation of the curve generated by repeatedly folding a point to a line. Let $P = (0, 1)$ and $L$ be the line $y = -1$ in the real plane. Suppose that we fold $P$ to the point $P' = (t, -1)$ on $L$. Following the same method used in analyzing the heptagon fold in Section 1.4, we see that our crease line will have slope $t/2$ and will pass through the point $(t/2, 0)$. Thus our crease line equation is

$$y = \frac{t}{2}x - \frac{t^2}{4}. \tag{2.1}$$

This is a parameterized family of lines, and we want to find the envelope of this family (a curve that is tangent to all the members of the family, see (Cox et al., 2005, Chapter 3.4)). Setting $F(x, y, t) = y - tx/2 + t^2/4$, the envelope is given by simultaneously solving

$$F(x, y, t) = 0 \quad \text{and} \quad \frac{\partial}{\partial t}F(x, y, t) = 0.$$

Thus we have $(\partial/\partial t)F(x, y, t) = x/2 - t/2 = 0$, or $x = t$. Plugging this into the line equation, we get $y = x^2/2 - x^2/4$, or $y = x^2/4$, a parabola.

Another way to obtain this equation would be while addressing a question posed in Diversion 1.6. That is, what choices for the point $P_2$ make operation O6 impossible? In other words, we are asking what region of the paper will not have any creases as we repeatedly fold $P_1$ to $L$? We answer this by supposing that we have a point $P_2 = (x, y)$ and we want to find a value of $t$ that makes our crease line in Equation (2.1) pass through $P_2$. All we have to do is solve Equation (2.1) with respect to $t$. The quadratic formula gives us

$$t = 2\left(\frac{x}{2} \pm \sqrt{\frac{x^2}{4} - y}\right).$$

This will not have a solution when the discriminant is less than zero: $x^2/4 - y < 0$. The boundary of this region is our parabola $y = x^2/4$, and all points above this curve are "bad" choices for $P_2$.

Seeing that operation O6 generates tangents to a parabola, one naturally wonders if other conic section tangents can be folded. The answer is yes, but before describing how, we'll give an informal motivation.

One can obtain a parabola from an ellipse in the following way: Recall that an ellipse is the set of all points $P$ such that the sum of the distances between $P$ and two fixed foci equals a constant called the major axis of the ellipse. Fix a focus of the ellipse as well as the vertex point closest to that focus and let the other focus slide off to infinity. The result will be a parabola. This process is equivalent to altering the angle at which a plane intersects a cone to go from elliptical intersection to a parabolic one. (This fact was known as far back as Kepler (1604). See also (Hilbert and Cohn-Vossen, 1956, pp. 3–4).)

Now consider an ellipse with foci $F_1$ and $F_2$ determined by the constant distance (major axis) $r$. Draw a circle $C$ with center $F_2$ and radius $r$. This circle has some useful properties, described in the following theorem.

**Theorem 2.2**    *Using the ellipse and circle described previously, let $P$ be a point on the ellipse and $Q$ be the point where the extended ray $F_2P$ meets the circle $C$. Then*

(i) *the shortest distance between $P$ and a point on the circle $C$ is given by the segment PQ, and*

(ii) *the tangent line to the ellipse at $P$ reflects $F_1$ onto the point $Q$.*

*Proof*    For (i), suppose that a different point $Q' \neq Q$ on $C$ has a length $PQ'$ less than that of $PQ$. Then we would have $|F_2P| + |PQ'| < r = |F_2Q|$, which violates the triangle inequality.

For (ii), notice that since $|F_2Q| = |F_2P| + |F_1P| = r$, we have $|F_1P| = |PQ|$. Also, a property of tangents to an ellipse is that they form equal angles with the lines from the point of tangency to the two foci (see (Hilbert and Cohn-Vossen, 1956, p. 4)). In the notation of Figure 2.2, this means that $\angle F_1PT_1 = \angle F_2PT_2$. But $\angle F_2PT_2$ and $\angle T_1PQ$ are vertical angles and thus are equal. Therefore $\angle F_1PT_1 = \angle T_1PQ$,

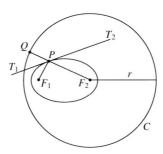

**Figure 2.2**  An ellipse and its folding circle.

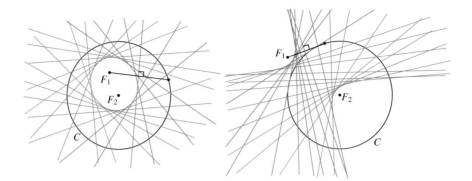

**Figure 2.3** Folding tangents to an ellipse and a hyperbola.

which means that the tangent at $P$ bisects $\angle F_1PQ$. This combined with $|F_1P| = |PQ|$ settles (ii). $\qquad\square$

We now ask what happens to the circle $C$ as we drag the focus $F_2$ to infinity, leaving the focus $F_1$ as well as the ellipse vertex closest to $F_1$ fixed. As the major axis of the ellipse grows, so will the radius of the circle, but one point of the circle, the point closest to $F_1$ where the circle intersects the line containing the major axis, will also remain fixed. Thus as $F_2$ moves farther and farther away, the circle will appear more and more like a straight line near the focus $F_1$. In the limit, we'll have that $F_1$ will be the focus of a parabola and the circle $C$ will have turned into the directrix of that parabola.

In other words, and in light of Theorem 2.2, the circle $C$ and focus $F_1$ play that same role for the ellipse as the directrix and focus play when folding tangents to a parabola. Thus if we draw a circle $C$ with center $F_2$ on our paper, and pick a point $F_1$ inside $C$, then repeatedly folding $F_1$ onto $C$ will generate the tangents to the ellipse with foci $F_1$ and $F_2$ and major axis constant equal to the radius of $C$, as shown in Figure 2.3 on the left. (In practice this is easier to do if one starts with a circular piece of paper, picks a point inside at random, and then folds the paper's circular boundary to the point over and over again.)

Similarly, the tangents to a hyperbola can be folded if we choose the focus $F_1$ to be **outside** the circle $C$. See the right side of Figure 2.3. In fact, a hyperbolic version of Theorem 2.2 can be proven using exactly the same argument as for the ellipse.

---

**Diversion 2.1** Derive the equations for an ellipse or hyperbola generated by these folding methods. For example, prove that if we fold the point $(1, 0)$ over and over again to the circle $(x + 1)^2 + y^2 = 16$, then the crease lines will be tangent to an ellipse (and find the equation of this ellipse). Hint: Parameterize the circle and then use the substitutions $\sin\theta = \frac{2t}{1+t^2}$ and $\cos\theta = \frac{1-t^2}{1+t^2}$ to obtain a trigonometry-less parameterization.

---

We have seen how folding a point to a line generates tangents whose envelope is a parabola and folding a point to a circle generates tangents whose envelope is an ellipse or hyperbola (or circle, if the point is the circle's center). In general, we can let $C$ be a curve in the plane and ask what curve will be the envelope of the crease lines made when folding a fixed point to $C$ repeatedly. Rupp (1924) addressed this question and showed that this process generates a copy of the negative pedal of the curve $C$ scaled down by a factor of 1/2.

The **negative pedal** of a curve $C$ is determined as follows: Given a fixed point $O$ (called the **pedal point**) and a point $P$ on the curve, let $L_P$ be the line through $P$ that is perpendicular to $OP$. The envelope of the lines $L_P$ over all $P \in C$ is the negative pedal curve of $C$ (Lockwood, 1967). To see why this relates to our operation of folding a point $O$ repeatedly to a curve $C$, notice that if $P \in C$ is the point to which we fold $O$, the locus of the midpoints $M$ of $OP$ (over all $P \in C$) forms a copy of the curve $C$ dilated toward the point $O$ by a factor of 1/2. The crease lines we fold would be perpendicular to the segments $OM$, and thus their envelope would form the negative pedal curve of this smaller copy of $C$. Therefore all the classic results for negative pedal curves would apply (Lockwood, 1967, pp. 156–159). For example, if $C$ were a cardioid and $O$ the cusp point, then folding $O$ to $C$ repeatedly will create creases tangent to a circle.

However, the major result from all this is that operation O6 is generating tangent lines to a parabola. That is, when we fold a point $P$ to a line $L$, the resulting crease will be tangent to the parabola with focus $P$ and directrix $L$. This hints at a connection between operation O6 and solving quadratic equations, which will be discussed in Section 2.2.

This also gives us a way to interpret origami operation O7, where we simultaneously fold two points to two lines. Each point folding to a line determines the focus and directrix of a parabola, and the crease line made will be tangent to both. Thus operation O7 is finding a common tangent to two parabolas. The implications of this will be explored in Section 2.3.

We end this section with some comments on the literature available for folding conic tangents. T. Sundara Row (1901, Note 235) seems to be the first to call attention to the fact that folding a point to a line produces a parabola tangent, although the proof given there is not the most elegant. Virtually all of the later publications, however, cite Row's work as their key influence. Lotka (1907) expanded Row's discovery to the case of folding a point to a circle, providing a very nice analytical method that simultaneously proves that the resulting envelopes are ellipses and hyperbolas. (His approach is similar to the one given here for the parabola, where we used the discriminant of a well-chosen quadratic to determine which points will not be hit by any crease lines.) Rupp (1924) seems to be the first to successfully analyze what happens when we fold a point to a general curve. Yates (1943), citing Row's and Lotka's work, provides proofs for the parabola, ellipse, and hyperbola that are the most concise and elegant that the author has seen.

Other publications have emphasized the educational potential of the origami-conic connection. The NCTM booklet *Paper Folding for the Mathematics Class*

(Johnson, 1957) presents minimal details and incomplete proofs. (Fehlen, 1975) presents the material through pedagogy in a style that leaves the proofs incomplete. (Bruckheimer and Hershkowitz, 1977) gives a very straightforward proof of the parabola case. (Scher, 1996) gives incomplete proofs but calls attention to how these folding activities can be dramatically modeled with computer software like Geometer's Sketchpad or GeoGebra. (Smith, 2003) offers alternate analytic proofs but doesn't cite any previous work. The author tried his hand at specifically framing this material for the classroom (Hull, 2012, Activity 6).

There are undoubtedly other references on operation O6 (we chose to not mention references that only state the folding exercise, without any attempt at a proof). In fact, it seems that numerous times people have completely re-invented the wheel on this topic. For example, during the years 1970–2000 there were at least four articles in *Mathematics Teacher* on folding tangents to conics, none of which reference each other or the previous work listed above. Despite the inherent simplicity and charm of exploring conics via operation O6, this has been, as Yates puts it, "unfortunately relegated to the limbo" (Yates, 1943, p. 230).

## 2.2    Solving Second Degree Equations

At this point it is fairly evident that origami should be able to solve second degree equations. The previous section provides good evidence for this. Also, the operations of a straightedge and compass seem to be mimicked by origami. Origami operation O1 simulates a straightedge, and O6 can be thought of as doing the work of a compass as follows: since the point $P_2$ lies on the crease line, it can be thought of as remaining fixed, like the center of a circle. Then $P_1$ is folded onto the line $L$, and in doing so the length of the segment $P_1P_2$ is preserved as $P_1$ is moved to its new location. Thus the image of $P_1$ under folding operation O6 is the same as finding a point of intersection between the line $L$ and a circle centered at $P_2$ with radius $|P_1P_2|$. This provides the intuition, at least, that origami constructions should be able to do everything that SE&C can do, and since the field of numbers constructible under SE&C is precisely those that are solutions to quadratic equations, origami should be able to find such solutions as well.

Actually doing this is another matter. As we shall see in Chapter 3, where we construct the field of origami constructible numbers, it is relatively easy to, given origami constructible nonzero lengths $a$ and $b$, construct any combination of these lengths using the operations addition, subtraction, multiplication, division, and square roots. Therefore the output of the quadratic formula can be constructed, giving us any real solutions to quadratics with rational coefficients.

However, in light of our work in Section 2.1, the fact that we can fold tangents to conics is too tempting to not apply to this problem.

Suppose that we want to find roots, if they exist, of $y = x^2 + ax + b$, where $a, b \in \mathbb{Q}$. By locating the focus and directrix of this parabola, we can try to pick a well-chosen tangent to fold and solve our problem. After enacting a change of coordinates to translate the parabola's vertex to the origin, we can use standard methods to see that the focus will be at the point $P_1 = (-a/2, b - a^2/4 + 1/4)$ and the directrix line $L$ will be $y = b - a^2/4 - 1/4$. By the same methods we used in Sections 1.4 and 2.1, we find that if we fold $P_1$ to an arbitrary point $(t, b - a^2/4 - 1/4)$ on $L$, then the resulting crease line will have equation $y = (2t + a)x - t^2 + b$. If we take $t = (-a + \sqrt{a^2 - 4b})/2$, then the crease will be tangent to the parabola at a root, and the intersection of the crease and the $x$-axis (which we assume is a constructed line) will locate the root for us. Substituting this value of $t$ into our line equation gives us the rather horrendous line equation

$$y = \sqrt{a^2 - 4bx} + \frac{a}{2}\sqrt{a^2 - 4b} + 2b - \frac{a^2}{2}.$$

But now we see that if we let $x = -a/2$ on this line, the square root terms will cancel and give us $y = 2b - a^2/2$. Therefore we can pick $P_2 = (-a/2, 2b - a^2/2)$, which is easy to construct from the coefficients of our original quadratic, and perform origami operation O6 to fold $P_1$ onto $L$ making the crease go through $P_2$; then, the intersection of the resulting crease and the $x$-axis will be a root.

The other root could have been found by taking $t = (-a - \sqrt{a^2 - 4b})/2$. Of course, if the vertex of the parabola, $(-a/2, b - a^2/4)$, is above the $x$-axis, then no real roots exist and the results of this origami method will be meaningless.

Figure 2.4 shows how this works to find a root of $x^2 - x - 1 = 0$. We pick $P_1 = (1/2, -1)$, $P_2 = (1/2, -5/2)$, and $L$ the line $y = -3/2$ and perform operation O6.

---

**Diversion 2.2**   Use the example in Figure 2.4 to solve Diversion 1.5 (constructing a regular pentagon via origami). The fact that $\cos(2\pi/5) = (1/2)/((1 + \sqrt{5})/2)$ might help.

---

Many other folding methods for solving quadratics exist, and readers are encouraged to discover their own. In fact, one way to develop origami constructions

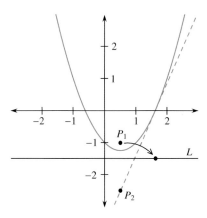

**Figure 2.4** Folding to find a root of $y = x^2 - x - 1$.

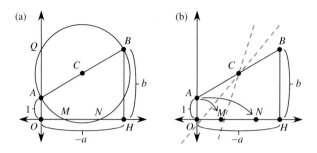

**Figure 2.5** Lill's construction for solving $x^2 + ax + b = 0$.

of various things is to find a non-origami construction (say, using SE&C or other tools) and then adapt them to paper folding. Alperin's origami heptagon construction in (Alperin, 2002) does just this, using Gleason's trisection construction in (Gleason, 1988) as a guide.

We can take a similar approach with an elegant SE&C construction for solving quadratics due to Lill and described in (Dickson, 1904). If we wish to solve $x^2 + ax + b = 0$ for $a, b \in \mathbb{Q}$, let $O$ be the origin and construct $A = (0, 1)$, $H = (-a, 0)$, and $B = (-a, b)$. Then draw a circle with diameter $AB$ centered at the midpoint $C$ of $AB$. If $M$ and $N$ are the points where this circle intersects the $x$-axis (see Figure 2.5(a)), then the lengths $|OM|$ and $|ON|$ (if they exist) will be solutions to $x^2 + ax + b = 0$.

---

**Diversion 2.3**   Prove that Lill's construction works. Hint: First prove that $|HB| = |OQ|$, $|OM| = |NH|$, and $|OM||ON| = |OA||OQ| = b$. Then note what $|OM| + |ON|$ is to get the result.

---

This method is easy to convert to paper folding by noting that the midpoint $C$ of $AB$ is obtained by folding $A$ to $B$ (operation O3). Then $M$ and $N$ can be found using O6, folding so that $A$ lands on the $x$-axis and the crease passes through the point $C$, as shown in Figure 2.5(b).

## 2.3   Solving Third Degree Equations

That origami can solve general cubic (third degree) equations was hinted at in the angle trisection and regular heptagon constructions of Chapter 1. Both of these constructions utilized origami operation O7, which we saw in Section 2.1 was equivalent to the geometric problem of finding a common tangent to two parabolas. Now, in the real projective plane, two conics will have at most four common tangents. In the case of two parabolas one of these tangents will be the line at infinity. Therefore there may be at most three common tangents that operation O7 could find for us, providing more evidence that such an origami operation is solving some kind of cubic equation.

(For more details of how origami constructions can be viewed on the projective plane, see Section 3.3.)

The purpose of this section is to present a general method for solving cubic equations via origami devised by Margherita Piazzolla Beloch (1936) in the 1930s and describe its relation to other origami methods for solving cubics that have been developed since.

Beloch's insight was to adapt an ingenious, non-origami construction for solving arbitrary polynomials with rational coefficients created by Lill in the 1860s (Lill, 1867). (See (Riaz, 1962) or (Hull, 2011) for a more easily obtainable, and in English, reference.) Lill's method proceeds as follows: Suppose that we want to locate a root of the equation

$$a_n x^n + a_{n-1} x^{n-1} + \cdots + a_1 x + a_0 = 0, \tag{2.2}$$

where the coefficients $a_i$ are rational. We will construct a real solution for this geometrically, if one exists, borrowing the imagery of "turtle graphics" from the computer language Logo. Imagine a turtle sitting at the origin, pointing in the direction of the $x$-axis, with its head in the positive direction. The turtle will move forward a distance of $a_n$, leaving a straight line trail as it does so and adopting the convention that if $a_n$ is negative then the turtle will move backward. Then the turtle will rotate by 90° counterclockwise and repeat this process with the coefficient $a_{n-1}$. Thus the turtle will travel up if $a_{n-1} > 0$ and down if $a_{n-1} < 0$. If $a_{n-1} = 0$ the turtle will remain in place. Then the turtle rotates by 90° counterclockwise again and we repeat. This continues for all the coefficients $a_i$. Let $T$ be the final location of the turtle. See Figure 2.6 for examples where all the coefficients are positive and some are negative.

More precisely, we can think of the turtle's **n-order path** as being made by connecting the following points in sequence:

$$O = (0,0),\ (a_n, 0),\ (a_n, a_{n-1}),\ (a_n - a_{n-2}, a_{n-1}),\ (a_n - a_{n-2}, a_{n-1} - a_{n-3}), \ldots,$$

$$\left( \sum_{k=0}^{\lfloor n/2 \rfloor} (-1)^k a_{n-2k},\ \sum_{k=0}^{\lfloor n/2 \rfloor} (-1)^k a_{n-2k-1} \right) = T,$$

where we might need to use $a_{-1} = 0$ to make the last summations end correctly.

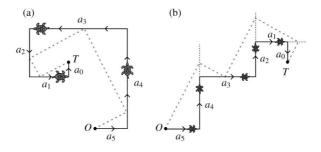

**Figure 2.6** Lill's method applied to quintics where (a) all coefficients are positive and (b) $a_3, a_2, a_0 < 0$ and the rest are positive.

We then construct an $(n-1)$-order path beginning at $O$ and ending at $T$ as follows: Position our turtle back at $O$, and this time point it at a yet-to-be-determined angle $-90° < \theta < 90°$ from the previous $a_n$ line. Let the turtle march in this direction until it comes to the infinite line containing the $a_{n-1}$ line segment. (Note that depending on $\theta$, the turtle might not hit the actual $a_{n-1}$ segment, so the extended line may be needed.) Then rotate the turtle by $\pm90°$ so that it faces the next $a_{n-2}$ line. Let it march again until it hits the line containing $a_{n-2}$. Then rotate by a right angle again and repeat this process. In the end the turtle will hit the line containing the $a_0$ segment, and now we get to see what our choice of $\theta$ should have been. Our aim is to choose a value $\theta$ that makes the turtle arrive in the end at point $T$. (This is illustrated by the dotted line paths in Figure 2.6.) If such a value of $\theta$ exists, then it will lead us to a solution of our original polynomial (2.2).

**Theorem 2.3** (Lill, 1867)  *In the above process, the value $x = -\tan\theta$ will be a root of $a_n x^n + a_{n-1} x^{n-1} + \cdots + a_0$.*

*Proof*  Since the turtle turns $90°$ at each corner, the $n$-order path and the $(n-1)$-order path form a sequence of right triangles all of which have an angle $\theta$ in common and are thus similar. Label the corners of the $n$-order path $P_n, P_{n-1}, \ldots, P_1$ and the corners of the $(n-1)$-order path $Q_{n-1}, Q_{n-2}, \ldots, Q_1$. An example where $n = 5$ is shown in Figure 2.7.

Let $x = -\tan\theta$ and consider the length of the side opposite $\theta$ in each of these similar triangles, $P_i Q_{i-1}$ for $i = n, \ldots, 2$ and $P_1 T$. We obtain

$$|P_n Q_{n-1}| = -a_n x, \ |P_{n-1} Q_{n-2}| = -x(a_{n-1} - |P_n Q_{n-1}|) = -x(a_{n-1} + a_n x),$$
$$|P_{n-2} Q_{n-3}| = -x(a_{n-2} - |P_{n-1} Q_{n-2}|) = -x(a_{n-2} + x(a_{n-1} + a_n x)), \ldots$$
$$|P_1 T| = -x(a_1 + x(a_2 + x(a_3 + \cdots + x(a_{n-1} + a_n x) \cdots ))).$$

Since $|P_1 T| = a_0$, we have that $a_0 = -a_1 x - a_2 x^2 - a_3 x^3 - \cdots - a_{n-1} x^{n-1} - a_n x^n$. Therefore $x = -\tan\theta$ is a solution to Equation (2.2). $\square$

Finding the proper angle $\theta$ to make this work is certainly not something that normal construction tools (compass, marked straightedge, what-have-you) can handle in general. As seen in Figure 2.5 in the previous section, the case where $n = 2$ can easily be modified for SE&C, but Lill did offer another suggestion for handling the cases

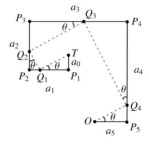

**Figure 2.7**  Proving Lill's method works in a case where $n = 5$.

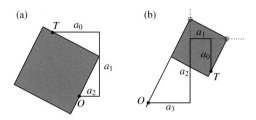

**Figure 2.8** Lill's use of transparent graph paper to find a solution square for the (a) $n = 2$ and (b) $n = 3$ cases.

where $n = 2$ or 3. He proposed, after drawing the initial 2- or 3-order path, overlaying it with translucent graph paper (with a fine mesh) and rotating it until the points $O$ and $T$ both lie on sides of a square (or on lines that contain a side of the square) and some of this square's corners lie on lines that contain a segment $a_i$. For the $n = 2$ case, we would want one corner of the square to lie on the $a_1$ segment (or a line that contains $a_1$); see Figure 2.8(a). In the cubic $n = 3$ case we would want one corner of the square to lie on the $a_1$ line and an adjacent corner to lie on the $a_2$ line. We'd also want, for the cubic case, the points $O$ and $T$ to lie on lines containing opposite sides of the square. See Figure 2.8(b) for an example. We call this square, in the proper position in the plane, the **solution square** for Lill's method in the $n = 2$ or $n = 3$ cases.

Of course, Lill's method won't always work. If no real solutions exist for the polynomial, then no angle $\theta$ will work, and no turning or positioning of our translucent graph paper will produce a solution square.

---

**Diversion 2.4**   Prove that the $(n - 1)$-order path generated by Lill's method to solve Equation (2.2) is a scalar multiple of the path made by the degree-$(n - 1)$ polynomial obtained by factoring $(x + \tan\theta)$ from Equation (2.2). In this way, Lill's method represents geometrically the process of successively factoring out the real roots from a polynomial.

---

It was the cubic case of Lill's method that Beloch discovered could be performed by paper folding. She states in (Beloch, 1936) that she developed this work while teaching a course on geometry, where she no doubt presented Lill's method, at the University of Ferarra, Italy, in the 1933–34 academic year. Beloch's technique is illustrated in Figure 2.9 and proceeds as follows: Given a cubic $a_3x^3 + a_2x^2 + a_1x + a_0$ with rational coefficients, we construct the 3-order path according to Lill's method, and our aim is to find the proper angle $\theta$, if it exists, to give us a 2-order path connecting $O$ and $T$. First construct a line $D_1$ at $x = 2a_3$ and another line $D_2$ at $y = a_2 + a_0$.

**Theorem 2.4** (Beloch, 1936)   *Using the above notation, if the origami operation O7, folding the point $O$ to the line $D_1$ and the point $T$ to the line $D_2$, is possible, then the line segment that the extended $a_2$ and $a_1$ lines cut out of the resulting crease will be one side of the solution square for Lill's method.*

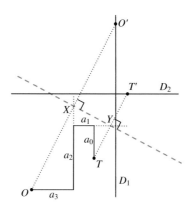

**Figure 2.9** Beloch's use of operation O7 to solve a cubic.

*Proof*   The proof is very simple once one understands the reasoning behind the location of the lines $D_1$ and $D_2$. Suppose that we wanted to consider a parabola with focus $O = (0,0)$ and vertex $P_3 = (a_3,0)$. Then the directrix of this parabola would be the line $x = 2a_3$, which is $D_1$. Therefore, any fold that places $O$ onto $D_1$ will be tangent to this parabola. Furthermore, if $O'$ is the image of $O$ on $D_1$ after the folding, then the midpoint $X$ of $OO'$ will be on the line containing the $a_2$ segment.

Similarly, consider the parabola with focus $T = (a_3 - a_1, a_2 - a_0)$ and vertex $P_1 = (a_3 - a_1, a_2)$. The directrix of this parabola will be the line $y = a_2 + a_0$, which is $D_2$. When we fold $T$ onto a point $T'$ on line $D_2$, the resulting crease will be tangent to this second parabola, and the midpoint $Y$ of $TT'$ will lie on the line containing the $a_1$ segment. See Figure 2.9.

When we fold $O$ onto $D_1$ and $T$ onto $D_2$ using operation O7, we have that the points $X$ and $Y$ will be on this crease line. Thus the segment $XY$ is perpendicular to $OX$ and $TY$, and thus $XY$ is the side of a square with two adjacent corners (the points $X$ and $Y$) on the $a_2$ and $a_1$ lines. Two (extended) opposite sides of this square contain the points $O$ and $T$, and therefore this is a solution square for Lill's method.   □

---

**Diversion 2.5**   Determine the operation O7 fold needed, and the 3-order path it generates for Lill's method, to find one of the three real solutions of $x^3 - 7x - 6 = 0$ using Beloch's approach. (Note that the lack of an $a_2$ term doesn't change the procedure; the $a_2$ segment will merely have zero length and the "line containing $a_2$" will still be perpendicular to the $a_3$ and $a_1$ lines. See (Riaz, 1962) or (Hull, 2012) for hints.)

---

It seems that Beloch's work on origami constructions laid in obscurity for many years. While attention was drawn to it in the early years of the origami mathematics community (see (Huzita and Scimemi, 1989) in particular), the methods of trisecting angles by Abe and Justin and Messer's cube root of two construction (Messer, 1986) were done without knowledge of Beloch's work. Furthermore, the reference

(Huzita and Scimemi, 1989) was in a very hard-to-find publication, and the more recent constructions of (Alperin, 2000) and (Hatori, 2003) for solving general cubics were made without any knowledge of Beloch. It is fascinating, then, to learn that both of these modern solutions are virtually identical to Beloch's approach, although without the reference to Lill.

In Hatori's construction (Hatori, 2003) we begin with an arbitrary cubic $x^3 + ax^2 + bx + c = 0$ and consider the parabola with focus $P_1 = (a, 1)$ and directrix $L_1: y = -1$ and the parabola with focus $P_2 = (c, b)$ and directrix $L_2: x = -c$.

---

**Diversion 2.6**   Prove that in Hatori's construction the slope of the crease line we obtain (if it exists) by folding $P_1$ onto $L_1$ and $P_2$ onto $L_2$ using origami operation O7 will be a real solution to $x^3 + ax^2 + bx + c = 0$.

---

Figure 2.10 shows how Hatori's construction can be interpreted as equivalent to Beloch's. Indeed, if we reflect this picture about the line $y = -x$ and translate $P_1$ to the origin, it is exactly what Beloch's application of Lill's method would generate.

Alperin's construction (Alperin, 2000) begins with the equivalent form of a general cubic $x^3 + ax + b = 0$. (See (Hull, 2012, Activity 6) or (Weisstein, n.d.) for an explanation.) He then asks us to consider the parabolas

$$y = \frac{1}{2}x^2 \quad \text{and} \quad \left(y - \frac{1}{2}a\right)^2 = 2bx.$$

The first one has focus $P_1 = (0, 1/2)$ and directrix $L_1: y = -1/2$. The second has focus $P_2 = (b/2, a/2)$ and directrix $L_2: x = -a/2$. These foci are the same that Beloch's method would give, except that they are rotated 90° and scaled down by 1/2, which is the equivalent of multiplying the cubic equation by 1/2. Figure 2.11 illustrates this, but the application of Lill's method can be hard to discern due to the lack of an $x^2$ term in the polynomial. One should think of the "line segment of length $a_2$" as being there, parallel to (actually, along) the $x$-axis but of zero length. (Thus the line "containing" the $a_2$ segment is the $x$-axis.)

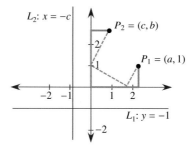

**Figure 2.10**  Hatori's cubic solution viewed through the lens of Beloch/Lill.

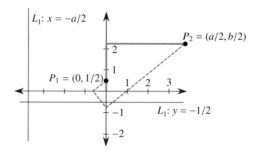

**Figure 2.11** Alperin's cubic solution viewed through the lens of Beloch/Lill.

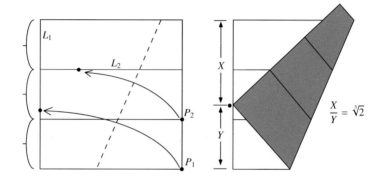

**Figure 2.12** Peter Messer's cube root of two construction.

---

**Diversion 2.7** Prove that in Alperin's construction the slope of the crease line we obtain (if it exists) by folding $P_1$ onto $L_1$ and $P_2$ onto $L_2$ using origami operation O7 will be a real solution to $x^3 + ax + b = 0$.

---

**Diversion 2.8** Use Beloch's method to devise a paper-folding way to construct $\sqrt[3]{2}$. (You're likely to come up with a similar folding procedure to that of Hatori (2003).)

---

Given how all these methods of solving cubics via origami are really the same at heart, it is surprising to then find one that is different. Peter Messer's cube root of two construction (Messer, 1986) does not fit as easily into the Beloch/Lill strategy; see Figure 2.12. Messer does utilize two parabolas whose directrices are perpendicular, as Beloch does. But if we were to use Beloch's method to solve $x^3 - 2 = 0$, our 3-order path would start with a horizontal segment of length 1, then two segments (vertical then horizontal) of length 0, and then a vertical segment of length 2. There is no way the points $O$ and $T$ would share the same $x$- or $y$-coordinate, as do Messer's points $P_1$ and $P_2$. However, Messer's method finds two lengths whose **ratios** give $\sqrt[3]{2}$, as opposed to a length obtained by a slope of the turtle path from Lill's method.

Nonetheless, there should be a way to interpret Messer's construction in terms of Beloch and Lill.

---

**Diversion 2.9** Prove that Messer's construction works, either using straightforward geometry or by finding an appropriate polynomial that Lill's method and Beloch's fold O7 can solve to generate Messer's construction. (Hint for the latter: Consider a cubic with $-(\sqrt[3]{2}+1)$ as a root.)

---

## 2.4 Cubic Curves and Beloch's Fold

A different way to see that BOO O7, which is sometimes called **Beloch's fold**, is solving a cubic equation is to actually compute the cubic curve it generates. To do this, start with a point $P_1$ and a line $L_1$. In fact, by applying a suitable affine transformation, we may assume that $P_1 = (0, 1)$ and $L_1$ is the line $y = -1$. As we saw in Section 2.1, if we fold $P_1$ to the point $(t, -1)$ on $L_1$, we find that $y = (t/2)x - t^2/4$ is the equation of our crease line.

Now let $P_2 = (a, b)$ be the second point in BOO O7. We wish to examine the image of $P_2$ after folding $P_1$ onto $L_1$.

---

**Diversion 2.10** Prove that if $(x, y)$ is the image of $P_2$ after folding $P_1$ onto $L_1$, then $(x, y)$ must satisfy the following cubic:

$$(y + b)(y - b)^2 = -(x^2 - a^2)(y - b) - 2(x - a)^2. \qquad (2.3)$$

---

The curve that the image of $P_2$ traces when $P_2 = (0.5, -0.5)$ is shown in Figure 2.13. If we were to then draw the line $L_2$, onto which $P_2$ is supposed to be folded in

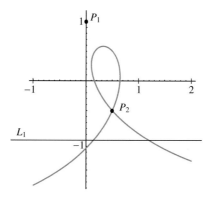

**Figure 2.13** The locus of the image of $P_2 = (0.5, -0.5)$ as $P_1$ is folded onto $L_1$ traces a cubic curve.

O7, we see that any point where $L_2$ intersects this cubic curve (2.3) will be a possible place to fold $P_2$ and still have $P_1$ folding onto $L_1$. In other words, O7 is finding a specific point on a cubic curve, which is solving a cubic equation.

**Remark 2.5**    In closing this section, we note that it is also possible to solve quartic equations using the BOOs O1–O7. This follows from the fact that any quartic equation can be factored into two quadratics or a cubic and a linear term. Thus, in theory origami can solve quartics, but this theory does not immediately provide instructions as to how it can be done. Indeed, solving quartics can be quite arduous. Robert Geretschläger performed this legwork and presented it at an AMS Special Session on mathematical origami at the Joint Mathematics Meetings in Baltimore, MD, 1998 (see (Geretschläger, 1998)). B. Carter Edwards and Jerry Shurman developed a more elegant method that takes advantage of projective geometry (Edwards and Shurman, 2001).

## 2.5    Historical Remarks

While the publication record seems to indicate that Eduard Lill developed his method by himself in the mid-1800s, it is quite amazing that a precursor of the cubic case of Lill's method was developed by the ancient Greeks. Eratosthenes of Cyrene (276–194 BC) developed several mechanical ways to approximate the solutions of cubic equations as a tool to help in the design of catapults, and one of his methods involves drawing the turtle path for the cubic and then using a wooden contraption made of three right angles, one side of which is expandable, in the same way as the solution square of Lill, as in Figure 2.8. See (Soedel and Foley, 1979) for more information. This method of Eratosthenes, however, was limited to solving cubics. Lill seems to be the first to generalize the idea to polynomials of arbitrary degree.

Interest in geometric constructions that can be performed by paper folding has a long and varied history. In nineteenth-century Europe, some people first encountered such folding geometry in the work of Friedrich Froebel (Heerwart, 1920; Liebschner, 1992), where he suggested using paper folding to educate young children on the topics of symmetry and geometry. In Japan several ancient **sangaku** (wooden tablets that contain geometry problems and were hung in Shinto shrines for visitors to ponder) dating back to the early 1800s have been found that deal with origami geometry; see (Fukagawa and Pedoe, 1989, p. 37). But the first serious work attempting to collect paper folding construction techniques for educational and research value was T. Sundara Row's book *Geometric Exercises in Paper Folding* (Row, 1901), first published in India in 1893 and then in the West in 1901. Felix Klein helped popularize Row's work by making explicit reference to it in a number of his books, such as (Klein, 1897, p. 42). Beloch (1936) suggested how popular paper folding geometry was becoming due to the "*autorevole giudizio*" (authoritative judgment) passed on by Klein.

In 1930 the Italian historian and mathematics devotee Giovanni Vacca wrote a paper summarizing some origami history and, apparently, the mathematics known about paper folding geometry at the time (Vacca, 1930). No mention is made in this paper about the possibility of origami being able to perform constructions beyond those of straightedge and compass. Row, in fact, states explicitly that it is impossible to "double the cube" (i.e., construct $\sqrt[3]{2}$) via origami (Row, 1901, Note 112). Therefore it does seem that Beloch was the first to realize that origami could perform operation O7 to construct the simultaneous tangent to two parabolas and thus solve general cubic equations, which is why it is referred to as Beloch's fold. To cap off this achievement, she includes in (Beloch, 1936) a paper folding construction method for the $\sqrt[3]{2}$, which is coincidentally virtually the same as the one independently discovered by Hatori (2003) nearly 70 years later.

If origami geometry enjoyed a period of popularity amongst mathematicians in 1930s Italy, it didn't seem to last long or receive much notice in the greater community. With the exception of mathematicians Benedetto Scimemi and Humiaki Huzita, no researchers in this area from the 1970s to the early 2000s seemed aware of Beloch's contributions.

More recent work has, however, distinguished itself by placing origami constructions in the context of algebra. This is the subject of the next chapter.

# 3    Origami Algebra

In this chapter we will look at the topic of origami geometric constructions from an algebraic viewpoint. In the previous chapter we saw how origami can find real roots of quadratic and cubic equations. However, if our paper is the complex plane $\mathbb{C}$, then we can locate complex roots as well. Our main goal is to determine what subfield of $\mathbb{C}$ is constructible via paper folding. For now, we continue to restrict ourselves to crease lines that are straight and to constructions where only one fold is made at a time.

## 3.1    Definitions and Origami Arithmetic

We let our sheet of paper be the complex plane $\mathbb{C}$, so that each point we construct will be a complex number. The choice of points and lines in $\mathbb{C}$ with which we begin our constructions is rather arbitrary. Following Alperin (2000), we let our initial points be 0 and 1, and to make things slightly more simple, we allow the real axis to be an initially constructed line.

**Definition 3.1**    A number $\alpha \in \mathbb{C}$ is called an **origami constructible point** if there is a finite sequence of origami folds (straight-crease, single-fold) that starts with the points 0 and 1 and the real axis line and ends with $\alpha$ being constructed. An **origami constructible line** is defined similarly. An origami constructible point is also called an **origami number**. We denote the set of all origami numbers by $\mathcal{O}$.

In view of Theorem 1.1, we know that the basic origami operations O1–O8 defined in Section 1.5 are all we need to consider in our finite sequence of origami folds for constructing points and lines. In fact, Theorem 1.2 tells us that all we really need are operations O2 and O7. However, it is often convenient to refer to the other operations rather than enunciate the special cases of O7.

That $\mathcal{O}$ is a subfield of $\mathbb{C}$ is a simple exercise to prove. However, it is the author's experience that readers who have not experimented sufficiently with paper folding will often be unconvinced that elementary complex arithmetic can be performed via origami. Thus, for completeness we include a full proof, via a sequence of lemmas, in this section. (Although a few details will be left as diversions for the reader.)

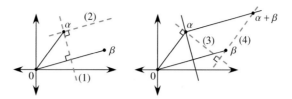

**Figure 3.1** Constructing $\alpha + \beta$.

**Lemma 3.2** *The origami numbers $\mathcal{O}$ form a group under complex addition.*

*Proof* We already have that the additive identity is in $\mathcal{O}$. For any nonzero $\alpha \in \mathcal{O}$, use BOO O1 to construct the line passing through the origin and $\alpha$. We then use O5 to fold a crease passing through the origin that is perpendicular to $\overline{0\alpha}$. The image of $\alpha$ under this fold will be $-\alpha$, which we therefore know is constructible by Remark 1.3.

Now let $\alpha, \beta \in \mathcal{O}$ such that they are not collinear with the origin. Figure 3.1 shows how to construct $\alpha + \beta$. First use O1 to construct lines $\overline{0\alpha}$ and $\overline{0\beta}$. Then use O5 to make a line, (1) in Figure 3.1, perpendicular to $\overline{0\beta}$ and passing through $\alpha$. Make a second crease (2) perpendicular to (1) passing through $\alpha$. Then repeat these last two steps but with the roles of $\alpha$ and $\beta$ switched. The end result are lines (2) and (4) in the figure, which complete the parallelogram with 0, $\alpha$, and $\beta$ as vertices. The fourth corner of this parallelogram, where (2) and (4) intersect, is $\alpha + \beta$.

---

**Diversion 3.1** Develop a construction for the case where $\alpha$, $\beta$, and the origin are collinear. □

---

**Lemma 3.3** *Let $\alpha = a + bi \in \mathbb{C}$, where $a, b \in \mathbb{R}$. Then $\alpha \in \mathcal{O}$ if and only if $a, b \in \mathcal{O}$.*

*Proof* Since the real axis is given to us and the imaginary axis is easy to construct, this proof is merely an application of operation O5. □

**Lemma 3.4** *Let $a, b \in \mathcal{O} \cap \mathbb{R}$ be nonzero. Then $ab$ and $1/a$ are also in $\mathcal{O}$.*

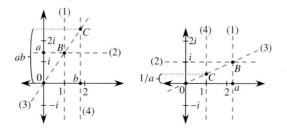

**Figure 3.2** Constructing $ab$ and $1/a$ for nonzero $a, b \in \mathcal{O} \cap \mathbb{R}$.

*Proof*  The constructions for $ab$ and $1/a$ are similar and shown in Figure 3.2. For $ab$, let $a$ be on the imaginary axis and $b$ be on the real axis. Construct a vertical line through the point 1 and a horizontal line through the point $a$ (lines (1) and (2) in the figure), letting the intersection of these two lines be $B$. Then fold the line through $B$ and the origin; the slope of this line (3) will be $a$. Thus, when we make a vertical line (4) through the point $b$, it will intersect the line (3) at the point $C = b + abi$, giving us $ab$.

For $1/a$, see the right part of Figure 3.2. We take $a$ on the real axis and construct a line with slope $1/a$. Where this line intersects the vertical line at the point 1 will be $C = 1 + (1/a)i$. $\qquad\qquad\square$

**Lemma 3.5**  *The set $\mathcal{O}$ is closed under multiplication and taking reciprocals of nonzero elements.*

*Proof*  This is merely a consequence of the previous two lemmas, for if $\alpha = a + bi$ and $\beta = c + di$ are in $\mathcal{O}$, then $a, b, c, d \in \mathcal{O}$ by Lemma 3.3. Thus $\alpha\beta$ is a complex number whose real and imaginary parts are products, sums, and differences of $a, b, c, d$, and thus are in $\mathcal{O}$, implying that $\alpha\beta \in \mathcal{O}$. Also, if $\alpha \neq 0$,

$$\frac{1}{\alpha} = \frac{1}{a + bi}\frac{a - bi}{a - bi} = \frac{a}{a^2 + b^2} + \frac{-b}{a^2 + b^2}i.$$

Thus the real and imaginary parts of $1/\alpha$ are in $\mathcal{O}$ by Lemma 3.4, so we have $1/\alpha \in \mathcal{O}$ as well. $\qquad\qquad\square$

Putting all this together, we conclude the following:

**Theorem 3.6**  *The set $\mathcal{O}$ is a subfield of $\mathbb{C}$.*

## 3.2  The Field of Origami Numbers

We now examine the structure of the subfield $\mathcal{O}$. Since any straightedge and compass construction can be performed by origami, we know that $\mathcal{O}$ must be closed under square roots. But we also know that things like $\sqrt[3]{2}$ can be constructed by origami as well. Our aim is to be very specific about what we can construct by describing what field extensions to the rational complex numbers are possible while still remaining in $\mathcal{O}$.

We begin with some details.

**Lemma 3.7**  *Let $\alpha \in \mathcal{O}$. Then $\sqrt{\alpha}$ and $\sqrt[3]{\alpha} \in \mathcal{O}$.*

*Proof*  Write $\alpha = re^{i\theta}$. First notice that the real number $r$ is in $\mathcal{O}$. (We can use O6 to fold the point $\alpha$ onto the real axis, making the crease pass through the origin. In doing this, $\alpha$ will be folded to the point $r$ on the real axis, and this is then a constructible point by Remark 1.3.) Next, we have that $e^{i\theta} \in \mathcal{O}$ by constructing $\alpha/r$. Also, we saw in Sections 2.2 and 2.3 that the operations O1–O8 can construct real solutions to quadratic and cubic equations, and thus $\sqrt{r}, \sqrt[3]{r} \in \mathcal{O}$. Angle bisection is easy by operation O4, and in Section 1.3 we saw that O7 allows us to perform angle trisections. Thus $e^{i\theta/2}, e^{i\theta/3} \in \mathcal{O}$, and we have, since $\mathcal{O}$ is a field, that $\sqrt{\alpha} = \pm\sqrt{r}e^{i\theta/2} \in \mathcal{O}$ and $\sqrt[3]{r}e^{i\theta/3} \in \mathcal{O}$. To get the other cube roots of $\alpha$, we notice that $e^{2\pi i/3} \in \mathcal{O}$ (we constructed $60°$ angles in Section 1.1), and thus $\sqrt[3]{r}e^{i\theta/3}e^{2\pi i/3}$ and $\sqrt[3]{r}e^{i\theta/3}e^{4\pi i/3}$ are in $\mathcal{O}$. $\qquad\square$

Having proven that $\mathcal{O}$ is closed under square and cube roots and that $\mathcal{O}$ is a subfield of $\mathbb{C}$, we then may use the quadratic formula and Cardano's formula to show that the complex roots of any second or third degree equation are also in $\mathcal{O}$. This gives us everything we need to describe the field extensions that make up $\mathcal{O}$.

The next definition follows (Cox, 2004).

**Definition 3.8**  A **2-3 tower** is a nested sequence of fields

$$\mathbb{Q} = F_0 \subset F_1 \subset \cdots \subset F_{n-1} \subset F_n \subset \mathbb{C}$$

such that $[F_i : F_{i-1}] = 2$ or $3$ for all $1 \le i \le n$.

Here $[K : F]$ denotes the **degree** of the field extension $K = F(\alpha)$ of $F$, which equals the degree of a minimal polynomial $f$ with coefficients in $F$ and $f(\alpha) = 0$.

We pause now to describe how this fits in with our origami constructions. We begin with our given points $0$ and $1$ (and our given line, the real axis). We know that, for starters, the origami operations can be used construct any point in the complex plane with rational coordinates, which is the field extension $\mathbb{Q}(i)$. With just this, our "tower" would be $\mathbb{Q} \subset \mathbb{Q}(i)$ with $[\mathbb{Q}(i) : \mathbb{Q}] = 2$ since $x^2 + 1$ is the minimal polynomial of $i$ over $\mathbb{Q}$. If we then construct $\sqrt[3]{2}$, we would conclude that $\sqrt[3]{2}$ is an origami number by realizing the 2-3 tower $\mathbb{Q} \subset \mathbb{Q}(i) \subset \mathbb{Q}(i, \sqrt[3]{2})$, which is contained in $\mathcal{O}$. (Here $[\mathbb{Q}(i, \sqrt[3]{2}) : \mathbb{Q}(i)] = 3$ because the minimal polynomial of $\sqrt[3]{2}$ over $\mathbb{Q}(i)$ is $x^3 - 2$.)

We will shortly be in need of the following lemma, which specifies how much "muscle" Beloch's fold O7 has for extending a field. In doing this, we momentarily leave $\mathbb{C}$ and return to the real plane.

**Lemma 3.9**  *When performing operation O7 using points in and lines with equations over a field $F$, the equation of the resulting crease line will have coefficients in a field extension of $F$ of degree 3 or less.*

Note that in Section 2.3 we saw how operation O7 can be used to solve an arbitrary cubic equation. This is different from Lemma 3.9, however, since now we need to prove that any application of O7 will do no more than solve a cubic.

*Proof*  Operation O7 requires two points, $P_1, P_2$ and two lines $L_1, L_2$. We first consider the case where the points are not contained in either of the lines. Using a suitable change in coordinates, we can assume that $P_1 = (0, 1)$ and $L_1$ is the line $y = -1$. We'll let $P_2 = (a, b)$ and $L_2$ be the line $a_1 x + b_1 y = c_1$, where $a, b, a_1, b_1, c_1 \in F$.

As seen in Section 2.1, if we fold $P_1$ to an arbitrary point $P'_1 = (t, -1)$ on $L_1$, the resulting crease line equation will be that shown in Equation (2.1): $y = (t/2)x - t^2/4$. Clearly if we can find a field extension $K$ of $F$ with $t \in K$ and $[K : F] \le 3$, then we'll be home free.

Let $P'_2 = (x, y)$ be the image of $P_2$ when folded (reflected) about the crease line. The slope of $P_1 P'_1$ must equal that of $P_2 P'_2$, which gives us $-2/t = (y - b)/(x - a)$, or

$$x = a + \frac{1}{2}(b - y)t. \tag{3.1}$$

Also, the crease line must pass through the midpoint of $P_2 P'_2$, which is $((x+a)/2, (y+b)/2)$. Plugging this point into the crease line Equation (2.1) and solving this with (3.1) gives us the following parameterization of the curve $P'_2$ traces as we vary $t$:

$$x = a + \frac{t^2 - 2at + 4b}{t^2 + 4}, \quad y = \frac{(b-2)t^2 + 4at - 4b}{t^2 + 4}. \tag{3.2}$$

For operation O7 to be successful, we need a value of $t$ that places this point $(x, y)$ on the line $L_2$. Plugging (3.2) into the line equation for $L_2$ and solving for $t$ yields a cubic equation with coefficients in terms of $a, b, a_1, b_1$, and $c_1$. Therefore the roots of this equation, and thus our desired values of $t$, will be contained in a field extension $K$ of $F$ of degree 3 or less.

We now turn to the case where some or all of the points $P_1$ and $P_2$ are on the lines $L_1, L_2$. As shown in the proof of Theorem 1.2, these degenerate cases of O7 give us operations O1, O3–O6, and O8. It is not difficult to show that the crease lines generated by these operations have equations with coefficients in a field extension of $F$ of degree 2 or less. For example, O1's equation is linear in terms of the coordinates of $P_1$ and $P_2$. We saw in Section 2.1 how O6 generates a crease line whose coefficients are found using a quadratic equation, thus needing an extension of degree 2.

---

**Diversion 3.2**   Prove that the other BOOs solve only linear and quadratic equations.

□

---

**Theorem 3.10**   *A number $\alpha \in \mathcal{O}$ if an only if there exists a 2-3 tower $\mathbb{Q} = F_0 \subset \cdots \subset F_n \subset \mathbb{C}$ such that $\alpha \in F_n$.*

*Proof*   Let $\mathbb{Q} = F_0 \subset \cdots \subset F_n \subset \mathbb{C}$ be a 2-3 tower with $\alpha \in F_n$. Our aim is to prove that $F_n \subset \mathcal{O}$, and we will do this by induction. When $n = 0$ there is nothing to prove, since $\mathbb{Q} \subset \mathcal{O}$. So assume that $F_{n-1} \subset \mathcal{O}$. Let $\alpha \in F_n$ and $f$ be the minimal polynomial of $\alpha$ over $F_{n-1}$, which is of degree 3 or less since $[F_n : F_{n-1}] = 2$ or 3. If $f$ has degree 1, then the extension $F_n$ was redundant and $\alpha \in \mathcal{O}$. If the degree of $f$ is 2 or 3, then we can construct all roots of $f$ using the quadratic formula or Cardano's formula (which use only square and cube roots) to obtain $\alpha \in \mathcal{O}$.

For the other direction, let $\alpha$ be an origami number. We wish to construct a 2-3 tower $\mathbb{Q} = F_0 \subset \cdots \subset F_n \subset \mathbb{C}$ such that $F_n$ contains the real and imaginary parts of all numbers that led up to the construction of $\alpha$. That is, if $\alpha = a + bi$, then if we could show that $a, b \in F_n$, we'd have that $\alpha \in F_n(i)$, and this last extension of degree 2 would complete our tower.

In light of Theorem 1.2, we know that $\alpha$ was constructed by a finite sequence of the operations O2 and O7. Our proof will be by induction on the number $N$ of times we used O2, the only operation that specifically generates new points. If $N = 0$ then $\alpha$ was in our initial given set of points and lines, and the trivial tower $\mathbb{Q} = F_0 \subset \mathbb{C}$ suffices.

Now assume that all origami numbers constructed by a finite sequence of O2 and O7 operations using $N - 1$ applications of O2 will have a 2-3 tower as per the hypothesis. Let $\alpha$ be an origami number whose sequence $S$ of operations for construction contains O2 $N$ times. We can assume that the last operation in $S$ was O2 that was used to find $\alpha$ as the intersection of lines $L_1$ and $L_2$. Then $L_1$ must have been constructed using O7 and two previously constructed points $\alpha_1, \beta_1$ (along with two other, previously constructed lines). Similarly, $L_2$ must have arisen from a use of O7 with previous points $\alpha_2, \beta_2$ (and two other lines). By our induction hypothesis, there exists a 2-3 tower $\mathbb{Q} = F_0 \subset \cdots \subset F_n \subset \mathbb{C}$ where the real and imaginary parts of $\alpha_1, \alpha_2, \beta_1, \beta_2$ are contained in $F_n$. By Lemma 3.9, this means that our 2-3 tower can be extended to one that includes the coefficients for the equations of lines $L_1$ and $L_2$. Finding the intersection point of these two lines is merely a linear process, and so we conclude that we have extended our 2-3 tower (again, adjoining $i$ at the end if needed) to one that contains $\alpha$, and we're done. $\square$

We may then give a Galois theory–based classification of origami numbers.

**Theorem 3.11**  *Let $\alpha \in \mathbb{C}$ be algebraic over $\mathbb{Q}$ and let $L$ be the splitting field of the minimal polynomial of $\alpha$ over $\mathbb{Q}$. Then $\alpha$ is an origami number if and only if $[L : \mathbb{Q}] = 2^a 3^b$ for some integers $a, b \geq 0$.*

*Proof*  Let $\alpha \in \mathcal{O}$. Then notice that $\mathcal{O}$ is a normal extension of $\mathbb{Q}$. This follows from our previous work: The folding operations that construct elements of $\mathcal{O}$ are O1–O8 (Theorem 1.1), we know that these operations generate points that are solutions of polynomial equations of degree 3 or less (Chapter 2 and Lemma 3.9), and $\mathcal{O}$ is closed under square and cube roots (Lemma 3.7). Thus $\mathcal{O}$ contains all roots of polynomials of degree 3 or less over $\mathbb{Q}$ by Cardano's formula for cubics and the quadratic formula. Thus every polynomial of degree 3 or less over $\mathbb{Q}$ splits over $\mathcal{O}$, and so $\mathcal{O}$ is a normal extension. This gives us that $L \subset \mathcal{O}$, and if we take a primitive element of $L$ and apply Theorem 3.10, the resulting 2-3 tower gives us that $[L : \mathbb{Q}] = 2^a 3^b$ for some integers $a, b \geq 0$.

For the other direction we know that the Galois group $\text{Gal}(L/\mathbb{Q})$ has $|\text{Gal}(L/\mathbb{Q})| = [L : \mathbb{Q}] = 2^a 3^b$. Burnside (1904) proved that such groups are solvable. Then the Galois correspondence can be used to construct a 2-3 tower for $\alpha$, proving by Theorem 3.10 that $\alpha \in \mathcal{O}$. $\square$

This can be used to determine if specific constructions will be possible with straight-crease, single-fold origami. For example, we can specify exactly what regular polygons will be foldable. We'll make use of the following definition, borrowed from Cox and Shurman (2005).

**Definition 3.12**   A prime $p$ is called a **Pierpont prime** if $p$ is greater than 3 and of the form $2^a 3^b + 1$ for some integers $a, b \geq 0$.

In the same way that Fermat primes (of the form $2^{2^a} + 1$) determine which regular polygons are SE&C constructible, Pierpont primes determine which are origami constructible.

**Theorem 3.13**   *A regular n-gon can be constructed by origami if and only if $n = 2^a 3^b p_1 \cdots p_k$ for some integers $a, b \geq 0$ and where $p_1, \ldots, p_r$ are distinct Pierpont primes.*

*Proof*   Let $\zeta_n = e^{2\pi i/n}$ be the $n$th root of unity. It is easy to see that the regular $n$-gon will be constructible if and only if $\zeta_n$ is constructible. Then $\mathbb{Q}(\zeta_n)$ is the splitting field of the separable polynomial $z^n - 1 \in \mathbb{Q}[z]$. Then by Theorem 3.11, $\zeta_n$ is origami constructible if and only if $[\mathbb{Q}(\zeta_n) : \mathbb{Q}] = 2^a 3^b$ for some integers $a, b \geq 0$.

Now, the minimal polynomial of $\zeta_n$ over $\mathbb{Q}$ is the $n$th cyclotomic polynomial

$$\Phi_n(z) = \prod_{\substack{0 \leq i < n \\ \gcd(i,n)=1}} (z - \zeta_n^i).$$

(This is proven in a number of Galois theory texts, for example (Cox, 2004, Section 9.1).) This means that $[\mathbb{Q}(\zeta_n) : \mathbb{Q}] = \deg(\Phi_n(z)) = \phi(n)$, where $\phi(n)$ is the Euler $\phi$-function, the number of positive integers $i$ less than $n$ with $\gcd(i, n) = 1$. Elementary number theory gives us many results for computing $\phi(n)$. In particular, we'll make use of $\phi(n) = n \prod_{p|n} (1 - 1/p)$ for integers $n > 1$, where this product is taken over all primes $p$ that divide $n$.

So suppose $n = 2^a 3^b p_1 \cdots p_k$, where $a, b \geq 0$ and $p_1, \ldots, p_r$ are distinct Pierpont primes. Then we have

$$
[\mathbb{Q}(\zeta_n) : \mathbb{Q}] = \phi(n)
$$

$$
= n \prod_{p|n} \left(1 - \frac{1}{p}\right) = 
\begin{cases}
2^a 3^{b-1}(p_1 - 1) \cdots (p_k - 1), & a, b > 0, \\
2^{a-1}(p_1 - 1) \cdots (p_k - 1), & a > 0, b = 0, \\
2 \cdot 3^{b-1}(p_1 - 1) \cdots (p_k - 1), & a = 0, b > 0, \\
(p_1 - 1) \cdots (p_k - 1), & a, b = 0.
\end{cases}
$$

Therefore $[\mathbb{Q}(\zeta_n) : \mathbb{Q}]$ is a power of 2 times a power of 3 since the $p_i$ are Pierpont primes. This implies that $\zeta_n \in \mathcal{O}$.

Conversely, suppose $\zeta_n \in \mathcal{O}$, so that $[\mathbb{Q}(\zeta_n) : \mathbb{Q}] = \phi(n)$ is a power of 2 times a power of 3. Let the factorization of $n$ into distinct powers of primes be $n = q_1^{a_1} \cdots q_r^{a_r}$. Then

$$
\phi(n) = n \prod_{p|n} \left(1 - \frac{1}{p}\right) = q_1^{a_1 - 1}(q_1 - 1) \cdots q_r^{a_r - 1}(q_r - 1).
$$

If any $q_i$ is odd, then either $q_i = 3$ or $a_i = 1$. Also, each factor $(q_i - 1)$ must be a power of 2 times a power of 3, and thus $q_i$ is a Pierpont prime. Thus each prime factor of $n$ is either 2, 3, or a Pierpont prime, as desired. ☐

As examples, we may now verify that the regular heptagon is constructible via origami since 7 is a Pierpont prime. Also, the regular nonagon (9-gon) and tridecagon (13-gon) are foldable. The smallest regular polygon not constructible by straight-crease, single-fold origami is the undecagon (11-gon).

## 3.3 Other Characterizations of Origami Numbers

There are other methods for describing the set of origami numbers $\mathcal{O}$. In this section we present a few equivalent construction tools and strategies, some of which can provide alternate analyses of origami constructions than that given in the previous section.

Andrew Gleason (1988) showed that the tools of straightedge, compass, and an arbitrary angle trisector will only construct numbers in $\mathbb{C}$ that are in a 2-3 tower of field extensions of $\mathbb{Q}$. Emert, Meeks, and Nelson also showed that the Mira construction tool, which is basically a mirror held perpendicularly to the plane on which points are being constructed, produces numbers that also must arise from a 2-3 tower (Emert et al., 1994). (For an earlier paper studying the Mira as a construction tool, see (Dayoub and Lott, 1977).) Thus these construction methods are equivalent to straight-crease, single-fold origami and produce the same subfield of constructible numbers $\mathcal{O} \subset \mathbb{C}$. Intuitively, this makes sense. Origami can trisect angles, and in doing so it utilizes the most powerful operation origami has to offer (O7). Also, the process of folding a piece of paper flat along a single crease is identical to reflecting half of the plane about a line, which is exactly what the Mira does.

The set $\mathcal{O}$ is also generated when using the tools of a **marked** straightedge and compass (Martin, 1985). This also makes intuitive sense. The side of a piece of paper functions as a straightedge, and if a fold or pinch-mark is made along it, then the fold mark on the side can be used to mark a unit length, essentially mimicking a marked straightedge.

Returning to the tools of paper folding, a completely different way to generate the subfield $\mathcal{O}$ was described by Alperin (2000). As mentioned in Section 2.3, the act of the basic origami operation O7 is finding a common tangent to two parabolas drawn in the plane. Therefore, one could take an approach of using conics to aid in geometric constructions. The idea of using conics in geometric constructions goes back to the Greeks. For example, Archimedes used conics to construct a regular heptagon, and Menaechmus used them to construct cube roots (Videla, 1997). However, such geometric constructions with conics typically involve locating points of intersection between two conics, which isn't exactly what Beloch's fold O7 is doing.

To see how such conic constructions relate to origami constructions, we need to move to the real projective plane and use the technique of duality. For a good introduction to projective geometry and the projective plane, see (Bumcroft, 1969). To review, we may define the real projective plane $\mathbb{RP}^2$ to be 3-dimensional space $\mathbb{R}^3$, where we define a "point" to be any line passing through the origin and a "line" to be any plane passing through the origin. Thus our "points" can be described by a point $(a, b, c) \in \mathbb{R}^3$, where any multiple of this point, like $(ta, tb, tc)$ for any $t \in \mathbb{R}$, is considered to be the same "point" in $\mathbb{RP}^2$ because $(a, b, c)$ and $(ta, tb, tc)$ both lie on the same line through the origin. In other words, the points of $\mathbb{R}^3$ are partitioned into equivalence classes determined by the lines through the origin, and each equivalence class is a "point" in $\mathbb{RP}^2$. Referring to a "point" by a 3-tuple $(a, b, c)$ as a representative from its equivalence class is called the **homogeneous coordinates** of the "point."

Similarly, our "lines" in $\mathbb{RP}^2$ are planes through the origin given by equations $ax + by + cz = 0$ for $a, b, c \in \mathbb{R}$. But any scalar multiple of such an equation will result in the same plane in $\mathbb{R}^3$, so our equations of planes through the origin form equivalence classes as well. Referring to a "line" $ax + by + cz = 0$ by a 3-tuple $[a, b, c]$, where we use square brackets to distinguish it from "points," is called the **homogeneous coordinates** of the "line."

These "points" and "lines" in $\mathbb{RP}^2$ can be made to look more familiar if we consider how they intersect the plane $z = 1$ in $\mathbb{R}^3$. Thus the "point" $(a, b, c)$ is the same as the point $(a/c, b/c, 1)$ on the plane $z = 1$ (provided $c \neq 0$), and the "line" $[a, b, c]$, whose equation is $ax + by + cz = 0$, intersects $z = 1$ at $ax + by + c = 0$, which is an honest-to-goodness line. It can be expressed in homogeneous coordinates as $[a, b, c]$ or $[a/c, b/c, 1]$.

The power of homogeneous coordinates is seen in the natural bijection that exists between "points" and "lines." Namely, given a "point" $(a, b, c) \in \mathbb{RP}^2$ we can associate with it the "line" $[a, b, c]$. This association is a bijection and is called **duality**, and it maps "points" to "lines" and vice versa.

Now, given a conic $C$ drawn in the plane $z = 1$, we can consider its **dual conic** $C^*$ to be as follows: Take all "points" $(a, b, 1)$ in $C$; their duals $[a, b, 1]$ will be a set of lines on the plane $z = 1$. The envelope of these lines (that is, the curve that is tangent to each of these lines) will be another conic $C^*$ in the plane $z = 1$.

Although we will not make use of it here, it is worth mentioning (see (Alperin, 2000)) that the duality operation for conics can be done analytically in a very elegant manner. If we define the real $3 \times 3$ symmetric matrix

$$A = \begin{pmatrix} a & b & c \\ b & d & e \\ c & e & f \end{pmatrix},$$

then the matrix equation $(x, y, z) \cdot A \cdot (x, y, z)^T = 0$ is the same as the homogeneous equation for a conic: $ax^2 + 2bxy + dy^2 + 2cxz + 2eyz + hz^2 = 0$. (Setting $z = 1$ gives a general conic in the plane.) The matrix equation for the dual conic is then $(x, y, z) \cdot Adj(A) \cdot (x, y, z)^T = 0$, where $Adj(A)$ is the adjoint matrix of $A$ (see (Andrilli and Hecker, 2003, p. 145)).

Returning to origami, suppose that we have two conics drawn on the plane $z = 1$ and we wish to find a common tangent for them, which is what Beloch's fold O7 tries to do. If such a common tangent exists, then we can take the duals of the two conics and the tangent line. The dual of the latter will be a point, and since the original line was tangent to the original conics, the dual point must lie on both of the dual conics. Therefore, finding our common tangent is equivalent to the dual problem of finding the points of intersection (if any exist) between two conics.

The set of numbers in $\mathbb{C}$ (considering, once again, the plane $z = 1$ to be the complex plane) constructible from the intersection of conics has been shown to be the smallest subfield of $\mathbb{C}$ closed under conjugation and square and cube roots. (Carlos R. Videla gives a nice proof of this in (Videla, 1997). He uses Galois theory in a similar spirit to the previous section.) That is, constructing numbers by intersecting conics gives us the origami numbers $\mathcal{O}$. Therefore the dual operation, finding common tangents to two conics, will also produce $\mathcal{O}$, giving us an alternate proof of the structure of $\mathcal{O}$ from that given in Section 3.2.

## 3.4 Other Work: Origami Rings

Of all the material covered in this book, origami geometric constructions is arguably the most well-trod ground. Nonetheless, mathematicians continue to find interesting new things to explore. One recent idea came from origami design: Box pleating (which we will encounter again in Section 6.6) is an origami design technique where the crease pattern vertices all lie on a square grid and only crease angles that are multiples of $\pi/4$ (45°) are allowed. One could then ask what set of origami constructible numbers can be generated only using angle multiples of $\pi/4$ for the creases. This turns out to not be very interesting; one gets merely finer and finer square lattice grids of points. But if we change the allowed angle multiple to $\pi/8$, then the kinds of crease patterns allowed include that of the classic origami bases (see Figure 8.5) and the flapping bird/crane (see Figure 1 in the Introduction). This constructibility question for the $\pi/8$ case was answered in (Tachi and Demaine, 2011). In (Buhler et al., 2012) the authors look at the general constructibility question where the allowed angles are multiples of $\pi/n$ and prove that in general one obtains a subring of $\mathbb{C}$ in such construction schemes. Further work on this can be found in (Kritschgau and Salerno, 2017; Nedrenco, 2019).

# 4    Beyond Classic Origami

Those who think the previous chapters encompass all the ways in which paper folding can perform geometric constructions are quite mistaken. Up till now we were very careful to include the stipulation "straight-crease, single-fold" when describing the type of origami under consideration. Relaxing these conditions leads to more unexplored realms of origami constructions. In this chapter we will examine the current state of such explorations.

## 4.1    Multifolds

In (Frigerio and Huzita, 1989) Humiaki Huzita first proposed that allowing ourselves to make more than one crease at the same time might lead to a bigger field of origami constructible numbers. However, he did not provide an example of such a fold. Indeed, it is rather difficult to imagine what such an origami operation would be that requires the simultaneous creation of multiple creases. We will nonetheless make this an official definition.

**Definition 4.1**    An origami operation is called a **multifold** if it requires the simultaneous creation of more than one crease. A multifold that creates $n$ different creases is called an $n$-**fold**.

---

**Example 4.2** (Thirds and trisections)    At least one example of such a folding maneuver is quite common. In Chapter 1 we saw the standard method of curling a piece of paper into an "S" shape in order to divide its length into thirds (see Figure 1.3). This is an example of a 2-fold, where the two crease lines, at 1/3 and 2/3 across the length of the paper, must be made at the same time in order for the operation to work. If a basic origami operation were to be created to encompass such folds, it might be written as follows:

**MO1:**  Given two lines $L_1$ and $L_2$, make exactly two creases $a$ and $b$ at the same time so that $L_1$ folds onto $b$ and $L_2$ folds onto $a$.

If the lines $L_1$ and $L_2$ are parallel, then this is the move shown in Figure 1.3, where $L_1$ and $L_2$ are the left and right sides of the paper. If $L_1$ and $L_2$ intersect, then MO1

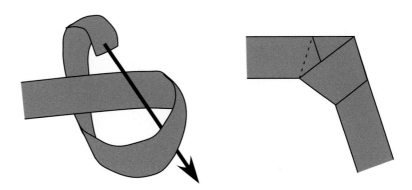

**Figure 4.1** Folding a pentagon knot from a strip of paper is a 3-fold.

trisects the angle between them. In other words, this multifold allows us to trisect line segments and angles in one step, just as the basic origami operation O4 in Section 1.5 allows us to bisect angles in one step.

Whether or not we should allow such origami moves involving multiple, simultaneous creases is a matter of some controversy. From an origami history perspective, the "fold into thirds" move in Figure 1.3 is commonly accepted amongst origamists, but the angle trisection version is never seen.

**Example 4.3** (Folding knots)   Another example of multifolds in the origami cannon is in the folding of polygonal knots from strips of paper. The most common example of this is twisting a strip of paper into a simple overhand knot and pulling it tight while flattening it. This is demonstrated in Figure 4.1. In order to make the pentagon regular the paper needs to tightened slowly and the three creases made at the same time.

The practice of making such knots goes back to ancient Japan. According to Fukagawa (Fukagawa and Pedoe, 1989), there is a **sangaku** tablet (geometry problems that were written on wooden tablets and hung in Shinto shrines in Edo-era Japan) from 1810 that depicts a pentagonal folded knot. In the Western world one of the earliest references seems to be a paper by F. V. Morley in the MAA *Monthly* from 1924 (Morley, 1924) that describes how to make knots in the form of regular pentagons, hexagons (this requires two strips of paper), and heptagons (Figure 4.2), as well as generalizations for any *n*-gon with $n > 6$.

When folding such a knotted polygon from a strip, the paper travels a path along a star polygon. For example, in the pentagon knot the paper touches every other side of the pentagon as it travels around the knot, following the path of a $\left\{\frac{5}{2}\right\}$ star polygon, where the 5 denotes the number of points we have (i.e., the order of the rotational symmetry group of the figure) and the 2 means we connect every second point as we go around, thus producing a star. (This is standard notation due to Coxeter (1973, pp. 93–94) .)

**Figure 4.2**  Folding a heptagon knot from a strip of paper is a 5-fold.

---

**Diversion 4.1**   Why is it that we can't make a knotted hexagon from a single strip of paper? Are there any other $n$-gons that can't be made in this way? A regular hexagon can be made, however, from two strips of paper knotted together like a square knot. Under what conditions can a regular $n$-gon be knotted from $k$ strips of paper? (See (Hull, 2012, Activity 10).)

---

A regular heptagon folded from a strip is shown in Figure 4.2. This is an example of a 5-fold and is very difficult to do. With practice, however, it can be done quite accurately. 9- and 11-gons can also be done, but they take a good deal of practice and patience. One could dispute that such folds are not "real" origami, since they are folded from strips of paper. However, these folds do create the proper angles for their respective $n$-gons, and they all could be made from a square piece of paper that was first pleated into a thin strip. The strip could then be knotted into the desired $n$-gon, which would create creases at the proper angles for the $n$-gon. The paper could then be unfolded and these creases used to make a more traditional-looking $n$-gon, if desired. This would certainly be a very awkward way to fold regular polygons, but it does demonstrate their theoretical possibility.

The fact that a regular 11-gon can be made with a multifold demonstrates how multifolds exceed the ability of standard straight-crease, single-fold origami constructions, since the construction of a regular 11-gon requires the solution of a quintic whose roots do not lie in the origami number field $\mathcal{O}$ from Chapter 3.

## 4.2    Lang's Angle Quintisection

Folding polygons from strips offers a good example of the power and feasibility of multifolds. However, the knotted strip of paper method of making polygons requires a 3-fold to make a pentagon, which is also possible with single-fold origami. If multifolds really do extend the geometric possibilities of paper folding in any kind of practical way, we would like an example of a 2-fold that does something beyond the realm of traditional folding.

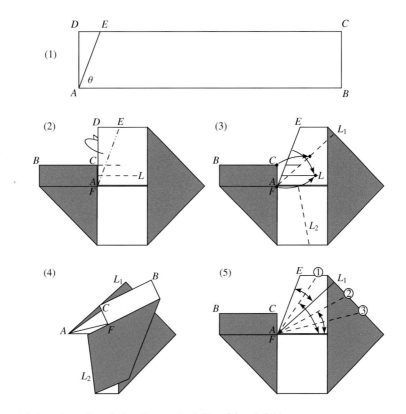

**Figure 4.3** Lang's angle quintisection method. Step 3 is a 2-fold.

For this we turn to Robert J. Lang's angle quintisection method (Lang, 2004). It is a 2-fold that, like the knotted polygons, requires a strip of paper. This method is quite difficult. Instructions follow and are depicted graphically in Figure 4.3.

(1) Label the corners of your strip $A$, $B$, $C$, and $D$, and make your angle $\theta$ to quinitsect at corner $A$, so that $\angle EAB = \theta$. Generally, a strip that is $1 \times 5$ in dimensions should work, but if $\theta$ is very acute then a longer strip will be needed.
(2) Fold the strip into the configuration shown in Figure 4.3(2). The exact proportions of this will depend on the size of $\theta$; the line $AE$ should be unobstructed, and $BC$ should protrude no more than half-way up the side $AD$. Pinch a small crease perpendicular to $AD$ that extends the line $BC$. Then make a longer crease $L$ by folding $A$ to the $BC$ line. Finally, mountain-fold $AE$ behind. (This makes the next step easier.)
(3) Now you're ready to do the tricky 2-fold. You'll make crease line $L_1$ by folding side $AE$ down and at the same time crease $L_2$ by folding point $F$ onto line $L$. By themselves, these two creases are not well-defined, but they can be made at the same time by having the folded side $AE$ cross line $L$ at the same point that $F$ folds to **and** by having point $C$ land on the crease $L_1$. To achieve all this, $AE$ needs to

be curled over and $F$ moved onto $L$ without making a crease, then everything can
be aligned and then the creases made simultaneously.

(4) After step (3) your paper might look something like Figure 4.3(4). Notice how $C$
is on $L_1$ and $F$ lands on the folded edge $AE$. Unfold step (3).

(5) We claim that the crease line $L_1$ makes an angle of $3\theta/5$ with the bottom edge
of the strip $AB$. So fold $AE$ to $L_1$ to make the $4\theta/5$ crease (labeled 1 in Figure
4.3). Then fold crease 1 to the bottom edge of the strip to make the $2\theta/5$ crease
(labeled 2). Lastly, fold crease 2 to the bottom edge to make $\theta/5$.

Proving that this method works would make a nice diversion, but the author's experience is that most people find it quite unbelievable. Incredulity over step (3) can,
perhaps, be a mental block, so we provide a proof here.

**Theorem 4.4**    *Lang's angle quintisection method works.*

*Proof*    Starting with our finished step (5) in Lang's method, draw $C'F'$ to be the image
of $CF$ under the folding of $L_2$. Draw a line $F'G$ perpendicular to the bottom edge of
the strip (what would be $AB$ if it were all unfolded). Also let $J$ be the point on $AE$
that landed on $F'$ under the folding of $L_1$. Let $H$ be the midpoint of $C'F'$ and $I$ be the
midpoint of $C'J$. (See Figure 4.4.) Our goal is to prove that $\angle C'AG = 3\theta/5$.

*Claim: $AH$ is perpendicular to $C'F'$.*

*Proof of claim:* Note that $A$ and $F$ are actually the same point and that $C'F'$ is the
reflection of $CF$ under the crease $L_2$. So $CF \cong C'F'$ and the pre-image of $H$ under
the $L_2$ fold is the midpoint of $CF$. Therefore the reflection of $AH$ (which is also $FH$)
about the crease $L_2$ is on the line $L$, since $F$ maps to $F'$ and $H$ maps to the midpoint of
$CF$, both of which are on $L$. Since $L \perp CF$, we have that $AH \perp C'F'$, which proves
the claim.

Also, we have that the length of $F'G$ is equal to one-half $CF$, which equals the
length of $F'H$. Thus $F'G \cong F'H \cong C'H$. Now, since $J$ maps to $F'$ under the fold $L_1$,
we have $C'J \cong C'F'$, and since $I$ is the midpoint of $C'J$, we get $IJ \cong C'I \cong C'H$ as
well.

In other words, $\triangle AC'J \cong \triangle AC'F'$, which means that $AI \perp C'J$. Thus we have
the set of five congruent right triangles $\triangle AIJ \cong \triangle AC'I \cong \triangle AC'H \cong \triangle AF'H \cong$

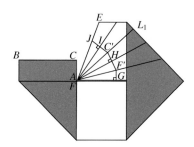

**Figure 4.4**  Proving Lang's angle quintisection method works

$\triangle AF'G$, which means that the angles they all make at $A$ are congruent. In other words, $\angle C'AG = 3\theta/5$.                                                                                    □

## 4.3  Describing Multifold Origami Operations

In order to study the possibilities of using multifolds in origami constructions, we would need to classify them. Unfortunately, this gets very unwieldy very fast.

Suppose, for instance, that we restrict ourselves to 2-folds. We would need to characterize all 2-folds with a list of basic folding moves like we did for single-fold origami constructions in Chapter 1. Recall that there are only seven single-fold origami operations (BOOs O1–O8 in Section 1.5 except we omit O2 since it only finds points of intersection and doesn't involve actual folding). A similar list of BOOs for 2-folds is too long to fully describe in this book. Alperin and Lang (2009) developed a notational scheme and computational method for enumerating all the distinct 2-fold BOOs and found that there are 489 of them. To help elucidate why there are so many, and to provide intuition on how multifolds work, we will summarize part of their results.

In Theorem 1.1 we proved that the list of BOOs O1–O8 was complete for single-fold origami constructions by paying careful attention to how points and lines can be aligned via folding to produce a single, straight crease line, and we used the notion of **degrees of freedom** to know when a crease was being well-defined or not. (By "well-defined" in this context we mean that there might be a finite number of crease lines that achieve the desired alignments, but not an infinite number of such creases.) We need to proceed similarly for 2-folds.

---

**Example 4.5** (Beloch revisited)   Single-fold operation O7, the Beloch fold, takes two points $P_1$ and $P_2$ and folds them, respectively and simultaneously, to two lines $L_1$ and $L_2$. When it is possible to do, this is a well-defined fold because aligning $P_1$ onto $L_1$, by itself, has one degree of freedom, and so we need another alignment to specify a specific crease. Having $P_2$ be folded onto $L_2$ does the job.

We could try something similar with a 2-fold in the following way: Create one crease, labeled $a$, by folding $P_1$ onto $L_1$ and a second crease, labeled $b$, by folding $P_2$ onto $L_2$. Since we are producing two creases here, the degrees of freedom are not reduced by these two alignments when combined, and therefore the two creases $a$ and $b$ are not well-defined. What is needed is another alignment (or two), and this can be done in many different ways. Two such ways are shown in Figure 4.5.

In Figure 4.5(a) we look at the image of a third point $P_3$ under the folding of crease $a$ and the image of a fourth point $P_4$ under the fold $b$. If we can find a point $P_5$ where these two images coincide, then this will make the creases $a$ and $b$ well-defined.

In Figure 4.5(b) we are given a third line $L_3$ and try to make the creases $a$ and $b$ intersect at a point on $L_3$ (labeled $P_3$ in the Figure). This, by itself, does not produce well-defined creases, since if one moved the image of $P_1$ on $L_1$ by a small amount from

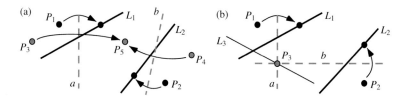

**Figure 4.5** Two different alignments for an AL6ab 2-fold.

that shown in Figure 4.5(b), then it would change the location of where $a$ intersects $L_3$. One could then move the image of $P_2$ on $L_2$, changing the slope of the crease $b$ so that it intersects $L_3$ at the same point as $a$. In this way, we may produce an infinite number of different candidates for the point $P_3$ and thus the creases $a$ and $b$.

In order to make this move well-defined, we need to stipulate another alignment. Figure 4.5(b) suggests that we intended the creases $a$ and $b$ to be perpendicular at the outset, and this is a legitimate alignment. Even though it may seem difficult to do in practice, it is not much different from the 1-fold BOO O5 (see Section 1.5) where we make a crease through a given point that is perpendicular to a given line. Thus we can fold the point $P_1$ onto $L_1$ to make a crease $a$ and fold $P_2$ onto $L_2$ to make a crease $b$ so that $a$ and $b$ are perpendicular and they intersect at a point on a third line $L_3$.

---

**Diversion 4.2**   In what other ways can the folds $a$ and $b$ in Example 4.5 be well-defined? That is, what additional alignments can be placed that determine the creases $a$ and $b$?

---

Alperin and Lang (2009) developed a notational system for denoting such 2-fold alignments. They label the two creases that a 2-fold will make as $a$ and $b$ (as we did in Figure 4.5) so that each crease can be specified as performing a certain alignment type, which they classify into ten different types, numbered 1–10 (see Figure 4.6). For instance, in Example 4.2 we align the line $L_1$ onto the crease $b$ and $L_2$ is aligned onto the crease $a$. Aligning an existing line onto a crease line that is being made is called alignment 4, or AL4 for short, in Alperin and Lang's notational scheme, and so the 2-fold move MO1 from Example 4.2 is called AL4ab, since both creases $a$ and $b$ are performing alignment AL4. This allows for no degree of freedom, so an AL4ab alignment produces a well-defined fold (assuming it's possible to do).

A fold that places a point onto a line is referred to as alignment 6, or AL6. A suffix of $a$ or $b$ is added to denote which crease line is following this alignment. So both of the examples in Figure 4.5 are notated AL6ab because both use alignment 6 for creases $a$ and $b$. Further alignments are specified by additional suffixes. In Figure 4.5(a) the additional alignment is used of folding a point onto the image of another point by the folds $a$ and $b$, respectively, and this is called alignment 8. So Figure 4.5(a) is an AL6ab8 fold.

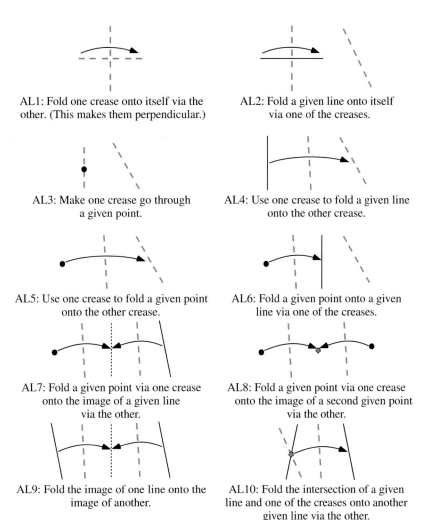

AL1: Fold one crease onto itself via the other. (This makes them perpendicular.)

AL2: Fold a given line onto itself via one of the creases.

AL3: Make one crease go through a given point.

AL4: Use one crease to fold a given line onto the other crease.

AL5: Use one crease to fold a given point onto the other crease.

AL6: Fold a given point onto a given line via one of the creases.

AL7: Fold a given point via one crease onto the image of a given line via the other.

AL8: Fold a given point via one crease onto the image of a second given point via the other.

AL9: Fold the image of one line onto the image of another.

AL10: Fold the intersection of a given line and one of the creases onto another given line via the other.

**Figure 4.6** Alperin and Lang's 10 distinct 2-fold alignments.

Figure 4.5(b) uses Alperin and Lang's alignment AL1, which makes the two crease lines be perpendicular, in addition to AL6ab. But we also want to make sure the intersection of $a$ and $b$ lies on the line $L_3$, and this uses their alignment AL10, which in this case would specify that crease line $a$ is folding $L_3$ onto itself, which then forces the crease $b$ to intersect $L_3$ at the same point as does $a$. It's a mouthful, but we can thus refer to this alignment as AL16ab10a.

Curious readers may like to know that the 2-fold used in step (3) of Lang's angle quintisection in Section 4.2 is notated by AL3a5b6b7b. However, since we won't be using this alignment notation elsewhere in this book, we won't give a full description of their alignment numbering scheme. See (Alperin and Lang, 2009) for further details.

With the brief description that we have provided, we can explain the method by which Alperin and Lang enumerated the 2-fold BOOs. Consider, again, the alignment shown in Figure 4.5(a), where we fold $P_1$ onto line $L_1$ and $P_2$ onto $L_2$ with two creases $a$ and $b$, respectively, in such a way that $a$ folds the point $P_3$ onto the same point that $b$ folds the point $P_4$. As we saw in Section 2.4, when we fold $P_1$ onto $L_1$, the image of the point $P_3$ will trace a cubic curve. Similarly, when we fold $P_2$ onto $L_2$, the point $P_4$ will trace another cubic curve. Candidates for the point $P_5$, where the images of $P_3$ and $P_4$ can coincide, will therefore lie on the intersection points of these two cubics.

In other words, determining whether or not the alignment AL6ab8 is a viable fold involves finding solutions to a cubic equation. The same thing can be done with any combination of alignments. Because of the degrees of freedom involved, one would need to consider all combinations of either 2, 3, or 4 of the the possible alignments, construct equations for them, and see if they can have solutions. Those that can will be viable BOOs for 2-folds. Those that can't will have an inconsistent or underdetermined set of equations to solve. Alperin and Lang wrote Mathematica code to examine all the possibilities, which is how they arrived at their list of 489 BOOs for 2-folds.

## 4.4    Solving Equations with Multifolds

In this section we will present some possibilities for solving equations via multifolds. We begin with a very straightforward method for solving general quartics with 2-folds.

Suppose that we have a quartic $a_4x^4 + a_3x^3 + a_2x^2 + a_1x + a_0 = 0$ and we consider using Lill's method to solve this. (See Section 2.3.) Our turtle path will start at the origin $O$ and travel to $P_4 = (a_4, 0)$, $P_3 = (a_4, a_3)$, $P_2 = (a_4 - a_2, a_3)$, $P_1 = (a_4 - a_2, a_3 - a_1)$, and finally $T = (a_4 - a_2 + a_0, a_3 - a_1)$.

To find the bullet path by which we may "shoot" the turtle, we employ a similar strategy to the one Beloch used to solve cubics with the 1-fold O7. Let $L_1$ be a line parallel to $\overleftrightarrow{P_1P_2}$ where $L_1$ is the same distance to, and on the opposite side of, $\overleftrightarrow{P_1P_2}$ as $O$. That is, $L_1$ is the line $x = 2a_4$. Similarly, let $L_2$ be defined by $x = a_4 - a_2 - a_0$ (which is parallel to $\overleftrightarrow{P_3P_4}$ and on the other side of, and equidistant to, $\overleftrightarrow{P_3P_4}$ than the point $T$).

We then perform the 2-fold AL16ab10a (the same move as shown in Figure 4.5(b)) by folding $O$ onto $L_1$ to produce a crease line $a$ and folding $T$ onto $L_2$ to produce a cease line $b$, so that $a \perp b$ and $a$ and $b$ will intersect each other at a point $Y$ located on the line $\overleftrightarrow{P_2P_3}$.

Figure 4.7 shows an example of this for the quartic polynomial $x^4 + 2x^3 + 3x^2 + x - 1$.

If we then let $X$ be the midpoint of $O$ and its image $O'$ under crease $a$ and we let $Z$ be the midpoint of $T$ and its image $T'$ under the crease $b$, then we have that $OX \perp XY$ and $XY \perp YZ$ and $YZ \perp ZT$. Furthermore, $X$ must lie on $\overleftrightarrow{P_1P_2}$ (by the construction of $L_1$) and $Z$ must lie on $\overleftrightarrow{P_3P_4}$ (by the construction of $L_2$). Therefore, $OXYZT$ is a bullet path and solution to Lill's method, which solves our quartic equation.

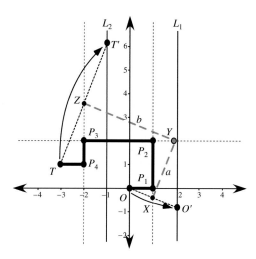

**Figure 4.7** Solving the quartic $x^4 + 2x^3 + 3x^2 + x - 1 = 0$ with a 2-fold.

Of course, since quartics can be reduced to cubics and quadratics, they can also be solved by 1-fold moves, indicating that the above construction isn't really showing any power of 2-folds over 1-folds. However, the 2-fold method for solving quartics is a very natural use of Lill's method, making it, theoretically, at least, much more elegant than the contortions one would have to go through to solve a quartic with 1-folds (as in (Geretschläger, 1998; Edwards and Shurman, 2001)).

It should be noted that to most people the example shown in Figure 4.7 will seem **impossible to perform** with a sheet of normal, convex paper. One thing that paper folding is never allowed to do is to make two points move farther away from each other than where they started, as this would require ripping the paper in some way. (This is only true for convex paper.) In Figure 4.7 we have the points $O$ and $T$ clearly being folded away from each other. However, if instead of folding $O$ onto the line $L_1$, we think of this alignment as being made by folding $L_1$ onto $O$ and similarly $L_2$ onto $T$, then instead of moving $O$ and $T$ farther apart, we will be folding parts of $L_1$ and $L_2$ closer together. Looked at this way, the 2-fold required in Figure 4.7 seems almost reasonable, even if incredibly difficult.

Next we will show how general quintics can be solved via 3-folds. The method presented here is due, again, to (Alperin and Lang, 2009) and is simply the same method used for the quartic but with an extra crease thrown in. (See Figure 4.8.)

To solve a quintic via Lill's method, a turtle path with six sides will be needed, call it $OP_1P_2P_3P_4P_5T$. As we did for the quartic case, we define a line $L_1$ to be equidistant from and on the opposite side of $\overleftrightarrow{P_1P_2}$ as the point $O$, and we define the line $L_2$ to be on the opposite side of $\overleftrightarrow{P_4P_5}$ as the point $T$ and equally spaced from $\overleftrightarrow{P_4P_5}$ as $T$. We then perform alignment AL6 twice to fold $O$ onto $L_1$ with a crease $a$ and $T$ onto $L_2$ with a crease $b$, but this time to make it well-defined we make a third crease $c$ that we want to have the following properties:

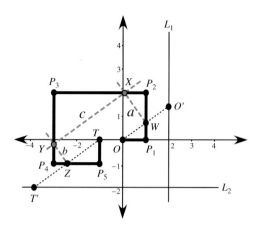

**Figure 4.8** Solving the quintic $x^5 + 2x^4 + 4x^3 + 3x^2 + 2x + 1 = 0$ with a 3-fold.

- $c$ should intersect $\overleftrightarrow{P_2P_3}$ at the same point $X$ as $a$ does.
- $c$ should intersect $\overleftrightarrow{P_3P_4}$ at the same point $Y$ as $b$ does.
- $a \perp c$ and $b \perp c$.

This makes the three creases $a$, $b$, and $c$ well-defined, assuming they're even possible. Making sure the creases are perpendicular is using AL1 again, and ensuring that they intersect on the required lines of the turtle path is using AL10. If we then let $W$ be where $a$ intersects $\overleftrightarrow{P_1P_2}$ and let $Z$ be where $b$ intersects $\overleftrightarrow{P_4P_5}$, we have that $OWXYZT$ is the bullet path that "shoots" the turtle, solving our quintic. An example of this is shown in Figure 4.8.

If the multifold solution for solving quartics seemed difficult to perform, then this method for quintics probably seems unfathomable and unfoldable. The folds $a$ and $b$ in Figure 4.8 aren't too bad if we think of $L_1$ being folded onto the point $O$ and $L_2$ being folded onto $T$. But the crease $c$ needs to be made at the same time in order to make it perpendicular to $a$ and $b$, and actually doing this will make the paper crumple and get incredibly messy. In theory it does work, but in practice it is next to impossible, even for the most experienced paper folder.

This raises a matter of some controversy concerning multifolds. Namely, should multifolds be considered legitimate origami operations even though they are often impossible to perform? Sometimes these folds can be made easier, as Lang did for his angle quintisection in Section 4.2, where he first folds the paper into a non-convex shape that facilitates the necessary 2-fold alignments. Certainly there are limits to such strategies, however, and as we consider $n$-folds for greater values of $n$, the folding moves quickly become impossible.

Nonetheless, from a theoretical point of view, multifolds can allow us to find real roots of any polynomial we desire.

**Theorem 4.6** (Alperin and Lang, 2009)   *Every polynomial equation of degree n with real solutions can be solved by an $(n - 2)$-fold.*

The proof is seen by generalizing the 3-fold solution for the quintic. A polynomial of degree $n$ will have a turtle path in Lill's method with $n + 1$ sides, and thus the bullet path for the solution will have $n$ sides. The first and last parts of the bullet path may be obtained by aligning $O$ and $T$ onto their proper directrices (lines $L_1$ and $L_2$ in the previous quartic and quintic examples), and thus the first and last segments won't need to be creases of our multifold. We will need creases for the remaining $n - 2$ sides of the bullet path. The second segment and second-to-last segment may be made by using AL6 to align $O$ onto $L_1$ and $T$ onto $L_2$, and the rest of the segments may be made with AL1 to make the creases perpendicular in sequence and with AL10 to make their intersections lie on the turtle path. In any case, this results in an $(n - 2)$-fold.

Theorem 4.6 does not give us the most efficient way to use multifolds to solve polynomial equations. In fact, finding efficient multifolds to do this sort of thing is an ongoing area of research. In 2013 Yasuzo Nishimura proved that arbitrary quintics can be solved by 2-fold origami (Nishimura, 2013), and in 2016 Joachim König and Dmitri Nedrenco proved 2-fold origami can solve arbitrary septic equations (König and Nedrenco, 2016). The latter is quite a surprise. It is currently unknown what the algebraic limit is of 2-fold origami, let alone 3-, 4-, or higher-fold origami.

## 4.5 Constructions with Curved Creases

Throughout this book thus far we have been careful to stipulate that the creases we've been folding are straight lines. Making **curved creases** is possible and has been studied extensively. See (Fuchs and Tabachnikov, 1999), for example. One way to make curved creases is to literally carve them into a sheet of paper. If one takes a ball-point pen and draws a curve with no self-intersections on a sheet of paper with a heavy hand, then this curve will be able to be folded. (This is sometimes more easily done if one cuts along a neighborhood of the curve to remove it from the rest of the paper, which can sometimes get in the way. But if the curve is simple enough, say with no points of inflection, then this won't be necessary.) With practice curved creases can be made without the aid of tools, and a growing number of origami artists are using curved creases to add dimension and realism to their artwork.

---

**Example 4.7** (Folding a circle)   Perhaps the most simple curve to fold would be a circle. Even with the "draw the curve with a pen" method, this is difficult to do, unless one allows the use of a compass. With practice, though, it can be done with good accuracy by the method shown in Figure 4.9

The procedure is to take a square piece of paper, crease it in half vertically and horizontally, and make four small 1/4 crease marks by folding the center of each side to the center of the square. Then fold the square into a cone by make a half-diagonal mountain crease as shown in step (2). In step (3) make a closed sink, which is a technical origami term for pushing the cone point down, inverting it, while not allowing the

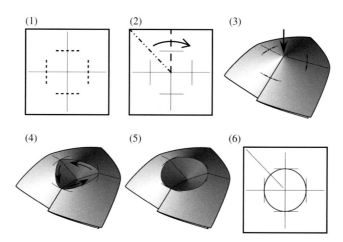

**Figure 4.9** One way to make a circle crease without tools. Step (2) is meant to fold the paper into a cone shape. Step (3) is a non-flat, closed sink, whose crease we smooth into a circle in step (4).

cone to unfold. The result will likely be more messy and chaotic than the illustration in step (4), but the point is that you want to start smoothing the sides of the closed sink, using the 1/4 marks as a guide to make the circle uniform, as shown in step (5). With the inverted part of the cone making a nice circle, crease it more firmly by running a fingernail along the inside of it. Unfolding the paper should reveal the crease pattern in step (6), complete with a creased circle.

Again, this method takes some practice to perform consistently and accurately.

While straight creases can be folded flat, curved creases cannot. Nonetheless, one may consider the possibility of using curved creases in origami geometric constructions.

It should come as no surprise, however, that if we allow the folding of circles, then we can construct $\pi$. For completeness, we describe how.

Consider the crease pattern on a strip of paper shown in Figure 4.10. First valley-fold a semicircle centered at a point $A$ on the strip's long side. Then make a valley crease from one end of the semicircle (point $B$) making an angle of 45° or less from the side of the paper.

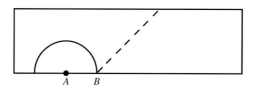

**Figure 4.10** A crease pattern for constructing $\pi$ from a strip of paper.

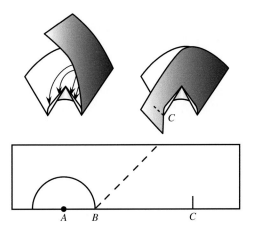

**Figure 4.11** Making the fold and marking the point $C$. $BC/AB = \pi$.

Fold the 45° crease flat and then fold the semicircle, which will make the paper form part of a cone. See Figure 4.11. The trick is then to slide the raw edge of the paper, brought into position by the 45° crease, into the semicircle fold. Mark with a crease where this raw edge meets the other end of the semicircle; this is the point $C$ in Figure 4.11. Unfolding the paper, we see that the length of the line segment $BC$ equals the perimeter of the semicircle. If the radius $AB$ of the semicircle equals 1, this means that $BC = \pi$.

Performing this origami maneuver is not easy. Even if you photocopy Figure 4.10 so as to make the semicircle crease more easily, it is still quite tricky to maneuver the raw edge of the paper into the semicircle crease. It should literally form part of a cylinder that will intersect the cone made by the rest of the paper along the semicircle. With practice, however, it can be done quite accurately. The author managed to fold one starting with a semicircle of radius 25 mm and creating a segment $BC$ of length 78.5 mm. If we assume that $AB$ is our unit length, this means that the relative length of $BC$ is 78.5/25 = 3.14, which is $\pi$ to two decimal places. For origami, that's pretty good accuracy.

Allowing curved creases therefore implies that non-algebraic numbers can be constructed. Determining exactly what transcendental numbers are possible is a wide-open question, and it likely depends on what kind of curves we deem to be foldable. The combination of folding curves and the ability of paper folding to measure lengths makes it seem that geometric constructions under curved-crease origami would be very powerful.

## 4.6    Open Problems

Multifolds have offered new areas in which to explore origami geometric constructions. We offer a few open problems that remain for multifold origami.

**Open Problem 4.1**   What is the full algebraic power of 2-fold origami? That is, what is the greatest $n$ for which 2-fold origami can always solve arbitrary polynomial equations of degree $n$?

**Open Problem 4.2**   Can a better bound be found on multifold origami than that given in Theorem 4.6? That is, can some $k_n$ be found so that any polynomial equation of degree $n$ with real solutions can be solved with a $k_n$-fold, where $k_n < n - 2$?

# Part II

# The Combinatorial Geometry of Flat Origami

The study of the mathematics of paper folding is inherently a modeling process. Folding is a physical action, and if we want to understand the rules at play, we need to decide on a way to model it.

The first part of this book focused on modeling the geometry of paper folding in terms of what it can construct. Such an approach clearly demonstrates the power and versatility of origami, but it tells us nothing about **how** paper can fold. For example, is there a way to look at a crease pattern and know what it could fold into? Or might some crease patterns not fold into anything?

A model of origami that includes combinatorics as well as geometry turns out to provide valuable insights into such questions. In Part II we will explore this model by considering **flat origami**, paper folds that can be pressed in a book without crumpling or adding new creases. First, local conditions and properties for flat-foldability will be explored, and then we will turn to the substantial challenges to extending these globally. We will also encounter applications to origami design and other classic folding problems.

# 5     Flat Vertex Folds: Local Properties

In this chapter we will begin our investigation of **flat origami**, that is, folded paper objects that lie flat, or can be pressed in a book without crumpling, by first considering the case where our crease pattern has only one vertex in the paper's interior. While this case, the local properties of flat origami crease patterns, is very well-understood, it still offers a number of surprises and open problems. This material also forms the foundation on which all other combinatorial work in origami is based.

## 5.1     Definitions and Flat-Foldability

Our first challenge is to provide a definition of an **origami fold** and in particular a **flat fold**. Note that in Part I we considered a "fold" to mean a single crease made on a piece of paper determined by aligning previously constructed points and lines. Now our interpretation of the word "fold" will change to mean any folded object, such as a folded crane or any origami model folded from a single sheet of paper that one might learn from an origami instruction book. Such folds are determined by their crease pattern. See Figure 5.1.

We start with our piece of paper, which is a closed region $R$ of the plane. It will be assumed that $R$ is also bounded and simply connected (no holes), unless otherwise noted. Typically in origami $R$ is a square, but it need not be.

**Definition 5.1**    Given a piece of paper $R \subset \mathbb{R}^2$, a **crease pattern on $R$** is a plane graph $G = (V, E)$ with $V \subset R$ and where each edge $e \in E$ is in the interior of $R$, except possibly its endpoints. Vertices on the boundary of $R$ are called **boundary vertices** and vertices in the interior of $R$ are called **interior vertices**. The **faces** of the crease pattern $G$ are the connected components of $R \setminus (V \cup (\cup E))$.

In cases where it is understood what $R$ is, or if it doesn't matter, we will simply refer to the plane graph $G$ as the crease pattern. The idea is that the edges of $G$ will be the individual folds of our origami model. Note that if a face $f$ of a crease pattern $G = (V, E)$ is not bounded by edges in $E$, then some of the boundary of $f$ will be part of the boundary of the paper $R$.

Our next task is to define what we mean by a **flat origami** model, and this is a tricky task to do precisely. Some basic assumptions about our folding medium and crease patterns need to be made:

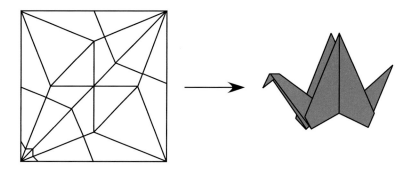

**Figure 5.1** A flat fold and its crease pattern.

- The edges in our crease patterns will be straight lines.
- The paper is assumed to be non-stretchable and cannot be torn.
- The paper cannot intersect itself.
- When we fold the paper into a shape, in the final result the paper should bend only along the edges of the crease pattern.

These assumptions do limit the kind of origami we are considering for now. For example, curved creases are possible but do not result in flat origami models.

Several researchers have tried to define flat origami in different ways. The definition we present here is a based on those given in (Bern and Hayes, 1996; Justin, 1997; Dacorogna et al., 2008).

**Definition 5.2**    Given a crease pattern $G = (V, E)$ on a region $R$, an **origami on $G$** is a continuous, one-to-one mapping $\sigma : R \to \mathbb{R}^3$ such that $\sigma$ is smooth (differentiable, say $C^\infty$ for purposes of simplicity) everywhere except along the creases $E$.

This definition is left intentionally flexible to cover the multitude of ways in which a sheet of paper can be contorted into a flat object. For example, nothing is said in this definition about not allowing the paper to stretch. We will insist, however, on more of our folding medium assumptions being met in the limit to a flat-folded state. For this we will need to define the **folding angle** at a crease.

**Definition 5.3**    Given two adjacent faces $f_1$ and $f_2$ in the crease pattern $G$ and an origami $\sigma$, if $\sigma$ restricted to $f_1$ and $f_2$ are both isometries, then we define the **folding angle** of the crease $e$ between these faces to be the signed angle of displacement from a flat plane exhibited by $f_1$ and $f_2$ under $\sigma$. If the folding angle is positive, we say the crease between $f_1$ and $f_2$ is a **valley crease**, and if the folding angle is negative, we say the crease is a **mountain crease**.

If $\sigma$ restricted to $f_1$ and $f_2$ are **not** both isometries, then we define the folding angle of the crease $e$ to be the signed angle of displacement along the crease with the smallest absolute value.

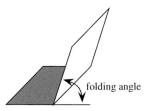

**Figure 5.2** The folding angle of a crease.

The folding angle is illustrated in Figure 5.2. Another way to think of the folding angle is that it is the supplement to the dihedral angle between the adjacent faces. Whether creases are mountains or valleys will depend, of course, on the general orientation of the origami fold, which corresponds to the fact that if a folded sheet of paper is turned over, all the valley creases will turn into mountain creases and vice versa. Note, however, that the generalness of our definition of **origami** allows for mappings $\sigma$ where the dihedral angle along a crease might not be constant. In such cases the folding angle along the crease will not be constant either, whereupon our folding angle is $\pi$ – the maximum dihedral angle present along the crease line.

**Definition 5.4**    Given a crease pattern $G = (V, E)$ on a region $R$, a **flat origami on** $G$ is an infinite sequence of origamis $\{\sigma_n\}_{n=1}^{\infty}$ on $G$ such that

- for each face $f$ of $G$, the images $\{\sigma_n(f)\}_{n=1}^{\infty}$ uniformly converge to a planar polygon congruent to $f$ and
- for each crease $l \in E$, the folding angles of the images $\{\sigma_n(l)\}_{n=1}^{\infty}$ converge to either $\pi$ or $-\pi$.

If there exists a flat origami on a given crease pattern $G$, then we say that $G$ is **flat-foldable** and that $G$ **folds flat**.

We may also define the limit map $\sigma : R \rightarrow \mathbb{R}^3$ to be $\sigma(x) = \lim_{n \to \infty} \sigma_n(x)$ for all $x \in R$, which we call the **folding map**. Note that we need the convergence in Definition 5.4 to be uniform to ensure that the limit map $\sigma$ is continuous. Also, $\sigma$ is an isometry on every face of $G$, but it does not fit the definition of an origami because it is not one-to-one. This is one of the problems with defining flat origami, in that we do not want the paper to intersect itself, but we also want the folding angles to all be $\pi$ or $-\pi$. Furthermore, $\sigma$ looses the information of which creases were mountains and which were valleys. Even though its codomain is $\mathbb{R}^3$, its range is planar, and so the only way to tell which creases were mountains and which were valleys would be to look at one of the maps $\sigma_n$.

This map $\sigma$ can be defined in another way that does not rely on limiting processes. We will see this in Sections 5.7 and 6.2.

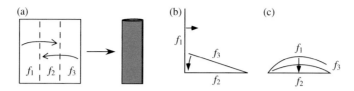

**Figure 5.3** Perfect thirds can't be folded flat without bending the faces.

**Example 5.5** (Folding thirds)  Let us motivate Definitions 5.2 and 5.4 by way of an example. Folding a square into thirds lengthwise, as in Figure 5.3(a), is a common origami move. If we try to perform this fold rigidly, keeping the faces $f_1, f_2$, and $f_3$ planar as the creases are folded, then we will only be able to fold one of the creases flat. As seen in Figure 5.3(b), the crease between faces $f_2$ and $f_3$ needs to be folded so that its folding angle approaches $\pi$, but as this limit is taken, the crease between $f_1$ and $f_2$ can only make a folding angle up to $\pi/2$.

If, however, we allow the faces $f_1$ and $f_2$ to smoothly bend, or even stretch, then faces $f_1$ and $f_3$ can both be folded down to $f_2$, as in Figure 5.3(c).

**Diversion 5.1**    Produce an infinite sequence of origamis $\sigma_n \colon [0, 1] \times [0, 1] \to \mathbb{R}^3$ to prove that the "folding thirds" crease pattern in Figure 5.3(a) is flat-foldable.

This example helps illustrate the way in which we can think and work with flat-foldable models. The sequence of origamis $\{\sigma_n\}_{n=1}^{\infty}$ is basically establishing a homotopy between the flat-folded state, described by the limit function $\sigma$, and a slightly unfolded state where the faces of the crease pattern might be slightly curved or stretched, but all of the folding angles are close to either $\pi$ or $-\pi$. Thinking of a flat origami model in this "slightly unfolded" state will be very useful in later proofs.

**Remark 5.6**    There are other ways in which we could model flat origami folds. The one presented here offers theoretical advantages as well as gives the impression of modeling how paper actually deforms into a flat-folded state. But computationally it leaves much to be desired. The example in Diversion 5.1 only has two creases, but creating a sequence of maps $\{\sigma_n\}_{n=1}^{\infty}$ for the flat origami definition is not too easy. Generating actual flat origami map sequences for crease patterns with more creases and vertices is very difficult to do in general, and it is best avoided.

A different approach to modeling flat origami that is more algorithmic can be found in Demaine and O'Rourke's text (Demaine and O'Rourke, 2007). The model they use defines origami as a non-stretching map from the paper into $\mathbb{R}^3$ together with a layering function for the parts that fold flat in order to distinguish what points of the paper lie on top of which others. They then impose restrictions, called non-crossing conditions, to ensure that the paper does not cross through itself. That is, they do not require that their folding map be one-to-one, and this allows them to consider folding angles of $\pi$ or $-\pi$ as routine, but it does mean that they have to perform other checks to make sure that the paper is not penetrating itself. The non-crossing conditions that Demaine and O'Rourke use were first developed by Justin (1997); we will encounter them in Chapter 6.

The major question of this part of the book is, "Given an origami crease pattern, can we fold it flat?" Definition 5.4 is far too cumbersome for practical use on general crease patterns, and so we strive to develop techniques and theorems that will aid us in answering this question.

**Definition 5.7**     A **mountain-valley assignment** (or **MV assignment**) for a crease pattern $G = (V, E)$ is a function $\mu : E \to \{-1, 1\}$ that assigns folding angles of $\mu(c)\pi$ to each crease $c \in E$. (So, $-1$ indicates a mountain and 1 a valley.) An MV assignment is called **valid** if it can be realized by a flat origami on the crease pattern.

In other words, an MV assignment $\mu$ is valid if there exists a flat origami $\{\sigma_n\}_{n=1}^{\infty}$ such that for every crease $l \in E$ the folding angles at $\sigma_n(l)$ converge to $\pi\mu(l)$. Also, any flat origami $\{\sigma_n\}_{n=1}^{\infty}$ generates a valid MV assignment by letting $\mu(l)$ equal the sign of the folding angles $\sigma_n(l)$ for all $l \in E$ and sufficiently large $n$.

Definition 5.7 can be generalized to non-flat origami models, which we will do in Part III of this book.

The following diversion is very useful for learning about nontrivial ways in which paper can fold flat. It is recommended, however, that readers approach it intuitively and by folding actual paper rather than applying Definition 5.4.

---

**Diversion 5.2**     Can the **square twist** crease pattern shown in Figure 5.4 be folded flat? If so, can it be folded flat in more than one way? (That is, how many different valid MV assignments are there for this crease pattern?)

---

Examples such as that in Diversion 5.2 show that there can be many different valid MV assignments for a given flat-foldable crease pattern, and therefore there can be many different flat origamis $\{\sigma_n\}_{n=1}^{\infty}$ that can be used to show that a given crease pattern folds flat.

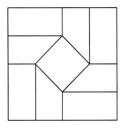

**Figure 5.4** The square twist crease pattern.

Our first theorem about flat folds was first observed by Toshiyuki Meguro in the 1980s. (Although it is also implied in Stewart Robertson's work (Robertson, 1977– 1978).)

**Theorem 5.8**    *The faces of a flat origami crease pattern are 2-colorable.*

*Proof*    We first give an informal proof that is short and very convincing. Given a flat origami crease pattern, choose one side of the paper to focus on and fold the crease pattern flat. Then hold the model in the outstretched palm of your hand and consider all the faces of the crease pattern in your folded model on the side of the paper on which you are focusing. Color green all such faces that are facing up, away from your palm, and all faces that are facing down into your palm color purple. Any two faces that are adjacent in the crease pattern will be facing in different directions when folded flat, so this results in a proper 2-coloring of the faces of the crease pattern.

A more rigorous proof is also useful in that it introduces some techniques that will be used later. Pick a face $f$ of the crease pattern $G = (V, E)$ and let $f'$ be any other face. Let $\gamma$ be any vertex-avoiding path (that is never tangent to a crease) on the crease pattern from a point in $f$ to a point in $f'$ and let $l_1, l_2, \ldots, l_k$ be the creases that $\gamma$ crosses, in order. Since we know that the crease pattern is flat-foldable, there exists a valid MV assignment $\mu : E \to \{-1, 1\}$. Then we define the function

$$Q(f') = \sum_{i=1}^{k} \mu(l_i) \quad \text{mod } 2.$$

Note that $Q(f')$ is merely measuring whether we've crossed an even or an odd number of creases when traveling from $f$ to $f'$.

We claim that $Q(f')$ is well-defined, that is, independent on the choice of the curve $\gamma$. To see this, let $\gamma'$ be a different vertex-avoiding path from a point on $f$ to a point in $f'$. Then if we follow $\gamma'$ and then $\gamma^{-1}$ (going backward along $\gamma$), we'll form a loop on the crease pattern from $f$ to $f$.

Let $f, f_1, f_2, f_3, \ldots, f_n, f$ be the faces, in order, that the loop $\ell = \gamma' \gamma^{-1}$ crosses. Then traveling along $\ell$ we have $Q(f) = 0$, $Q(f_1) = 1$, $Q(f_2) = 0$, $Q(f_3) = 1$, and $Q(f_i)$ will be 1 if $i$ is odd and 0 is $i$ even. This can be interpreted as the faces being flipped "up" ($Q(f_i) = 0$) or "down" ($Q(f_i) = 1$) as we follow this loop on the flat-folded piece of paper. However, when we complete the loop we should still have $Q(f) = 0$, and so $Q(f_n) = 1$, which means $n$ is odd. Therefore the loop $\gamma' \gamma^{-1}$ crosses an even number of faces, and computing $Q(f')$ based on $\gamma$ or $\gamma'$ will give the same even/odd parity, or they wouldn't add to an even number. Thus $Q(f')$ results in the same value whether we travel along $\gamma$ or $\gamma'$.

Now we can establish our proper 2-face coloring as follows: color the face $f$ green, and for any other face $f'$, color $f'$ green if $Q(f') = 0$ and purple if $Q(f') = 1$.    □

**Definition 5.9**    A **single-vertex fold** is a crease pattern $G = (V, E)$ with only one interior vertex. A **flat vertex fold** is a single-vertex fold that folds flat.

As stated previously, flat vertex folds are the subject of this chapter. The following is a direct result of Theorem 5.8.

**Corollary 5.10** *The degree of the interior vertex in a flat vertex fold is even.*

This result can be proved in a number of ways, but the 2-colorability of the faces of flat folds seems the most natural way to describe why flat vertex folds have even degree.

As a final comment for this section, we note that while the definitions for flat origamis specify that our piece of paper is a closed bounded region $R$ of the plane, they only use the flatness of $R$ in the limiting process of the sequence of origamis $\{\sigma_n\}_{n=1}^{\infty}$. This concept can be generalized to other 2-manifolds, and some of these will be explored in Chapter 10. But it is very useful, and necessary, for us to expand the concept of single- and flat vertex folds to ones where the region $R$ is a **cone** with the sole interior vertex at the cone's apex. In fact, many (but not all!) of the results we will see for flat vertex folds will also hold for cone-shaped paper.

**Definition 5.11** A single-vertex crease pattern on a **cone** is a single-vertex crease pattern $G$ on a closed bounded region $R$ on the surface of a cone that includes the apex and where the interior vertex of $G$ is placed at the apex of the cone. The **cone angle** of the cone is the angle measure around the apex of the cone. If the cone angle equals $2\pi$, then we are folding **flat paper**. If the cone angle is greater than $2\pi$, then we are folding **hyperbolic paper**.

Being able to consider single-vertex crease patterns on cones is important for some recursive folding processes that we will encounter.

## 5.2 Mountain-Valley Parity

One of the most basic results for flat vertex folds is known as Maekawa's Theorem ((Kasahara and Takahama, 1987), although this book originally appeared in Japanese in 1985), named after the Japanese physicist and origami artist Jun Maekawa. Jacques Justin (1986a) independently discovered this same result.

**Maekawa's Theorem** *The difference between the number of mountain and valley creases in a flat vertex fold on a cone with cone angle $\leq 2\pi$ is 2.*

*In other words, if $\mu$ is a valid MV assignment for a flat vertex fold on a cone with creases $l_1, \ldots l_{2n}$ and cone angle $\leq 2\pi$, then $\sum_{i=1}^{2n} \mu(l_i) = \pm 2$.*

*Proof* Fold the vertex flat and imagine cutting the vertex off with scissors, leaving a flat polygonal cross section. (See Figure 5.5.) Imagine a monorail traveling along this cross section in a counterclockwise manner. Then, assuming that we're looking at the cross section from above, every time the monorail gets to a mountain crease, it

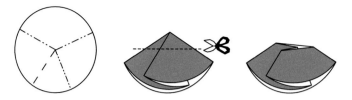

**Figure 5.5** The cross section of a flat vertex fold.

will rotate $\pi$ radians, and every time it gets to a valley crease, it'll rotate $-\pi$ radians. When it gets back to where it started, it will have rotated a full $2\pi$ radians. If $M$ and $V$ denote the number of mountain and valley creases, respectively, then we have

$$\pi M - \pi V = 2\pi \;\Rightarrow\; M - V = 2. \tag{5.1}$$

That is, $-\sum_{i=1}^{2n} \mu(l_i) = 2$. If we had looked at the vertex "from below," we would have gotten $-2$. □

**Remark 5.12** The proof of Maekawa's Theorem may seem too informal, but the definition of flat origami makes the argument completely valid. We're not really cutting the vertex off of the flat-folded state (the image of the limiting map $\sigma$) of the flat vertex fold, for if we did then the "cross section" would be just a line segment. Rather, since the vertex is flat-foldable, we know that there exists an infinite sequence of origamis $\{\sigma_n\}_{n=1}^{\infty}$ that converges uniformly to $\sigma$, and so we can take a specific $\sigma_n$ for large enough $n$ that will approximate the flat-folded state but will have the cross section we seek for our proof. Technically, the folding angles will not be exactly $\pi$ or $-\pi$, but they'll be very close and in the limit will give us Equation (5.1).

In general, whenever we consider the cross section of a flat origami, we will assume that we are actually considering the cross section of one of the maps $\sigma_n$ as close to the limit map $\sigma$ as we wish. This technique will be used without mention in several of the proofs that follow in this chapter.

Also, the assumption in Maekawa's Theorem that the cone angle of the paper is less than or equal to $2\pi$ is what allows us to cut away the folded vertex as shown in Figure 5.5. This subtlety will be explored more later on, in Theorem 5.17.

Note that Maekawa's Theorem is only a necessary condition for flat vertex folds.

---

**Diversion 5.3**    Find a flat vertex fold for which not all MV assignments that satisfy Maekawa's Theorem are valid. Are there any single-vertex crease patterns for which **all** MV assignments with $M - V = \pm 2$ are valid?

---

The case of flat vertex folds on hyperbolic paper is more complicated. Such folds could collapse in a manner similar to flat paper, giving $M - V = \pm 2$ **or** they could collapse in a different way that gives $M - V = 0$. (See Figure 5.6.) To help distinguish between these two cases, we introduce some terminology.

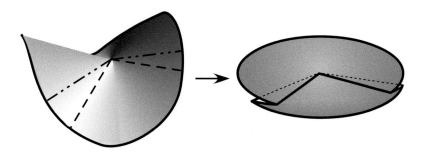

**Figure 5.6** A different way to fold hyperbolic paper flat.

**Definition 5.13** Let $v$ be an interior vertex in the crease pattern $G$ of a flat origami. Let $R_\varepsilon$ be a circle of radius $\varepsilon$ centered at $v$ where $\varepsilon$ is taken small enough so that the only creases of $G$ that intersect the boundary of $R_\varepsilon$ are those adjacent to $v$. Then the subset $G'$ of the embedded crease pattern $G$ containing only the vertex $v$ and its adjacent edges intersected with $R_\varepsilon$ is called the **local flat vertex fold of $G$ at $v$ on the disc $R_\varepsilon$.**

**Definition 5.14** Given a crease pattern $G = (V, E)$ on a region $R$ of a flat origami $\{\sigma_n\}_{n=1}^\infty$ converging to the flat-folded state map $\sigma$, let $x \in \sigma(R)$ be a point that is not the image of a crease line or a vertex of $G$. Then we say that the **number of layers at** $x$, denoted $L(x)$, is $|\sigma^{-1}(x)|$ (the size of the pre-image of $x$).

**Theorem 5.15** *Let $v$ be a vertex of a flat-foldable crease pattern $G$ with limit flat-folded state map $\sigma$, and consider the local flat vertex fold at $v$ on $R_\varepsilon$. Then $L(x)$ will either always be even or always be odd for all $x \in \sigma(R_\varepsilon)$ that are not the image of a crease or vertex.*

*Proof* Let $P_1$ and $P_2 \in \sigma(R_\varepsilon)$ be points that are not the image of a crease point or a vertex of the local flat vertex fold of $G$ at $v$. Our goal is to show that $L(P_1)$ and $L(P_2)$ have the same even/odd parity. Let $\gamma$ be a vertex-avoiding curve from $P_1$ to $P_2$ on $\sigma(R_\varepsilon)$ parameterized by the variable $t$ for $0 \le t \le 1$. We start at $P_1 = \gamma(0)$. As we travel along $\gamma$ the value of $L(\gamma(t))$ will not change until we come to a point $\gamma(a)$ that is the image of a point on a crease line (possibly more than one crease line). Crossing this point, we'll have that $L(\gamma(a + h))$ (for small $h$) will be $L(P_1)$ plus or minus 2 for every crease whose image under $\sigma$ contains $\gamma(a)$. Continuing along $\gamma$ we may cross more creases and either add or subtract 2 to $L(\gamma(t))$ each time, resulting in $L(P_2)$ having the same parity as $L(P_1)$. $\qquad\square$

A version of Theorem 5.15 holds for multiple-vertex flat origamis.

**Definition 5.16** Let $v$ be an interior vertex in the crease pattern of a flat origami fold $G$, and consider the local flat vertex fold of $G$ at $v$ on a disc $R_\varepsilon$. If, when we fold $R_\varepsilon$ flat according to the crease pattern, we have an even number of layers at the

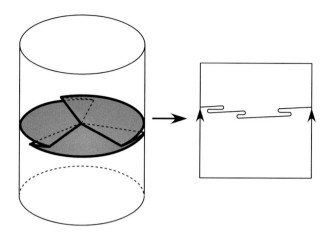

**Figure 5.7** The boundary of a folded disc flat vertex fold projects onto a cylinder.

non-crease points on the folded image of $R_\varepsilon$, then we say that the flat vertex fold is **pointy**. Otherwise, if there are an odd number of layers, then we say that the flat vertex fold is a **folded disc**.

The flat vertex folds shown in Figures 5.6 and 5.7 are examples of folded discs. The ones shown in Figures 5.5 and 5.8 are pointy.

**Theorem 5.17**  *The difference between the number of mountain and valley creases in a flat vertex fold on hyperbolic paper is 2 if the fold is pointy and 0 if it is a folded disc.*

*Proof*  If the flat vertex fold is pointy, then the proof proceeds exactly as for Maekawa's Theorem, although perhaps a better way to summarize the proof is that in the pointy case, the cross section of the folded vertex can be projected (from the vertex) down onto a plane, and so will be a loop whose turning number is 1, thus giving us either $\pi M - \pi V = 2\pi$ or $\pi V - \pi M = 2\pi$. (We describe this in terms of turning numbers because they will be used in what follows. The **turning number** of a curve is the total rotation of its tangent vector as we travel around the curve divided by $2\pi$. For background on turning numbers, see (Whitney, 1937).)

However, if the flat vertex fold is a folded disc, then the cross section (i.e., boundary of an $R_\varepsilon$ local fold of the vertex) of the folded paper will project onto a cylinder from the vertex. (See Figure 5.7.) Topologically, this is very different from the previous case in that this projected curve on the cylinder does not surround a region homeomorphic to an open disc. Rather, it divides the cylinder into two half-cynilders. In other words, the turning number of this curve on the cylinder is 0, giving $\pi M - \pi V = 0$ (or $\pi V - \pi M = 0$).  □

The difference between a flat vertex fold being pointy or a folded disc is an important one that we could define differently. We will resume this discussion in Section 5.4

## 5.3      Necessary and Sufficient Conditions

Whether or not a single-vertex crease pattern is flat-foldable is completely determined by the angles between the creases. Interestingly, the specific MV assignment does not matter! If the angles between the creases meet the proper condition, then a valid MV assignment will be guaranteed to exist. This result is known as Kawasaki's Theorem (Kasahara and Takahama, 1987), although it was also discovered independently by Robertson (1977–1978) and Justin (1984). Robertson's version, although first to appear, only proves the necessary direction. He does, however, prove it in much more generality than Kawasaki and Justin, placing the result in the context of isometric foldings of Riemannian manifolds of arbitrary dimension. We will explore Robertson's approach later, in Chapter 10.

We will be greatly aided by a result concerning the partial alternating sums of certain finite sequences. If $(\alpha_0, \alpha_1, \ldots, \alpha_{2n-1})$ is a finite sequence of positive real numbers, define for $0 \leq i, j \leq 2n - 1$,

$$S(i, j) = \alpha_i - \alpha_{i+1} + \alpha_{i+2} - \cdots \pm \alpha_j,$$

where the subscripts are taken mod $2n$.

**Lemma 5.18**    *Let* $(\alpha_0, \ldots \alpha_{2n-1})$ *be a sequence of* $2n$ *positive real numbers with the property that* $\alpha_0 - \alpha_1 + \alpha_2 - \cdots - \alpha_{2n-1} = 0$. *Then there exists a* $k$ *with* $0 \leq k \leq 2n - 1$ *such that* $S(k, i) \geq 0$ *for all* $0 \leq i \leq 2n - 1$.

*Proof*    Suppose that there exist $i$ such that $S(0, i) < 0$ (otherwise we could take $k = 0$ and be done), and let $k - 1$ be the index that achieves $\min\{S(0, i) : 0 \leq i \leq 2n - 1\}$. If there is more than one index that achieves this (negative) minimum, then $k - 1$ can be any one of them. Also, since these partial sums $S(0, i)$ start with $\alpha_0$ being positive, the minimum must occur on one of the subtracted odd-indexed terms, so $k - 1$ is odd.

We claim that $S(k, i) \geq 0$ for all $0 \leq i \leq 2n - 1$. To see this, first notice that for $k \leq i \leq 2n - 1$ we have $S(0, k - 1) \leq S(0, i)$, and subtracting the terms on the left gives $0 \leq S(k, i)$.

If $0 \leq i \leq k - 2$ then we still have $S(0, k - 1) \leq S(0, i)$, but now this implies $S(i + 1, k - 1) \leq 0$. But $S(i + 1, k - 1) + S(k, i) = S(0, 2n - 1) = 0$, thus $S(k, i) \geq 0$.

If $i = k - 1$, then $S(k, k - 1) = \alpha_k - \alpha_{k+1} + \cdots - \alpha_{2n-1} + \alpha_0 - \cdots - \alpha_{k-1} = S(0, 2n - 1) = 0$.      $\square$

**Kawasaki's Theorem**    *Let* $G$ *be a single-vertex crease pattern on a cone with cone angle* $\leq 2\pi$ *and with consecutive angles between the creases* $\alpha_0, \ldots, \alpha_{2n-1}$. *Then* $G$ *is flat-foldable if and only if*

$$\alpha_0 - \alpha_1 + \alpha_2 - \cdots - \alpha_{2n-1} = 0.$$

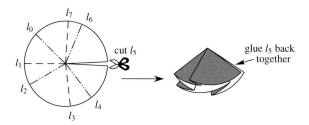

**Figure 5.8** Creating a valid MV assignment for a flat vertex fold.

*Proof*   Let $G$ be a flat vertex fold with consecutive angles $\alpha_0, \ldots, \alpha_{2n-1}$ and suppose that $G$ is embedded on a piece of paper $R$ that is either a circle or cone of radius 1 with the interior vertex at the circle's center (or cone's apex). Then the angles $\alpha_i$ in radian measure are equal to the arc-lengths, they subtend on the circle/cone. Let $\gamma$ be the oriented curve on the boundary of $R$, and assume that $\gamma$ starts at the crease between angles $\alpha_{2n-1}$ and $\alpha_0$ and then travels through the arcs $\alpha_1, \alpha_2$, and so on. Fold this embedding of $G$ on $R$ flat and consider the image $\sigma(\gamma)$ under this folding. It will follow an arc of length $\alpha_0$ in one direction, and then an arc of length $\alpha_1$ in the opposite direction, and then an $\alpha_2$-arc in the $\alpha_0$-direction, and so on until traversing an arc with length $\alpha_{2n-1}$ to get back to where it started. The (oriented) distance traveled can thus be expressed as $\alpha_0 - \alpha_1 + \alpha_2 - \cdots - \alpha_{2n-1} = 0$.

For the other direction we are given a single-vertex fold $G$ with angles $\alpha_i$ that satisfy $S(0, 2n-1) = 0$, using the notation of Lemma 5.18, which then tells us that there exists an index $k$ such that $S(k, i) \geq 0$ for all $0 \leq i \leq 2n - 1$. Let $l_i$ denote the crease line between angles $\alpha_{i-1}$ and $\alpha_i$ in $G$, where the indices are taken mod $2n$. Consider the following MV assignment for $G$:

$$\mu(l_{k+i}) = \begin{cases} -1 & \text{for } i \text{ odd,} \\ 1 & \text{for } i \text{ even, } i \neq 0, \\ -1 & \text{for } i = 0 \end{cases}$$

(where the index $k + i$ is taken mod $2n$). We claim that $\mu$ is a valid MV assignment for $G$, for if we cut our paper along the crease $l_k$, then the other creases $l_i$ ($i \neq k$) can be folded using $\mu$, which is simply folding them with alternating mountains and valleys into a zig-zag shape (called an **accordion pleat** by origamists), and can be made flat without any self-intersections of the paper. Furthermore, since $S(k, i) \geq 0$ for all $0 \leq i \leq 2n - 1$, we know that the cut ends of crease $l_k$ won't have any layers of paper in between them after this accordion pleat is folded. Thus they can be glued back together, making $l_k$ a mountain crease and proving that $\mu$ is a valid MV assignment. (An example of this process is shown in Figure 5.8.)    $\square$

The alternating sum of the angles condition in Kawasaki's Theorem is sometimes referred to as the **Kawasaki condition**. If one adds this condition to $\alpha_0 + \cdots + \alpha_{2n-1} = $ the cone angle of the paper, we obtain the following version of Kawasaki's Theorem:

**Corollary 5.19** *Let G be a single-vertex crease pattern on a cone with cone angle $A \leq 2\pi$ and with consecutive angles between the creases $\alpha_0, \ldots, \alpha_{2n-1}$. Then G is flat-foldable if and only if*

$$\alpha_1 + \alpha_3 + \cdots + \alpha_{2n-1} = \alpha_0 + \alpha_2 + \cdots + \alpha_{2n-2} = \frac{A}{2}.$$

Like Maekawa's Theorem, there is a slightly different version of Kawasaki's Theorem for flat vertex folds with cone angle greater than $2\pi$. As far as we know, Demaine and O'Rourke (2007) were the first to specify this generalization, but their proof is different from the one we present here.

**Theorem 5.20** *Let G be a single-vertex crease pattern on a cone with cone angle $> 2\pi$ (on hyperbolic paper) and with consecutive angles between the creases $\alpha_0, \ldots, \alpha_{2n-1}$. Then G is flat-foldable if and only if*

$$\alpha_0 - \alpha_1 + \alpha_2 - \cdots - \alpha_{2n-1} \in \{-2\pi, 0, 2\pi\}.$$

*Proof* If a vertex fold on hyperbolic paper is flat-foldable, then it is either pointy or a folded disc. If it is pointy, then the alternating angle sum will equal 0 as in the non-hyperbolic case. If it is a folded disc, then the alternating angle sum will be $\pm 2\pi$ because the flat-folded disc still has $2\pi$ radians around it. That is, instead of the alternating angle sum "coming back to where we started" and giving us 0, we will come back to where we started after traveling around the folded disc, giving us $2\pi$ (or $-2\pi$ if the initial angle causes the traveling to go in the opposite direction).

For the converse direction, if the alternating sum of the angles is 0, then the proof used in Kawasaki's Theorem can be used to prove that the vertex folds flat. If the alternating angle sum is $\pm 2\pi$, then we need to show that the vertex can be folded into a folded disc. To do that we will create an MV assignment that will fold the boundary of an $R_\varepsilon$ local fold of this vertex into a curve that can be projected from the vertex onto a cylinder without self-intersections.

This, it turns out, is very easy to do; just assign the creases to be mountains and valleys alternating around the vertex. To see this, label the creases $l_i$ as in the proof of Kawasaki's Theorem, and cut the paper (our $R_\varepsilon$ region) along crease $l_1$. Then define the MV assignment:

$$\mu(l_i) = \begin{cases} -1 & \text{for } i \text{ even,} \\ 1 & \text{for } i \text{ odd.} \end{cases}$$

We can then apply this MV assignment to the crease lines $l_2, \ldots, l_{2n}$, making an accordion pleat. Because the alternating sum of the angles is $\pm 2\pi$, the loose ends of the cut crease $l_1$ will line up with each other after the paper wraps around the folded disc (or as the boundary of $R_\varepsilon$ wraps around on the projected cylinder, if you prefer). Since the MV assignment is only alternating mountains and valleys, there won't be any layers of paper in between these loose ends of $l_1$ and they can be glued together, making $l_1$ a valley. The only possible obstacle to this would be if the paper was made to "spiral" in the wrong direction, so that the loose ends of $l_1$ were made to be opposite instead

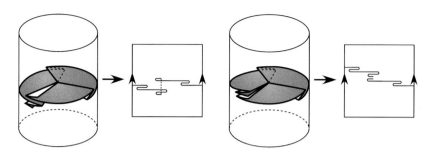

Figure 5.9 Changing the "spiraling" when folding a folded disc on hyperbolic paper.

of facing each other. But this is easily fixed by changing the direction of this spiraling (see Figure 5.9).                                                                                          □

Kawasaki's Theorem and Theorem 5.20 tell us exactly when a single-vertex crease pattern without an MV assignment will fold flat. As we will see in Chapter 6, determining whether a multiple-vertex crease pattern is flat-foldable is much more difficult.

But what if an MV assignment were included with a flat vertex fold? As you may have discovered in Diversion 5.3, not all MV assignments that satisfy Maekawa's Theorem will be valid for a flat vertex fold. Determining when an MV assignment will be valid is very much related to the problem of counting how many valid MV assignments there are for a given flat vertex fold, and this is the subject of Section 5.5.

First, however, we will consider a different way to view what we've learned about folding hyperbolic paper.

## 5.4     Cone Folds

Theorems 5.17 and 5.20 might seem like anomalies, contradictions to Maekawa's and Kawasaki's Theorems that only happen when folding hyperbolic paper. Yet one can view these cases as being more natural if we expand our concept of flat origami when it comes to single-vertex folds. Let us denote by $C_A$ a cone with cone angle $A$.

**Definition 5.21**    A **cone fold** is a single-vertex fold where the region $R = C_A$ is the bounded surface of a cone with the interior vertex of the fold located at the apex of $C_A$.

For convenience, we will assume that the region $C_A$ of a cone fold is the set of all points on the cone no more than a fixed radius away from the vertex, like a region $R_\varepsilon$ but on a cone.

We want to consider origamis and flat origamis on cone folds. The definition of an origami may be used as is, but let us modify our notion of flat folds in this context.

**Definition 5.22**  Given a cone fold $G = (V, E)$ on a cone $C_A$, a **flat cone fold on** $G$ is an infinite sequence of origamis $\{\sigma_n\}_{n=1}^{\infty}$ on $G$ and another cone $C_B$ such that

- for each face $f$ of $G$ (which is a sector of a cone), the images $\{\sigma_n(f)\}_{n=1}^{\infty}$ uniformly converge to a sector of $C_B$ congruent (intrinsically) to $f$ and
- for each crease $l \in E$, the folding angles of the images $\{\sigma_n(l)\}_{n=1}^{\infty}$ converge to either $\pi$ or $-\pi$.

By two cone sectors being **congruent** we mean relative to their intrinsic geometry, so that if the sectors were unrolled into a plane, then these unrolled images would be congruent sectors of a circle.

Note that Definition 5.22 does not specify that the limit map $\sigma = \lim_{n\to\infty} \sigma_n$ be surjective. Indeed, if our domain region is $C_{2\pi}$, then our "cone" is just flat paper, and if $\sigma$ is not surjective, then our flat cone fold is pointy. This case is just a flat vertex fold, so flat cone folds are a generalization of flat vertex folds.

**Theorem 5.23**  *Let $G$ be a flat cone fold with limit folding map $\sigma : C_A \to C_B$ and let $\alpha_0, \alpha_1, \ldots, \alpha_{2n-1}$ be the angles, in order, surrounding the interior vertex of $G$. Then $G$ is pointy if and only if $\alpha_0 - \alpha_1 + \cdots - \alpha_{2n-1} = 0$ and $G$ is a folded disc if and only if $\alpha_0 - \alpha_1 + \cdots - \alpha_{2n-1} = \pm B$.*

*Proof*  The proof is very similar to that of Theorem 5.20, but we need to provide a little more care because the codomain of our flat fold is a cone and it might be possible for the folded paper to cover the surface of the cone (i.e., project onto a cylinder) and still have turning number 1. (That is, it is possible to have $G$ be pointy and still have $\sigma$ be surjective.)

Let $\gamma$ be an oriented circle drawn on $C_A$ centered at the apex. We consider traveling along the image curve $\sigma(\gamma)$, in particular how it passes through the image of a point $x \in \gamma$ that is not on a crease of $G$. Using the orientation of $\gamma$, let $W_L(x)$ and $W_R(x)$ be the number of times $\sigma(\gamma)$ passes through $\sigma(x)$ going left to right and right to left, respectively. Then $W_L(x) - W_R(x)$ is the (oriented) turning number of $\sigma(\gamma)$ around the apex of $C_B$. (Any difference between $W_L(x)$ and $W_R(x)$ represents a trip made around the vertex, which is also the degree of the tangent map $\sigma'(\gamma(t))/|\sigma'(\gamma(t))|$, which equals the turning number. See (Whitney, 1937).) Note that since $\gamma$ has turning number 1 around the apex of $C_A$, $\sigma$ is continuous, and the sequence of origamis $\sigma_n$ are all one-to-one, we have that the turning number of $\sigma(\gamma)$ can be at most 1 in absolute value. (Winding twice or more around the apex of $C_B$ would require the paper to cross itself in order for $\sigma(\gamma)$ to come back to where it started.) That is, $W_L(x) - W_R(x)$ is either 0 or $\pm 1$.

With all that, we have that $G$ is pointy $\Leftrightarrow$ there are an even number of layers at $\sigma(x) \Leftrightarrow W_L(x) + W_R(x)$ is even $\Leftrightarrow W_L(x) - W_R(x)$ is even $\Leftrightarrow \sigma(\gamma)$ has turning number $0 \Leftrightarrow \alpha_0 - \alpha_1 + \cdots - \alpha_{2n-1} = 0$, since the alternating sum of the angles represents the distance traveled along $\sigma(\gamma)$, and thus how many times we've wound around the vertex.

For the other part of the theorem, $G$ is a folded disc $\Leftrightarrow$ there are an odd number of layers at $\sigma(x)$ $\Leftrightarrow$ the turning number of $\sigma(\gamma)$ is odd, so $\pm 1$ $\Leftrightarrow$ our oriented angle sum $\alpha_0 - \alpha_1 + \cdots - \alpha_{2n-1}$ will make one full revolution around the apex of $C_B$ in either the positive or negative direction, giving a sum of $\pm B$. $\qquad \square$

We combine what we're learned in this and the previous two sections into a nice, convenient theorem on cone folds.

**Theorem 5.24**   *Let $G$ be a flat cone fold with limit folding map $\sigma : C_A \rightarrow C_B$, $\alpha_0, \alpha_1, \ldots, \alpha_{2n-1}$ the angles, in order, surrounding the interior vertex of $G$, and $M$ and $V$ the number of mountain and valley creases, respectively, of $G$. Then the following are equivalent:*

- *$G$ is pointy.*
- *$\alpha_0 - \alpha_1 + \cdots - \alpha_{2n-1} = 0$.*
- *$M - V = \pm 2$.*

*Also the following are equivalent:*

- *$G$ is a folded disc.*
- *$\alpha_0 - \alpha_1 + \cdots - \alpha_{2n-1} = \pm B$.*
- *$M - V = 0$.*

This illustrates a strong connection between Maekawa's and Kawasaki's Theorems. Basically, each of them are related to the turning number of the flat vertex, and this establishes a link between the mountain-valley difference and the alternating angle sum at the vertex. This connection is not apparent at all in the flat vertex fold case, where we are simply folding flat paper, because the paper cannot "wrap around" the folded vertex in the image. Generalizing to cones allows this to happen and thus illustrates the connection.

Maekawa's and Kawasaki's Theorems do not fully generalize to multiple-vertex crease patterns. They are, at heart, local conditions of flat-foldability. Amazingly, their connection to each other via the turning number does generalize, giving us a powerful result that we will call Justin's Theorem, since Jacques Justin was the first (and only) person to discover it. We will present this in Chapter 6.

## 5.5   Counting Valid Mountain-Valley Assignments

Much of this section follows references (Hull, 2002, 2003). We begin with a highly recommended diversion.

---

**Diversion 5.4**   Figure 5.10 displays three degree-4 flat vertex folds. In each case, how many valid MV assignments are there for the crease pattern?

---

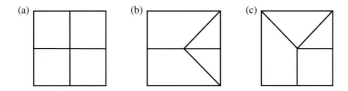

**Figure 5.10** Three different degree-4 flat vertex folds.

Note that since each crease pattern is an embedding of a planar graph, each crease is considered different, and so MV assignments that assign different values to some of the creases but that are identical under some symmetry of the crease pattern are considered to be different when we count them. For example, in Figure 5.10(a), we could have the crease in the "north" position be a valley and all the others mountain, or we could have the crease in the "south" position be a valley and the rest mountains. These are considered two different valid MV assignments.

Given a flat vertex fold $G = (V, E)$, let $E = \{l_0, l_2, \dots, l_{2n-1}\}$ be the creases meeting at the vertex and let $\alpha_i$ be the angle between the creases $l_i$ and $l_{i+1}$ (and $\alpha_0$ is between $l_{2n-1}$ and $l_0$). As Diversion 5.4 illustrates, the number of valid MV assignments for $G$ will depend on the circular sequence of angles $\alpha_i$. Therefore, we denote

$$C(\alpha_0, \dots, \alpha_{2n-1}) = \text{the number of valid MV assignments for } G.$$

Computing $C(\alpha_0, \dots, \alpha_{2n-1})$ is an exercise in understanding when certain sequences of crease lines can be assigned specific combinations of mountains and valleys without resulting in the folded paper intersecting itself.

For example, consider the degree-4 vertex in Figure 5.10(c), where our sequence of angles is $(\alpha_0, \alpha_1, \alpha_2, \alpha_3) = (90°, 45°, 90°, 135°)$. Notice that in this example any MV assignment $\mu$ cannot have $\mu(l_1) = \mu(l_2)$, for otherwise we would have two 90° angles trying to cover a 45° angle on the same side of the paper, which would force a self-intersection (or force a new crease to be made). This phenomenon occurs whenever a small angle is surrounded by larger angles in a single-vertex crease pattern, and it is sometimes referred to as the **Big-Little-Big Lemma**.

**Lemma 5.25** (Big-Little-Big Lemma)   *Let G be a flat vertex fold with angle sequence $\alpha_i$ and a valid MV assignment $\mu$. If $\alpha_{i-1} > \alpha_i < \alpha_{i+1}$ for some i, then $\mu(l_i) \neq \mu(l_{i+1})$. (That is, $\mu(l_i) + \mu(l_{i+1}) = 0$.)*

On the other extreme, what if all the angles $\alpha_i$ were equal? Then it wouldn't matter what MV assignment we picked. As long as it satisfied Maekawa's Theorem, the angles would not interfere with each other, so the MV assignment would be valid.

**Theorem 5.26**   *Let G be a flat vertex fold with angle sequence $(\alpha_0, \dots, \alpha_{2n-1})$. Then*

$$2^n \leq C(\alpha_0, \dots, \alpha_{2n-1}) \leq 2\binom{2n}{n-1}$$

*are sharp bounds.*

*Proof*  Both bounds will be proven by induction on the number of creases, using the fact that our paper can be a cone. Let $A$ be the cone angle of our paper.

For the upper bound, we want to prove that in the crease pattern with all angles equal, any MV assignment that satisfies $M - V = \pm 2$ will be valid. Given any such MV assignment $\mu$ on $2n$ crease lines with all angles $\alpha_i$ equal, there must exist an angle $\alpha_k$ with $\mu(l_k) \neq \mu(l_{k+1})$. Then fold these two creases according to $\mu$ to get a new single-vertex crease pattern on $2n - 2$ creases with angles $(\alpha_0, \ldots, \alpha_{k-1}, \alpha_{k+2}, \ldots, \alpha_{2n-1})$ on a cone with cone angle $A - 2\alpha_k$. The MV assignment $\mu' = \mu$ restricted to the creases $l_0, \ldots, l_{k-1}, l_{k+2}, \ldots, l_{2n-1}$ will still satisfy Maekawa's Theorem, so by the induction hypothesis $\mu'$ is a valid MV assignment. This immediately implies that $\mu$ is valid on the original set of $2n$ creases.

Therefore, given a flat vertex fold of degree $2n$ with all angles equal, we can pick any $n - 1$ of them to be mountains and the rest valleys, or we could pick any $n - 1$ creases to be valleys and the rest mountains. Each one of these possibilities is a distinct, valid MV assignment, so the total number of ways to fold such a vertex flat is $2\binom{2n}{n-1}$. Since this is merely enumerating all the possible MV assignments that satisfy Maekawa's Theorem, it must be a sharp upper bound for $C(\alpha_0, \ldots, \alpha_{2n-1})$.

For the lower bound, imagine that we have a flat vertex fold of degree $2n$ on a cone with cone angle $A$, and suppose that $\alpha_i$ is the smallest angle surrounding the vertex (or one of the smallest, if there is a tie). Then we have **at least two** possibilities for the MV assignment of $l_i$ and $l_{i+1}$, which are the two ways in which we could have $\mu(l_i) \neq \mu(l_{i+1})$. (Of course, there might be other possibilities as well.)

Thus if we fold $l_i$ and $l_{i+1}$ using one of these two possibilities and fuse, or identify, the layers of paper together, then the paper will turn into a cone with cone angle $A - 2\alpha_i$ and angle sequence

$$(\alpha_0, \ldots, \alpha_{i-2}, \alpha_{i-1} - \alpha_i + \alpha_{i+1}, \alpha_{i+2}, \ldots, \alpha_{2n-1}).$$

Note that the new angle $\alpha_{i-1} - \alpha_i + \alpha_{i+1}$ will be positive because $\alpha_i$ was chosen to be one of the smallest angles. By the induction hypothesis, this new flat vertex fold on $2n - 2$ creases will have at least $2^{n-1}$ valid MV assignments, and adding the two choices we have for $\mu(l_i)$ and $\mu(l_{i+1})$ gives us $2^n \leq C(\alpha_0, \ldots, \alpha_{2n-1})$.  □

The lower bound in Theorem 5.26 becomes equality for **generic** flat vertex folds, which are those where none of the angles are consecutively equal and none of the combined angles are equal to their neighbors throughout the inductive process outlined above except for the $n = 2$ base case. (The use of the adjective "generic" will be justified in Section 5.6.) As a six-crease example, notice that

$$C(100°, 70°, 50°, 40°, 30°, 70°) = 2C(100°, 70°, 50°, 80°)$$
$$= 2^2 C(100°, 100°)$$
$$= 2^3 = 8.$$

The bounds in Theorem 5.26 show the wide range of possibilities for $C(\alpha_0, \ldots, \alpha_{2n-1})$ for a fixed $n$ between the generic case and the all-angles-equal case. A next logical question is, "What values between $2^n$ and $2\binom{2n}{n-1}$ can $C(\alpha_0, \ldots, \alpha_{2n-1})$

achieve?" In Diversion 5.4 we see that when $n = 4$, $C(\alpha_0, \ldots \alpha_3)$ can equal 4, 6, or 8. (Notice that we'll only get even numbers, since the mountains and valleys can always be switched.) The degree-6 case becomes less predictable:

$$C(\alpha_0, \ldots, \alpha_5) \in \{8, 12, 16, 18, 20, 24, 30\}.$$

These values can be computed using recursive formulas for $C(\alpha_0, \ldots, \alpha_{2n-1})$, but in order to develop these, we will need a stronger version of the Big-Little-Big Lemma.

Before we do this, however, note that if $l_i, \ldots, l_{i+k}$ are crease lines and $\mu$ is an MV assignment, then the quantity $\sum_{j=i}^{i+k} \mu(l_j)$ counts the difference between the number of mountains and valleys among these creases.

**Theorem 5.27** (Hull 2002, 2003)  *Let $G$ be a flat vertex fold with angle sequence $(\alpha_0, \ldots, \alpha_{2n-1})$, and suppose that we have $\alpha_i = \alpha_{i+1} = \alpha_{i+2} = \cdots = \alpha_{i+k}$ and $\alpha_{i-1} > \alpha_i$ and $\alpha_{i+k+1} > \alpha_{i+k}$ for some $i$ and $k$ (where the indices are taken mod $2n$). Then an MV assignment $\mu$ for $G$ will be valid among the creases $l_i, \ldots, l_{i+k+i}$ if and only if*

$$\sum_{j=i}^{i+k+1} \mu(l_j) = \begin{cases} 0 & \text{if } k \text{ is even,} \\ \pm 1 & \text{if } k \text{ is odd.} \end{cases}$$

*Proof*  The result follows from Maekawa's Theorem. If $k$ is even, then the cross section of the paper around the creases in question might look as shown in Figure 5.11. (We say "might" because the equal angles may be twisted among themselves in a number of different ways.) If we consider this sequence of angles by itself and add a section of paper with angle $\beta = \alpha_{i-1} - \alpha_i + \alpha_{i+k+1}$ to connect the loose ends at the left and right (see Figure 5.11), then we'll have a flat-folded, pointy cone that must satisfy Maekawa's Theorem. The angle $\beta$ added two extra creases, which must be both mountains or both valleys. Thus we may either add or subtract two from the result of Maekawa's Theorem to get $\sum_{j=i}^{i+k+1} \mu(l_j) = 0$.

If $k$ is odd (Figure 5.12), then this angle sequence, if considered by itself, will have the loose ends from angles $\alpha_{i-1}$ and $\alpha_{i+k+1}$ pointing in the same direction. If we glue these together, possibly extending one of them if $\alpha_{i-1} \neq \alpha_{i+k+1}$, then Maekawa's Theorem may be applied. After subtracting (or adding) one to the result of

**Figure 5.11**  A case when $k$ is even.

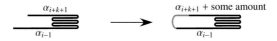

**Figure 5.12**  A case when $k$ is odd.

Maekawa's Theorem, because of the extra crease made when gluing the loose flaps, we get $\sum_{j=i}^{i+k+1} \mu(l_j) = \pm 1$.  □

**Theorem 5.28** (Hull, 2002, 2003)  *Let G be a flat vertex fold with angle sequence* $(\alpha_0, \ldots, \alpha_{2n-1})$, *and suppose that we have* $\alpha_i = \alpha_{i+1} = \alpha_{i+2} = \cdots = \alpha_{i+k}$ *and* $\alpha_{i-1} > \alpha_i$ *and* $\alpha_{i+k+1} > \alpha_{i+k}$ *for some i and k. Then*

$$C(\alpha_0, \ldots, \alpha_{2n-1}) = \binom{k+2}{\frac{k+2}{2}} C(\alpha_0, \ldots, \alpha_{i-2}, \alpha_{i-1} - \alpha_i + \alpha_{i+k+1}, \alpha_{i+k+2}, \ldots, \alpha_{2n-1})$$

*if k is even, and*

$$C(\alpha_0, \ldots, \alpha_{2n-1}) = \binom{k+2}{\frac{k+1}{2}} C(\alpha_0, \ldots, \alpha_{i-1}, \alpha_{i+k+1}, \ldots, \alpha_{2n-1})$$

*if k is odd.*

*Proof*  If $k$ is even, then Theorem 5.27 gives us $\sum_{j=i}^{i+k+1} \mu(l_j) = 0$, which means that among the $k+2$ creases $l_i, \ldots, l_{i+k+1}$, any $(k+2)/2$ of them can be valleys, and the rest mountains, since all the angles are the same. If we take one of these possibilities and fuse the layers of paper around these angles together, then angles $\alpha_{i-1}, \ldots, \alpha_{i+k+1}$ will be replaced with one angle with measure $\alpha_{i-1} - \alpha_i + \alpha_{i+k+1}$. This gives us the stated recursion.

If $k$ is odd, then $\sum_{j=i}^{i+k+1} \mu(l_j) = \pm 1$. Thus we could pick any $(k+1)/2$ of the $k+2$ creases $l_i, \ldots, l_{i+k+1}$ to be mountains and the rest valleys, or vice versa. Thus there are $2\binom{k+2}{(k+1)/2}$ MV assignments for these creases. However, because $k$ is odd, fusing all these layers together will create a new crease line whose mountain-valley assignment will be forced and ruin our hopes of recursion. To avoid this, we allow one of the crease lines to remain **unassigned** and divide the number of MV assignments by two. When the folded layers of paper are fused together, the angles $\alpha_i, \ldots, \alpha_{i+k}$ will be absorbed by the angles $\alpha_{i-1}$ or $\alpha_{i+k+1}$, which gives the stated recursion.  □

These recursions first appeared in (Hull, 2002) and (Hull, 2003), but the basic idea for how they work can also be found in (Justin, 1997).

As described in (Demaine and O'Rourke, 2007), these recursions allow one to compute any $C(\alpha_0, \ldots, \alpha_{2n-1})$ in linear time.

---

**Example 5.29**  Justin (1997) gave the following example with eight crease lines: $20°$, $10°$, $40°$, $50°$, $60°$, $60°$, $60°$, $60°$; see Figure 5.13(a). Here we find that

$$C(20, 10, 40, 50, 60, 60, 60, 60) = \binom{2}{1} C(50, 50, 60, 60, 60, 60)$$

$$= \binom{2}{1}\binom{3}{1} C(60, 60, 60, 60)$$

$$= \binom{2}{1}\binom{3}{1} 2\binom{4}{1} = 48.$$

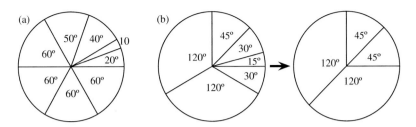

**Figure 5.13** The flat vertex folds from Example 5.29.

One should note, however, that identifying the sequences of equal angles that are used in the recurrence of Theorem 5.28 can be difficult. Figure 5.13(b) shows such an example. In the first step of the recursion, we identify the smallest angle 15° and apply the Big-Little-Big Lemma to replace the 15° angle and its two 30° neighbors with a sector with angle $30° - 15° + 30° = 45°$. This gives us two consecutive 45° angles, giving us a $k = 1$ case of the recursion in Theorem 5.28. Thus, this "two consecutive equal angles" recursive step was hidden; that such a case existed in this crease pattern was not apparent until we performed the first recursive step.

The recursions in Theorem 5.28 can be used to generate all possible values of $C(\alpha_0, \ldots, \alpha_{2n-1})$ for a fixed $n$. Let $SC(n)$ denote the set of all possible values of $C(\alpha_0, \ldots, \alpha_{2n-1})$. Then we have that $SC(1) = \{2\}$ (where we have the degenerate case of a degree-2 vertex whose two creases must be either both mountains or both valleys).

We also know from Diversion 5.4 that $SC(2) = \{4, 6, 8\}$, but this could also have been determined by our recurrences. Any degree-4 flat vertex fold that does not have all angles equal will either have one smallest angle surrounded by bigger angles or two smallest angles in a row surrounded by bigger angles. These will give $k = 0$ and $k = 1$ cases in Theorem 5.28, respectively, and both will recurse to the degree-2 case. When $k = 0$ we get $\binom{2}{1}C(\alpha_0, \alpha_1) = 4$, and when $k = 1$ we get $\binom{3}{1}C(\alpha_0, \alpha_1) = 6$. The all-angles-equal case gives us the upper bound in Theorem 5.26, which is 8, so $SC(2) = \{4, 6, 8\}$.

In other words, $SC(2) = \binom{2}{1}SC(1) \cup \binom{3}{1}SC(1) \cup \{2\binom{4}{1}\}$, where $k$ times $SC(n)$ is all the elements of $SC(n)$ multiplied by $k$. For $SC(3)$ we have

$$SC(3) = \binom{2}{1}SC(2) \cup \binom{3}{1}SC(2) \cup \binom{4}{2}SC(1) \cup \binom{5}{2}SC(1) \cup \left\{2\binom{6}{2}\right\}$$

$$= \{8, 12, 16\} \cup \{12, 18, 24\} \cup \{12\} \cup \{20\} \cup \{30\}$$

$$= \{8, 12, 16, 18, 20, 24, 30\}.$$

In general, we have the following theorem:

**Theorem 5.30**  *If $SC(n)$ is the set of possible values for $C(\alpha_0, \ldots, \alpha_{2n-1})$, then*

$$SC(n) = \left( \bigcup_{k=1}^{n-1} \left( \binom{2n-2k}{n-k} SC(k) \cup \binom{2n-2k+1}{n-k} SC(k) \right) \right) \cup \left\{ 2 \binom{2n}{n-1} \right\}$$

*for $n \geq 2$ and $SC(1) = \{2\}$.*

---

**Diversion 5.5**  Prove Theorem 5.30.

---

Thus we can compute

$$SC(4) = \{16, 24, 32, 36, 40, 48, 54, 60, 70, 72, 80, 90, 112\},$$
$$SC(5) = \{32, 48, 64, 72, 80, 96, 108, 120, 140, 144, 160, 162, 180, 200, 210,$$
$$216, 224, 240, 252, 270, 280, 300, 336, 420\},$$

and so on. The size of these sets, $|SC(n)|$, generates the sequence

$$1, 3, 7, 13, 24, 39, 62, 97, 147, 215, 312, 440, 617, 851, 1161, \ldots, \tag{5.2}$$

which does not (as of this writing) exhibit any discernible pattern.

The different values for $C(\alpha_0, \ldots, \alpha_{2n-1})$ indicate the different combinations of mountains and valleys that can be chosen for certain collections of creases around the vertex. However, these combinations of creases also force symmetry on the flat vertex fold, requiring that some angles be equal and others not. These kinds of distinctions are highlighted when we consider the configuration space in which flat vertex folds live, which we will explore next.

## 5.6　The Configuration Space of Flat Vertex Folds

Determining configuration spaces is a popular activity in computational and combinatorial geometry. The classic example is that of a robot arm with two joints, each of which can rotate in the plane a full $360°$. The range of positions for the tip of the arm can be described by the amount of rotation at each joint, and so its configuration space can be described as the surface of a torus, $[0, 2\pi] \times [0, 2\pi]$.

Similar, but much more complicated results have been found for closed chains of robot arm lengths (see (Milgram and Trinkle, 2004)), and such work has been applied to origami (see (Balkcom, 2002)). The idea is that if we have a single-vertex fold and we consider the vertex to be at the center of a sphere, then the folded paper would intersect the sphere to form a closed chain of spherical arcs, and the folding and unfolding of the vertex would be equivalent to this closed chain flexing like a robot arm, albeit a closed loop robot arm. The resulting configuration space from this approach describes the range of motions that a specific closed chain can achieve, capturing **how** a single-vertex fold can fold up, not just its final folded state. Such configuration spaces will fit more properly in the context of non-flat folding and rigid origami, which we will cover in Part IV of this book.

In this section we seek to find the configuration space for **all flat vertex folds of a given degree**. That is, for a fixed degree $2n$, how can we visualize the space of possible angles $\alpha_1, \ldots, \alpha_{2n}$ that could form a flat vertex fold of degree $2n$? The exposition presented here follows that of (Hull, 2009).

Consider the case of degree-4 flat vertex folds, where the angles are $\alpha_1, \ldots, \alpha_4$. Kawasaki's Theorem tells us that $\alpha_3 = \pi - \alpha_1$ and $\alpha_4 = \pi - \alpha_2$. All four angles are determined by $\alpha_1$ and $\alpha_2$, so $\alpha_1$ and $\alpha_2$ can be the parameters of our configuration space.

Assign $\alpha_1$ to our first coordinate and $\alpha_2$ to our second coordinate. Notice that the range for these parameters is $0 < \alpha_1, \alpha_2 < \pi$, since if either were 0, we wouldn't have four creases, and if either were $\pi$ then one of $\alpha_3, \alpha_4$ would be zero. Furthermore, if we pick any $\alpha_1$ and $\alpha_2$ between 0 and $\pi$, we can let $\alpha_3 = \pi - \alpha_1$ and $\alpha_4 = \pi - \alpha_2$ to obtain angles for a degree-4 flat vertex fold, showing that $(\alpha_1, \alpha_2)$ must be in our configuration space. Therefore the configuration space for degree-4 flat vertex folds, which we'll denote $P_4$, is the open square (see Figure 5.14)

$$P_4 = (0, \pi) \times (0, \pi).$$

Now, within $P_4$ there exist subsets for the different values of $C(\alpha_1, \ldots, \alpha_4)$. The maximal $C(\alpha_1, \ldots, \alpha_4)$ is 8, which corresponds to all the angles being equal. This is the point $(\pi/2, \pi/2)$ in $P_4$.

Next is $C(\alpha_1, \ldots, \alpha_4) = 6$, and this occurs when two adjacent angles are equal and different from the other pair. For example, we could have $\alpha_1 = \alpha_2$ (which implies that $\alpha_3 = \alpha_4$). This corresponds to the line $y = x$ in $P_4$, for $0 < x < \pi/2$ and $\pi/2 < x < \pi$.

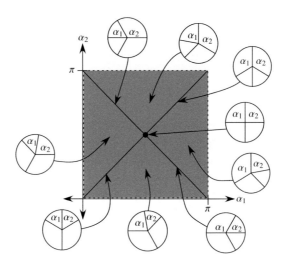

**Figure 5.14** Decomposing $P_4$ into different angle configuration subsets.

Or we could have $\alpha_2 = \alpha_3$, which implies that $\alpha_2 = \pi - \alpha_1$ (which forces $\alpha_1 = \alpha_4$) and gives us the line $y = \pi - x$ in $P_4$ for $0 < x < \pi/2$ and $\pi/2 < x < \pi$.

The remaining regions of $P_4$ are open right triangles, and these correspond to $C(\alpha_1, \ldots, \alpha_4) = 4$. For example, the region bounded by the y-axis, $y = x$, and $y = \pi - x$ has $\alpha_1 < \alpha_2$, $\alpha_1 < \pi/2$, and $\alpha_2 < \pi - \alpha_1$. Kawasaki's Theorem then gives us that $\alpha_3 > \pi/2$ and $\alpha_1 < \alpha_4$. In other words, $\alpha_1$ is the unique smallest angle, and we have a flat vertex fold similar to the one in Diversion 5.4(c) where $C(\alpha_1, \ldots, \alpha_4) = 4$. The other three triangular regions are similar.

This decomposition of $P_4$ into subsets gives us a complete classifications of all the possibilities for $C(\alpha_1, \ldots, \alpha_4)$ and is illustrated in Figure 5.14.

The configuration spaces $P_{2n}$ for flat vertex folds of degree-$2n$ quickly become very difficult to visualize for $n > 2$, as they are bounded, open sets in $\mathbb{R}^{2n-2}$. As another example, consider the $n = 3$ case. Letting $\alpha_1, \ldots, \alpha_6$ be the angles, we can express $\alpha_5$ and $\alpha_6$ in terms of the other angles (using Kawasaki's Theorem), and thus we may parameterize $P_6$ by the angles $\alpha_1, \ldots, \alpha_4$. That is, $P_6 \subset \mathbb{R}^4$. Our reasoning from the $n = 2$ case as well as the Kawasaki conditions $\alpha_1 + \alpha_3 + \alpha_5 = \alpha_2 + \alpha_4 + \alpha_6 = \pi$ give us the following restrictions on the angles:

$$0 < \alpha_i < \pi \text{ for all } i, \quad 0 < \alpha_1 + \alpha_3 < \pi, \quad \text{and} \quad 0 < \alpha_2 + \alpha_4 < \pi. \quad (5.3)$$

This means that the 2-dimensional cross section of $P_6$ along the $\alpha_1\alpha_2$-coordinate plane will be an open square, as in the $n = 2$ case. However, the 2-dimensional cross section along the $\alpha_1\alpha_3$-plane will be an open triangle bounded by $\alpha_1 > 0$, $\alpha_3 > 0$, and $\alpha_3 < \pi - \alpha_1$.

In fact, any point $(\alpha_1, \alpha_2, \alpha_3, \alpha_4)$ satisfying (5.3) will be part of a viable degree-6 flat vertex fold (along with the proper angles $\alpha_5$ and $\alpha_6$ given by Kawasaki's Theorem) and thus be in $P_6$. That is, $P_6$ is an open set. The closure of this set, $\overline{P_6}$, will have as extreme points (vertices) all angle configurations that give equality for the Equations (5.3) and that are the most degenerate, where one of the angles $\alpha_1, \alpha_3, \alpha_5$ equals $\pi$, one of the angles $\alpha_2, \alpha_4, \alpha_6$ equals $\pi$, and the rest equal 0. Thus $\overline{P_6}$ is the polytope formed by the convex hull of the points

$$(0, 0, 0, 0), (\pi, 0, 0, 0), (0, \pi, 0, 0), (0, 0, \pi, 0), (0, 0, 0, \pi),$$
$$(\pi, \pi, 0, 0), (\pi, 0, 0, \pi), (0, \pi, \pi, 0), (0, 0, \pi, \pi).$$

(This can also be seen by viewing the inequalities in (5.3) as defining the supporting hyperplanes for the polytope $\overline{P_6}$.)

There is enough information at hand to determine all the facets of $\overline{P_6}$, but instead we will turn to the general case $P_{2n}$.

If our angles are, in order, $\alpha_1, \ldots, \alpha_{2n}$, we know by Kawasaki's Theorem that the space can be parameterized by $\alpha_1, \ldots, \alpha_{2n-2}$. In other words, $P_{2n} \subset \mathbb{R}^{2n-2}$.

We say that a point $x = (\alpha_1, \ldots, \alpha_{2n-2}) \in \mathbb{R}^{2n-2}$, where $\alpha_i \geq 0$, **corresponds to a set of angles** if there exists $\alpha_{2n-1}, \alpha_{2n} \geq 0$ such that $(\alpha_1, \ldots, \alpha_{2n})$ satisfy the Kawasaki condition. (That is, if $\alpha_{2n-1} = \pi - (\alpha_1 + \alpha_3 + \cdots + \alpha_{2n-3})$ and $\alpha_{2n} = \pi - (\alpha_2 + \alpha_4 + \cdots + \alpha_{2n-2})$.) Note that this corresponding set of angles might not be a degree-$2n$ flat vertex fold, since the definition allows some of the angles to be zero or $\pi$.

**Theorem 5.31** *The configuration space $P_{2n}$ is an open set. Furthermore, if $x \in \overline{P_{2n}} - P_{2n}$ (the boundary of $P_{2n}$), then $x$ corresponds to a degenerate set of angles where at least one of the angles $\alpha_i$ equals 0 or $\pi$.*

*Proof* The fact that all angles in a degree-$2n$ flat vertex fold must be nonzero and less that $\pi$, together with the Kawasaki condition (Corollary 5.19), give us that every point in $P_{2n}$ must satisfy the inequalities

$$0 < \alpha_i < \pi \text{ for all } i, \quad 0 < \alpha_1 + \alpha_3 + \cdots + \alpha_{2n-3} < \pi,$$

$$\text{and} \quad 0 < \alpha_2 + \alpha_4 + \cdots + \alpha_{2n-2} < \pi. \tag{5.4}$$

Furthermore, any point satisfying these equations must be in $P_{2n}$, which proves that $P_{2n}$ is open. Any point $x$ on the boundary of $P_{2n}$ must also satisfy the Equations (5.4) but have at least one of the inequalities being an equality. Thus $x$ corresponds to a set of angles $\alpha_1, \ldots, \alpha_{2n}$ where either at least one of the $\alpha_i$ is 0 or $\pi$ for some $1 \leq i \leq 2n - 2$ (in which case, we're done) or $\alpha_1 + \alpha_3 + \cdots + \alpha_{2n-3}$ equals 0 or $\pi$ or $\alpha_2 + \alpha_4 + \cdots + \alpha_{2n-2}$ equals 0 or $\pi$. These latter two cases imply that either $\alpha_{2n-1}$ or $\alpha_{2n}$ equals 0 or $\pi$. Thus every case results in $x$ corresponding to a set of angles where at least one of the $\alpha_i$ equals 0 or $\pi$. $\square$

We can use Theorem 5.31 to examine more carefully the faces of $\overline{P_{2n}}$. The vertices of $\overline{P_{2n}}$, for example, will correspond to the most extreme degenerate degree-$2n$ flat vertex folds, where two angles are equal to $\pi$ and the rest are equal to 0. In order for such a case to satisfy Kawasaki's Theorem, one of the $\pi$ angles must be an even-indexed angle and the other an odd-indexed angle. This is illustrated in the left side of Figure 5.15.

Thus we have that $\overline{P_{2n}}$ has $n^2$ vertices whose coordinates are $(\alpha_1, \ldots, \alpha_{2n-2})$ where at most one of the $\alpha_{2i} = \pi$, at most one of the $\alpha_{2i+1} = \pi$, and the remaining $\alpha_i = 0$. (If all the $\alpha_i = 0$ then we have $\alpha_{2n-1} = \alpha_{2n} = \pi$ in the corresponding set of angles.)

An edge (1-face) of $\overline{P_{2n}}$ will be a line segment of points $E(u, v) = \{\lambda u + (1 - \lambda)v : 0 \leq \lambda \leq 1\}$ connecting two vertices $u$ and $v$ where the points of $E(u, v)$, aside from the endpoints, correspond to slightly less extreme degenerate degree-$2n$ flat vertex folds than those of the vertices. That is, instead of having an even-indexed angle and an odd-indexed angle equaling $\pi$ as we did for the vertices, each point in the relative

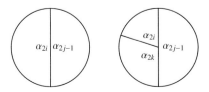

**Figure 5.15** Degenerate angle configurations for a vertex (left) and an edge (right) of $\overline{P_{2n}}$.

interior of $E(u, v)$ will correspond to a set of angles with either one even-indexed angle equaling $\pi$ and two odd-indexed angles adding to $\pi$, or vice versa (one odd-indexed angle is $\pi$ and two even-indexed angles sum to $\pi$). All the other angles would have to be 0; see the right side of Figure 5.15. Thus, if the nonzero corresponding set of angles for the vertex $u$ are at coordinate positions $2i$ and $2j - 1$ and those for $v$ are at coordinate positions $2s$ and $2t - 1$, then either $i = s$ or $j = t$ must be true in order for $E(u, v)$ to be an edge of $\overline{P_{2n}}$. That is, $u$ and $v$ must have a $\pi$ in a common coordinate so that their other $\pi$ coordinates can switch places as we travel along the edge $E(u, v)$.

The number of edges of $\overline{P_{2n}}$ will therefore be $\binom{n}{1}\binom{n}{2} + \binom{n}{2}\binom{n}{1}$, because in the corresponding set of angles $(\alpha_1, \ldots, \alpha_{2n})$ we could choose one of the $n$ even-indexed angles to be $\pi$, two of the $n$ odd-indexed angles to sum to $\pi$, and the rest to be 0, or we could pick two even-indexed angles to sum to $\pi$, one of the odd-indexed angles to be $\pi$, and the rest to be 0.

The 2-faces of $\overline{P_{2n}}$ follow similarly. In the corresponding set of angles for any point of a 2-face, we could have one even-indexed angle $\alpha_{2i} = \pi$ (and the rest = 0) and three odd-indexed angles $\alpha_{2j-1}, \alpha_{2k-1}$, and $\alpha_{2l-1}$ being nonzero but adding up to $\pi$ (and the rest = 0). This gives us two parameters (say $\alpha_{2j-1}$ and $\alpha_{2k-1}$, which then determine $\alpha_{2l-1}$) and thus will span a 2-face. Or we could have chosen two even-indexed angles and two odd, or three even-indexed angles and one odd. Thus there are $\binom{n}{1}\binom{n}{3} + \binom{n}{2}\binom{n}{2} + \binom{n}{3}\binom{n}{1}$ 2-faces total.

Thus we obtain the following:

**Theorem 5.32**   *The number of $k$-cells in $\overline{P_{2n}}$ is*

$$f_k = \sum_{i=0}^{k} \binom{n}{i+1}\binom{n}{k-i+1} = \binom{2n}{k+2} - 2\binom{n}{k+2}.$$

*Proof*   The previous arguments illustrate how we obtain the summation, and the summation identity can be obtained via standard combinatorial methods such as generating functions. We also offer a different combinatorial reasoning: To count $f_k$ we want to pick $k+2$ angles from the $2n$ corresponding angles to be nonzero in order to create our degenerate flat vertex fold. But we don't want all of the angles to be even-indexed or all odd-indexed, so we subtract the $2\binom{n}{k+2}$ ways in which this can happen. The result is all the ways to have all angles 0 except for $k + 2$ of them, where some are even-indexed and some odd-indexed. The even-indexed angles must sum to $\pi$, and so must the odd-indexed angles. This means that to parameterize these degenerate cases, we don't need all of the $k + 2$ angles; we can eliminate one of the even-indexed angles and one of the odd-indexed angles, leaving us with $k$ parameter coordinates for this face, thus creating a $k$-face.   □

The arguments given for Theorem 5.32 provide everything needed to calculate the coordinates for the vertices, edges, etc. of $\overline{P_{2n}}$, which can then be generated using Mathematica or other visualization software.

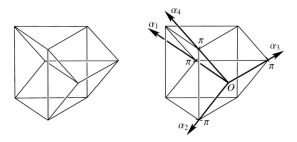

**Figure 5.16** A projection of the 4-dimensional polytope $\overline{P_6}$.

Figure 5.16 shows a projection of $\overline{P_6}$. We can try to compare our general calculations with the intuition developed earlier for the degree-6 flat vertex fold case. For example, recall that slicing $P_6$ along the $\alpha_1\alpha_3$-plane gives a right triangle. To make such a slice a 2-face of $\overline{P_6}$, we'd need the other angles (the even-indexed ones) to be extreme, either 0 or $\pi$, while still obeying Kawasaki's Theorem. So we could have

$$(\alpha_1, 0, \alpha_3, 0) \quad \text{where } 0 \le \alpha_1 + \alpha_3 \le \pi \text{ and } \alpha_6 = \pi,$$

$$(\alpha_1, \pi, \alpha_3, 0) \quad \text{where } 0 \le \alpha_1 + \alpha_3 \le \pi \text{ and } \alpha_6 = 0,$$

$$(\alpha_1, 0, \alpha_3, \pi) \quad \text{where } 0 \le \alpha_1 + \alpha_3 \le \pi \text{ and } \alpha_6 = 0.$$

The same reasoning applies to slices along the $\alpha_2\alpha_4$-plane, giving $\overline{P_6}$ six faces that will be 45° right triangles. Careful examination of Figure 5.16 reveals these faces.

In fact, going back to the general case, we can be more specific about the structure of $\overline{P_{2n}}$. Let $e_i \in \mathbb{R}^{2n-2}$ be the point with 0 for every coordinate except the $i$th, which is $\pi$. Let $o$ denote the origin. We denote the convex hull of a finite set of points $x_i$ by $\text{conv}(x_1, \ldots, x_n) = \{\lambda_1 x_1 + \cdots + \lambda_n x_n : \lambda_i \ge 0, \sum_{i=1}^{n} \lambda_i = 1\}$. Define

$$EP_{2n} = \text{conv}(o, e_2, e_4, \ldots, e_{2n-2}) \quad \text{and} \quad OP_{2n} = \text{conv}(o, e_1, e_3, \ldots, e_{2n-3}).$$

Then $EP_{2n}$ and $OP_{2n}$ are both $(n-1)$-simplices in $\mathbb{R}^{2n-2}$.

**Lemma 5.33** *A point $x \in EP_{2n}$ (resp. $OP_{2n}$) if and only if $x = \lambda_1 e_2 + \lambda_2 e_4 + \cdots + \lambda_{n-1} e_{2n-2}$ (resp. $x = \lambda_1 e_1 + \lambda_2 e_3 + \cdots + \lambda_{n-1} e_{2n-3}$) where $\lambda_i \ge 0$ and $\sum_{i=1}^{n-1} \lambda_i \le 1$.*

*Proof* If $x \in EP_{2n}$ or $OP_{2n}$ then certainly $x$ can be written as described in the lemma, since $\lambda_0 o$ is just the zero vector. For the other direction, if we write $x = \lambda_0 o + \lambda_1 e_2 + \cdots + \lambda_{n-1} e_{2n-2}$ where $\lambda_0 = 1 - \sum_{i=1}^{n-1} \lambda_i$, then we have that $x \in EP_{2n}$. The same argument with the points $e_{2i}$ switched to $e_{2i-1}$ handles the $OP_{2n}$ case. $\square$

Recall that if $A$ and $B$ are sets of points, then their **Minkowski sum** is $A + B = \{x + y : x \in A, y \in B\}$.

**Theorem 5.34** $\overline{P_{2n}} = OP_{2n} + EP_{2n}$.

*Proof*   Note that $x = (\alpha_1, \ldots, \alpha_{2n-2}) \in \overline{P_{2n}}$ if and only if

$$x = \frac{\alpha_1}{\pi}e_1 + \frac{\alpha_2}{\pi}e_2 + \cdots + \frac{\alpha_{2n-2}}{\pi}e_{2n-2}$$

$$= \left(\frac{\alpha_1}{\pi}e_1 + \frac{\alpha_3}{\pi}e_3 + \cdots + \frac{\alpha_{2n-3}}{\pi}e_{2n-3}\right) + \left(\frac{\alpha_2}{\pi}e_2 + \frac{\alpha_4}{\pi}e_4 + \cdots + \frac{\alpha_{2n-2}}{\pi}e_{2n-2}\right),$$

where $0 \leq \alpha_i \leq \pi$ for all $i$ and the coordinates of $x$ correspond to a set of angles that satisfy Kawasaki's Theorem. These conditions on $x$ are satisfied if and only if $0 \leq \alpha_i/\pi \leq 1$, $\sum_{i=1}^{n-1} \alpha_{2i-1} \leq \pi$, and $\sum_{i=1}^{n-1} \alpha_{2i} \leq \pi$, in other words,

$$\frac{\alpha_1}{\pi} + \frac{\alpha_3}{\pi} + \cdots \frac{\alpha_{2n-3}}{\pi} \leq \frac{\pi}{\pi} = 1 \quad \text{and} \quad \frac{\alpha_2}{\pi} + \frac{\alpha_4}{\pi} + \cdots \frac{\alpha_{2n-2}}{\pi} \leq \frac{\pi}{\pi} = 1.$$

Thus by Lemma 5.33 we have that $x \in \overline{P_{2n}}$ if and only if $x \in OP_{2n} + EP_{2n}$.    □

In other words, $\overline{P_{2n}}$ is the Minkowski sum of two $(n-1)$-simplices.

The question then is, "Can the configuration spaces $P_{2n}$ tell us anything about $C(\alpha_1, \ldots, \alpha_{2n})$?" In Figure 5.14 we see that $P_4$ can be decomposed into subsets of specific dimension corresponding to the different values of $C(\alpha_1, \ldots, \alpha_4)$. (The 0-dimensional center point corresponds to the only way to get $C(\alpha_1, \ldots, \alpha_4) = 8$, the 1-dimensional lines $y = x$ and $y = \pi - x$ correspond to $C(\alpha_1, \ldots, \alpha_4) = 6$ (excluding the center point), and the remaining 2-dimensional triangular regions correspond to $C(\alpha_1, \ldots, \alpha_4) = 4$.)

This nice correspondence between different dimensional subsets and distinct values of $C(\alpha_1, \ldots, \alpha_{2n})$ does not persist, however.

---

**Diversion 5.6**   Show that there are three different configurations of six angles that will give $C(\alpha_1, \ldots, \alpha_6) = 12$. Then show how two of them correspond to subsets of $P_6$ of dimension 2 and one of them corresponds to a subset of $P_6$ of dimension 3.

---

Diversion 5.6 illustrates how subsets of different shapes and dimensions can achieve the same $C(\alpha_1, \ldots, \alpha_{2n})$ value. This is just a consequence of different ways that the binomial coefficients in the Theorem 5.28 recursion can multiply together to get the same value.

However, these configuration spaces justify our use of the term "generic" for flat vertex folds with $C(\alpha_1, \ldots, \alpha_{2n}) = 2^n$, as mentioned in Section 5.5. Let us consider connected subsets $A_k \subset P_{2n}$ of constant $k = C(\alpha_1, \ldots, \alpha_{2n})$. We call such a subset **maximal** if no more points could be added to $A_k$ while preserving connectivity.

We would like to discuss the dimension of maximal subsets $A_k$, where by **dimension** we mean the size of a basis describing the points in, say, an $\varepsilon$-neighborhood of a point $x$ in the interior of $A_k$. However, as seen in Diversion 5.6, there can be regions of $P_{2n}$ of different dimension but with the same value of $C(\alpha_1, \ldots, \alpha_{2n})$, and if two such regions were adjacent in $P_{2n}$, then the corresponding maximal $A_k$ set would contain points $x$ with different dimension than that of the $\varepsilon$-ball $B(x, \varepsilon)$.

Nonetheless, this is not a problem for maximal subsets $A_k$ of codimension zero. Furthermore, this case gives us the motivation for the term "generic" flat vertex folds.

**Theorem 5.35**  *Let $A \subset P_{2n}$ be maximal with constant $C(\alpha_1, \ldots, \alpha_{2n}) = k$. Then the dimension of $A_k$ is $2n - 2$ (that is, has codimension zero) if and only if $k = 2^n$.*

*Proof*  The first key idea of the proof is, that $C(\alpha_1, \ldots, \alpha_{2n}) = 2^n$ if and only if each step in the recursive calculation of $C(\alpha_1, \ldots, \alpha_{2n})$ from Theorem 5.28 gave the binomial coefficient $\binom{2}{1}$, and thus had an angle $\alpha_i$ with $\alpha_{i-1} > \alpha_i$ and $\alpha_{i+1} > \alpha_i$ (the Big-Little-Big configuration).

Second, in this situation where each step in the Theorem 5.28 recursion is a Big-Little-Big configuration, we can modify any pair of angles of the form $\alpha_i, \alpha_{i+2}$ by a small amount while still maintaining flat-foldability and $C(\alpha_1, \ldots, \alpha_{2n}) = 2^n$. The reason for this is because if $C(\alpha_1, \ldots, \alpha_{2n}) = 2^n$ then no sector angles of the vertex are constrained to be equal to their neighbors or otherwise tied to other angle values farther into the recursion, such as those as in Figure 5.13(b). Thus, given any sector angle $\alpha_i$, we can find an $\varepsilon > 0$ such that replacing $\alpha_i$ and $\alpha_{i+2}$ with $\alpha_i \pm \varepsilon$ and $\alpha_{i+2} \mp \varepsilon$ will still satisfy Kawasaki's Theorem and maintain any Big-Little-Big relationships. Furthermore, the $C(\alpha_1, \ldots, \alpha_{2n}) = 2^n$ case is the **only** $C(\alpha_1, \ldots, \alpha_{2n})$ case that has this property, for otherwise there would be sector angles that are constrained to be equal or tied together and thus cannot be individually modified without changing $C(\alpha_1, \ldots, \alpha_{2n})$.

With this in mind, let $x = (\alpha_1, \ldots, \alpha_{2n-2}) \in P_{2n}$ and let $A_k$ be the maximal subset of $P_{2n}$ with constant $k = C(\alpha_1, \ldots, \alpha_{2n})$ containing $x$. By the above argument, $k = 2^n$ if and only if for each $i = 1, \ldots, 2n-2$ we can find an $\varepsilon_i > 0$ such that $C(\alpha_1, \ldots, \alpha_i \pm \varepsilon_i, \alpha_{i+1}, \alpha_{i+2} \mp \varepsilon_i, \ldots, \alpha_{2n}) = 2^n$. We take $\varepsilon = \min\{\varepsilon_1, \ldots, \varepsilon_{2n-2}\}$.

Then let $b_i \in \mathbb{R}^{2n-2}$ be the vector whose coordinates are all 0 except for the $i$th, which is 1, and the $(i+2)$-coordinate, which is $-1$, for all $i = 1, \ldots, 2n - 4$. We also take $b_{2n-3}$ to be all 0 except coordinate $2n - 3$, which is a 1, and $b_{2n-2}$ to be all 0 except the $2n - 2$ coordinate, which is a 1. Then the set of vectors $\{b_1, \ldots, b_{2n-2}\}$ are linearly independent and form a basis for the open ball $B(x, \varepsilon)$ that is totally contained in $A_k$. Since this holds for any $x \in A_k$, we have that $A_k$ has dimension $2n - 2$.  □

Theorem 5.35 implies that flat vertex folds of degree-$2n$ with exactly $2^n$ ways to fold flat are the only degree-$2n$ flat vertex folds that fill up regions of $P_{2n}$ with $(2n-2)$-dimensional volume. In other words, if we were to pick a point $x \in P_{2n}$ uniformly at random, then with probability 1 we would have that $x$ corresponds to a flat vertex fold with $C(\alpha_1, \ldots, \alpha_{2n}) = 2^n$. Therefore we are justified in making the following definition.

**Definition 5.36**  A flat vertex fold with sector angles $\alpha_1, \ldots, \alpha_{2n}$ is called **generic** if $C(\alpha_1, \ldots, \alpha_{2n}) = 2^n$.

It remains to be seen whether or not the configuration spaces $P_{2n}$ can be of further assistance in determining the patterns of $C(\alpha_1, \ldots, \alpha_{2n})$.

## 5.7    Matrix Model for Flat Vertex Folds

A computationally efficient way of describing flat vertex folds is with a matrix model. In fact, modeling paper folding with matrix transformations turns out to be a very powerful tool. It can be extended to multiple-vertex crease patterns and even non-flat folding, and it has proven useful for engineers wishing to solve folding problems in real-world applications. We will see these other uses in later chapters. For now we will merely introduce the single-vertex flat matrix model.

In any flat origami fold, each crease can be viewed as reflecting part of the paper across the crease line. It would thus be natural to model flat-folded creases by reflection matrices. Let $G$ be a flat vertex fold with crease lines $E = \{l_1, l_2, \ldots, l_{2n}\}$, in order around the vertex. We will assume that our graph $G$ is embedded in the plane with the vertex at the origin. Denote

$$R(l_i) = \text{the matrix that reflects the plane about the line } l_i.$$

The first people to develop this model seem to have been Toshikazu Kawasaki (Kawasaki and Yoshida, 1988) and Jacques Justin (1989), independently and at roughly the same time, in the mid-1980s. Both of them can be credited with discovering the following theorem. (We let $I$ denote the identity matrix, in this case a $2 \times 2$ matrix.)

**Theorem 5.37**    *Let $G = (V, E)$ be a single-vertex fold with $E = \{l_1, \ldots, l_{2n}\}$ the creases of the vertex, in order. Then $G$ will fold flat if and only if*

$$R(l_1)R(l_2)\cdots R(l_{2n}) = I. \tag{5.5}$$

This can be viewed as a matrix form of Kawasaki's Theorem. In fact, if we assume Kawasaki's Theorem, the proof is very simple.

*Proof*    Since the product of two reflections is a rotation by twice the angle between the two lines of reflection, the condition $R(l_1)R(l_2)\cdots R(l_{2n}) = I$ is equivalent to the equation

$$2\alpha_1 + 2\alpha_3 + 2\alpha_5 + \cdots + 2\alpha_{2n-1} = 2\pi,$$

where $\alpha_i$ is the angle between creases $l_i$ and $l_{i+1}$. Dividing by two gives us the angle conditions needed for Corollary 5.19 of Kawasak's Theorem, which then proves Theorem 5.37.    □

It will be useful to see a proof that does not rely so heavily on Kawasaki's Theorem. Denote all the faces of $G$ by $F = \{F_1, F_2, \ldots, F_{2n}\}$, where $F_i$ is the face between $l_i$ and $l_{i+1}$. Fix the face $F_{2n}$ and let us consider how all the other faces will fold. We do this by defining a **folding map** $\sigma : F \to \mathbb{R}^2$ by

$$\sigma(F_i) = R(l_1)R(l_2)\cdots R(l_i)[F_i],$$

where by $R(l_i)[F_i]$ we mean apply all points in $F_i$ to the matrix $R(l_i)$. Thus $\sigma(F_i)$ is the image of the face $F_i$ after we've folded creases $l_i, l_{i-1}, \ldots, l_2$, and $l_1$.

If the vertex $G$ folds flat, then clearly $\sigma(F_{2n}) = F_{2n}$, which implies Equation (5.5). Conversely, if we're given Equation (5.5), then $\sigma(F_{2n}) = F_{2n}$ and so we have that the folding map is continuous and well-defined on all parts of the paper $R$ (or $R_\varepsilon$, if you prefer). All that remains to be shown is that a valid MV assignment can be found for the creases of $G$, and to do this Lemma 5.18 can be employed as in the proof of Kawasaki's Theorem.

The definition of the folding map used above is basically the limit map $\sigma$ for a flat origami $\{\sigma_n\}_{n=1}^{\infty}$, only it leaves the face $F_{2n}$ fixed and is defined with reflection matrices instead of a limiting process. (We alluded to this alternate way to compute $\sigma$ in Section 5.1.)

The concept of a folding map can be extended to multiple-vertex crease patterns, and doing this will be important for generalizing results like Kawasaki's Theorem. We will explore this in Section 6.2.

---

**Diversion 5.7**   Verify Theorem 5.37 for the degree-4 crease patterns in Figure 5.10.

---

## 5.8   Open Problems

It may be clear to the reader that the single-vertex case of flat-foldable crease patterns has a surprising amount of complexity. We summarize in this section a few of the open problems for flat vertex folds, with the caveat that this is still an active area of research.

**Open Problem 5.1**   What is the asymptotic nature of the sequence $|SC(n)|$, given in (5.2)? It appears to be exponential, and a first attempt at bounds is given in (Chang and Hull, 2011), but improvements should be possible.

**Open Problem 5.2**   Are there other ways to dissect or label points in the configuration spaces $P_{2n}$ to make them correspond to the values and recursive structure of $C(\alpha_1, \ldots \alpha_{2n})$?

## 5.9   Historical Remarks

While studies in origami geometric constructions go back to the 1800s, research in flat-foldability is much more recent. The earliest reference to results such as those given in this chapter is in a pair of papers from 1966 by Saburo Murata (1966a, 1966b), where Maekawa's Theorem is stated, as is the necessary direction of Kawasaki's Theorem for the degree-4 and -6 cases, all without proof. Independently, it seems, in the 1970s Huffman (1976) and Husimi (1979) provided (and proved) the degree-4 case, necessary direction case of Kawasaki's Theorem. The first to state this result for arbitrary-degree vertices was Robertson (1977–1978), although as previously stated, he proved it for general surfaces but only as a necessary condition. (See Chapter 10.) Later the result was independently re-discovered by Kawasaki (Kasahara and Takahama, 1987) (this book appeared in Japanese under the name *Top Origami* in 1985) and Justin (1986a). At the same time, Maekawa (Kasahara and Takahama, 1987) and Justin (1986a) discovered Maekawa's Theorem, again independently of Murata. It seems, however, that full proofs of these results for single-vertex folds did not appear in print until (Hull, 1994). In the origami community, these results were named after Kawasaki and Maekawa due to the influence of (Kasahara and Takahama, 1987), which was a very popular origami book at the time. The author, in (Hull, 2002) and (Hull, 2003), chose to name them the "Kawasaki–Justin" and "Maekawa–Justin" Theorems because of the simultaneous nature of their discovery. In this book, however, we revert back to their more common names and adopt the name "Justin's Theorem" for a result from (Justin, 1997) that was presented in 1994 at the Second International Meeting of Origami Science and Scientific Origami in Otsu, Japan. This result concerns multiple-vertex crease patterns and contains Maekawa's and Kawasaki's Theorems as special cases. It will be presented in Chapter 6.

The idea of modeling flat-folded creases with reflections was discovered by Kawasaki (Kawasaki and Yoshida, 1988) and Justin (1989). Kawasaki further developed this idea and specifically used matrices in subsequent publications, like (Kawasaki, 1989a, 1997); see Chapter 12 for more details.

# 6 Multiple-Vertex Flat Folds: Global Properties

In this chapter we will explore flat origamis with more than one interior vertex in their crease pattern. We will see straight away that the results from Chapter 5 do not all extend to multiple-vertex flat folds, and thus we will seek to explore the differences between the local and global behavior of flat origami. It turns out that keeping track of the different possible layerings and potential self-intersections of the paper in multiple-vertex flat folds is combinatorially quite complex.

As was mentioned in Chapter 5, there is more than one way to model flat origami folds. We continue to use the model developed in Chapter 5, where a flat origami is a sequence of folds $\{\sigma_n\}_{n=1}^{\infty}$, where each $\sigma_n$ does not necessarily possess all the properties that a folded sheet of paper should possess (like non-elasticity) but does possess others (like continuity and being one-to-one), and the limit function $\sigma$ is a continuous piecewise isometry. Otherwise the approach followed here will be very similar to that of Justin (1997).

## 6.1 Impossible Crease Patterns

We begin by demonstrating that while Kawasaki's Theorem is a necessary condition for each vertex in a flat-foldable crease pattern, it is by no means a sufficient condition for multiple-vertex crease patterns to be flat-foldable. Figure 6.1 shows two multiple-vertex crease patterns that satisfy the Kawasaki angle condition at each vertex but that are impossible to fold flat. The reasons for why these two crease patterns do not work, however, are very different from each other.

---

**Diversion 6.1**   Prove that the crease patterns in Figure 6.1 cannot fold flat.

---

We will describe solutions to Diversion 6.1 shortly, but it would be very worthwhile for the reader to make copies of these crease patterns and try to fold them. It should be clear that while each vertex is perfectly flat-foldable, the multiple vertices won't all fold flat at the same time without the paper either ripping or intersecting itself. We capture this local but possibly not global flat-foldability in a definition.

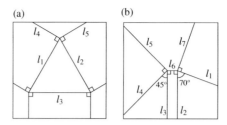

**Figure 6.1** Two crease patterns that are locally, but not globally, flat-foldable.

**Definition 6.1**     A crease pattern $G$ is called a **phantom fold** if each interior vertex satisfies the condition in Kawasaki's Theorem.

Thus a phantom fold will always fold flat locally at each vertex, but it may or may not be globally flat-foldable.

The crease pattern in Figure 6.1(a) won't fold flat because of a **mountain-valley contradiction**. Suppose that it did fold flat. Then there would exist a valid MV assignment $\mu : E \to \{-1, 1\}$ for the edges $E$ of the crease pattern. By Lemma 5.25 (the Big-Little-Big Lemma), we have that $\mu(l_1) \neq \mu(l_2)$ (the angle between $l_1$ and $l_2$ is 60°, whereas the angles to the left and the right of this angle are both 90°). But then the same argument also shows that $\mu(l_2) \neq \mu(l_3)$ and $\mu(l_3) \neq \mu(l_1)$, which is impossible.

In other words, flat-foldable crease patterns often contain "chains" of creases that determine their MV parity. If such chains form a triangle or some other non-2-colorable configuration (thinking of the assignments "mountain" and "valley" as two different colors), then these MV chains will force a contradiction and the crease pattern will be impossible to fold flat. We capture this by defining the **origami line graph** (modified from the one given in (Hull, 1994)) for a phantom fold crease pattern and considering if it is properly 2-vertex colorable.

Before we do this, recall that in graph theory, $P_2$ denotes a **path of length 2**, which has three vertices $a$, $b$, and $c$ and edges $\{a, b\}$ and $\{b, c\}$. The vertices $a$ and $c$ are called the **ends of** $P_2$ and must get the same color when we 2-color the vertices of $P_2$. In fact, the end vertices must always be the same color when 2-coloring the vertices of any even-length path $P_{2n}$.

**Definition 6.2**     Given a phantom fold $G = (V, E)$, we define the **origami line graph** $G_L = (V_L, E_L)$ to be a graph produced as follows: Let our initial set of vertices $V_L$ be the creases $\{c_1, \dots, c_n\}$ in $G$. Then perform the following steps:

(i) For all pair of creases $c_i, c_j \in E$, if they are forced to have different MV parity, then let $\{c_i, c_j\} \in E_L$.

(ii) For all pair of creases $c_i, c_j \in E$, if they are forced to have the same MV parity and $c_i$ and $c_j$ are not already the ends of a path of even length from performing step (i), then add a new vertex $v_{i,j}$ to $V_L$ and let $\{c_i, v_{i,j}\}, \{v_{i,j}, c_j\} \in E_L$.

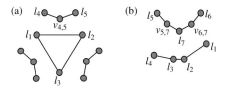

**Figure 6.2** The origami line graphs of the crease patterns in Figure 6.1.

**Theorem 6.3** *If the origami line graph $G_L$ of a phantom fold $G$ is not 2-vertex colorable, then $G$ is not flat-foldable.*

Figure 6.2 shows the origami line graphs of the phantom folds from Figure 6.1. The mountain-valley contradiction that emerges in Figure 6.1(a) is due to the presence of a triangle in Figure 6.2(a), which is not 2-colorable. Notice, however, that the origami line graph in Figure 6.2(b) is 2-colorable, and so the fact that the crease pattern in Figure 6.1(b) is not flat-foldable has nothing to do with mountain-valley contradictions.

What does make the 2-colorable crease pattern in Figure 6.1(b) impossible to fold flat, then? A reader who has not tried folding this crease pattern for herself might actually think it is possible, and we suggest again that you photocopy the crease pattern and try to fold it. There is something more nefarious than mountain-valley contradictions going on here.

At this stage, the only way to prove that this example is impossible to fold flat is to check all the possible layerings of the paper and argue that each case forces a self-intersection. We will describe one way to do this systematically; see the illustration in Figure 6.3.

Imagine that we have cut along the edges of the crease pattern to separate all the faces. Fix one of the faces $f$ and then reflect all of the other faces relative to $f$ as they would if the paper were being folded flat. (This is easily done using the folding map described in Section 5.1 relative to the fixed face $f$, which we will define formally in the next section.) We then impose a layer ordering of these faces to simulate how the paper would actually stack up in layers. Finally, we check to see if our layer ordering

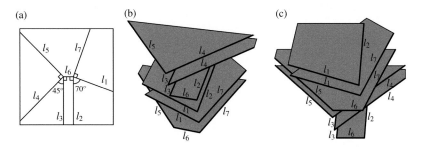

**Figure 6.3** Two different layering possibilities for the crease pattern in Figure 6.1(b).

is feasible by trying to glue the creases back together, which will only work if there are no other layers of paper in between the loose ends of the creases.

For example, the layering shown in Figure 6.3(b) won't work because the layers at crease line $l_1$ will keep crease $l_5$ from being glued back together. In Figure 6.3(c) we see a different arrangement of the layers, but this time the crease $l_2$ can't be glued back together because the $l_4$ and $l_7$ layers are in the way. One would need to check all such possible layerings to firmly prove that this crease pattern is impossible to fold flat.

Please note, however, that this examination of the Figure 6.1(b) crease pattern is very informal. It is only included as a way to build background and intuition for what is to follow. The simple layer ordering done above has a major pitfall and will not work for all phantom crease patterns. That is, it can be used to erroneously claim that some flat-foldable crease patterns are not flat-foldable. In the next section we will create the terminology and tools needed to analyze multiple-vertex flat origami properly.

---

**Diversion 6.2**    Prove that the two crease patterns in Figure 6.4 are phantom folds but are not flat-foldable. (Crease pattern (a) is derived from an example of Kawasaki (1989b) and (b) is from (Hull, 1994).)

---

## 6.2    Generalized Kawasaki: Necessary Conditions and the Folding Map

The reflections-based version of Kawasaki's Theorem (Theorem 5.37) gives us a necessary condition for flat-foldability that can be extended to crease patterns with more than one vertex.

**Theorem 6.4**    *Let $G = (V, E)$ be a flat-foldable crease pattern on a region R. Let $\gamma$ be a vertex-avoiding, simple closed curve drawn on R that is not tangent to any creases and that crosses, in order, the crease lines $l_1, l_2, \ldots, l_{2n}$. Then we have*

$$R(l_1)R(l_2) \cdots R(l_{2n}) = I. \tag{6.1}$$

Intuitively, this theorem makes a lot of sense, since the only way such a product of reflections could **not** be the identity would be if the paper ripped somewhere and made a discontinuity. In fact, the proof used in Theorem 5.37 can be modified to make

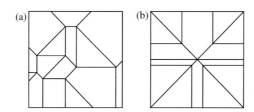

**Figure 6.4** Two more impossible crease patterns.

a convincing, informal proof. But the real reason why Theorem 6.4 is true is because of an inductive process, which we will employ in our proof.

*Proof* We proceed by induction on the number of vertices in the crease pattern that are in $\text{int}(\gamma)$, the interior of the closed curve $\gamma$.

If $\text{int}(\gamma)$ contains no vertices of the crease pattern, then either $\gamma$ does not cross any crease lines or, if it does cross a crease line $l$, it must cross $l$ an even number of times. This implies that Equation (6.1) must be true. Also note that if $\text{int}(\gamma)$ contains only one vertex, then we are done by Theorem 5.37.

Now suppose that $\text{int}(\gamma)$ contains $n + 1$ vertices of the crease pattern. Then we can find a vertex of the crease pattern $v$ in $\text{int}(\gamma)$ such that $\gamma$ can be bent into a new curve $\gamma'$ that is exactly the curve $\gamma$ except that it "misses" the vertex $v$ so that $v$ is **not** in $\text{int}(\gamma')$.

Specifically, $\gamma'$ will cross all of the crease lines that $\gamma$ crosses except for those creases adjacent to $v$. If $l_1, l_2, \ldots, l_{2n}$ are the crease lines adjacent to $v$, and $\gamma$ crosses $l_1, \ldots, l_i$, in that order, then $\gamma'$ will cross $l_{2n}, l_{2n-1}, \ldots, l_{i+1}$, in that order, in order to go around $v$ the other way. See Figure 6.5.

Since we can assume that Equation (6.1) holds for $\gamma'$, we have that

$$R_1 R(l_{2n}) R(l_{2n-1}) \cdots R(l_{i+1}) R_2 = I,$$

where $R_1$ and $R_2$ are the products of reflections from the creases before and after, respectively, those incident to $v$. We also know from Theorem 5.37 that $R(l_1) \cdots R(l_{2n}) = I$. Therefore,

$$R(l_1) \cdots R(l_i) = R(l_{2n}) R(l_{2n-1}) \cdots R(l_{i+1})$$
$$\Rightarrow R_1 R(l_1) \cdots R(l_i) R_2 = I,$$

as desired. $\qquad\qquad\qquad\qquad\qquad\qquad\qquad\qquad\qquad\qquad\qquad\qquad\qquad\qquad\square$

Theorem 6.4 is sometimes called the **generalized Kawasaki's Theorem**, although this can be misleading since Theorem 6.4 is only a necessary result for flat-foldable crease patterns, not a sufficient one.

Theorem 6.4 gives us what we need to create a more general definition of a folding map. We encountered the single-vertex version in Section 5.7.

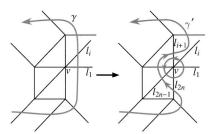

**Figure 6.5** How to make $\gamma$ avoid a vertex $v$.

**Definition 6.5**  Let $G = (V, E)$ be a flat-foldable crease pattern on a region $R$ and fix a face $f$ of $G$. We define the **folding map** $\sigma_f \colon R \to \mathbb{R}^2$ (also sometimes called the folding map of $G$ **toward** $f$) as follows: For all $x \in R$ let $\gamma$ be a simple curve in $R$ from a point in $f$ to $x$ that avoids all vertices of $G$ (and isn't tangent to any creases). Let $l_1$, $l_2, \ldots, l_k \in E$ be the crease lines, in order, that $\gamma$ crosses. Then

$$\sigma_f(x) = R(l_1)R(l_2) \cdots R(l_k)(x),$$

where $R(l_i)$ is the reflection transformation about the line $l_i$.

**Theorem 6.6**  *A folding map $\sigma_f$ of a crease pattern $G$ on a region $R$ is well-defined, that is, it is independent of the choice of the curve $\gamma$.*

*Proof*  Pick a point $x \in R$ and let $\gamma$ and $\gamma'$ be two different vertex-avoiding paths in $R$ from inside our fixed face $f$ to $x$. Let $\gamma$ cross, in order, $l_1, \ldots, l_k \in E$ and $\gamma'$ cross, in order, $l'_1, \ldots, l'_j \in E$. If we follow $\gamma'$ and then $\gamma^{-1}$ (going backward along $\gamma$), we'll form a closed curve from the face $f$ to itself. Because $G$ folds flat, we have that

$$R(l'_1)R(l'_2) \cdots R(l'_j)R(l_k) \cdots R(l_2)R(l_1) = I$$
$$\Rightarrow R(l'_1)R(l'_2) \cdots R(l'_j) = R(l_1)R(l_2) \cdots R(l_k).$$

Therefore, $\sigma_f$ gives the same result using the curve $\gamma$ as it does for $\gamma'$.    □

The folding map will prove to be very useful in what follows. It is used heavily in the work of Jusitn and Kawasaki, for instance. Also, notice the subtly in our development of the folding map: To prove Theorem 5.37 we needed a very restricted folding map for folding a single vertex, and thus did not need a curve $\gamma$ to define it. And to prove that the folding maps for multiple-vertex crease patterns are well-defined, we needed the generalized Kawasaki's Theorem (Theorem 6.4), which therefore needed to be proven without relying on the folding map.

---

**Diversion 6.3**  Prove that a crease pattern $G$ is a phantom fold if and only if Equation (6.1) holds for all closed curves $\gamma$ drawn on the crease pattern.

---

---

**Diversion 6.4**  Notice that when generalizing the necessary direction of Kawasaki's Theorem, we had to rely on the reflections model of flat origami. Why couldn't we have generalized it in terms of the angles between the creases crossed by the curve $\gamma$?

---

## 6.3    Generalized Maekawa

The idea of having the quantity $M - V$ from Maekawa's Theorem be some kind of characteristic invariant beyond the local, single-vertex results of Chapter 5 is, unfortunately, too much to hope for. A quick look at some examples reveals that

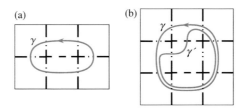

**Figure 6.6** Examples of closed curves drawn on multiple-vertex, flat-foldable crease patterns.

$M - V$ can equal just about anything for multiple-vertex flat-foldable crease patterns. The simple crease patterns in Figure 6.6, for example, give $M - V = 5$ and $M - V = 8$.

Attempts to find patterns in $M - V$ for general, multiple-vertex crease patterns have only revealed straightforward results. However, absolutely nothing can be said unless we are careful about how we think of crease lines with respect to the paper's boundary. All crease lines in an origami model come in three types:

(1) Crease lines that have both endpoints in the interior of the paper (called **interior creases**)
(2) Crease lines that have one interior endpoint and one endpoint on the boundary
(3) Crease lines that have both endpoints on the boundary of the paper

Crease lines of type (3) turn out to be the fox in the henhouse. Indeed, many multiple-vertex folds can be made to have $M - V$ equal to whatever we wish by adding many rows of, say, parallel mountain creases of type (3) to a suitable corner of the paper where no other creases lie. If we hope to have some kind of connection between $M - V$ and the interior vertices of our crease pattern, we need to consider any crease line with both endpoints on the boundary of the paper to actually be **two** crease lines with an interior vertex of degree 2 separating them. Such degree-2 vertices will still satisfy Maekawa's Theorem and thus will contribute to $M - V$ globally in the proper way.

In (Hull, 1994) a global version of Maekawa's Theorem is presented for multiple-vertex flat folds. Let us define interior vertices in a multiple-vertex flat fold with a valid MV assignment to be **up vertices** and **down vertices** if they locally have $M - V = 2$ or $-2$, respectively. Given a multiple-vertex crease pattern with a valid MV assignment, let $U$ be the number of up vertices and $D$ the number of down vertices. Also, let $IV$ be the number of internal valley creases and $IM$ the number of internal mountain creases. Then computing $M - V$ will involve adding 2 for every up vertex and $-2$ for every down vertex, except that every internal crease will be counted twice. Thus we obtain the next theorem:

**Theorem 6.7**   *Let $G = (V, E)$ be a flat origami crease pattern with mountain-valley assignment $\mu$. Then using the above notation, we have*

$$M - V = 2U - 2D - IM + IV.$$

## 6.4    Justin's Theorem and Paper with Holes

Astute readers may recall that in Section 5.4 we hinted at a combined generalization of Maekawa's and Kawasaki's Theorems to multiple-vertex flat origami crease patterns. The idea that there is a connection between the mountain-valley condition of Maekawa's Theorem and the angle relation of Kawasaki's Theorem was seen when considering flat cone folds. Similar results can be seen when considering a closed, vertex-avoiding path $\gamma$ on a multiple-vertex flat origami crease pattern, counting the mountains and valleys as we go as well as calculating the change in angle between the consecutive creases crossed by $\gamma$.

For example, consider the flat-foldable crease patterns, with valid MV assignments, shown in Figure 6.6. In part (a) we have that along the curve $\gamma$, $M - V = 6$ and the alternating sum of the angles between the consecutive creases that $\gamma$ crosses is 0. If we split these angles into alternate sums $\alpha_1 + \alpha_3 + \cdots$ and $\alpha_2 + \alpha_4 + \cdots$, as in Corollary 5.19, we get that the odd-indexed set of angles adds up to $180°$ and so does the even-indexed set of angles.

In Figure 6.6(b), however, we have that, along $\gamma$, $M - V = 8$ and the sum of the odd-indexed angles is $360°$ while the sum of the even-indexed angles is $0°$ (labeled appropriately, of course). Notice, though, that along $\gamma'$ we have $M - V = 6$ and the odd- and even-indexed sets of angles both add up to $180°$ once again.

The reader is encouraged to try more examples. One discovers that the odd- and even-indexed sets of angles always sum up to either $\pi$ or $2\pi$, and they do this depending on the parity of $(M - V)/2$.

Jacques Justin is the only person, it seems, to have discovered this relation. It appears in (Justin, 1997), but it was presented at a conference, the Second International Meeting of Origami Science and Scientific Origami in Otsu, Japan, in 1994. What's more, Justin realized that this result holds for flat-foldable crease patterns on paper that is not simply connected (i.e., has holes in it).

Having holes in one's paper poses further challenges, however. Consider the flat-folded ring of paper and crease pattern in Figure 6.7(a) and (b). It has two creases, one mountain and one valley. The angle between these creases is $\pi$ (this is the amount of turning a curve $\gamma$ would have to do on the paper to go from one crease to the other), and so we have that $\alpha_1 = \alpha_2 = \pi$ is the sum of the odd- and even-indexed set of angles. But $M - V = 0$, which is not the correct parity that our previous examples

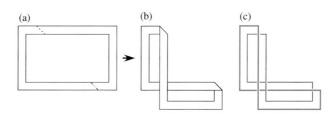

(a)          (b)          (c)

**Figure 6.7**  A twisted, flat-folded ring of paper.

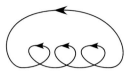

**Figure 6.8** A generic closed plane curve $\gamma$ with $turn(\gamma) = 4$ and the number of self-intersections $p = 3$, the minimum in its homotopy equivalence class.

were showing (we should have gotten $2\pi$). This is good! The fold shown in Figure 6.7(b) is impossible; the ring of paper has been given an illegal "twist" even though its crease pattern is valid, aside from the MV assignment. Another way to see this is to follow the boundaries of this ring of paper. The inner and outer boundaries of the ring form two linked closed curves, shown in Figure 6.7(c). If this fold were possible, the two boundary curves would have to be unlinked, since in the unfolded state they are unlinked.

**Diversion 6.5** Compute the linking number of the folded boundary curves in Figure 6.7(c) to prove that they are, indeed, linked.

Because of examples such as that in Figure 6.7, Justin refers to his discovery as the **non-twist relation**, and the proof that he provides, which we reproduce here, uses some elementary knot theory to capture the non-twisting nature of the paper's boundaries. We shall refer to his result as **Justin's Theorem**, however, because of the connection it makes between Maekawa's and Kawasaki's Theorems, which Justin also independently discovered.

First, however, we will need a lemma concerning turning numbers. Recall that the **turning number** of a closed plane curve $\gamma$, which we will denote $turn(\gamma)$, is the total rotational angle at which $\gamma'(t)$ turns in the range of $t$ divided by $2\pi$. That is, if $d\theta$ is the differential element of rotation, then

$$turn(\gamma) = \frac{1}{2\pi} \int_{\gamma} d\theta.$$

We also say that a closed plane curve $\gamma$ is **generic** if it has a finite number of self-intersections, $\gamma$ does not pass through any point in the plane more than twice, and if $\gamma(t_1) = \gamma(t_2)$ is an intersection point, then $\gamma'(t_1)$ and $\gamma'(t_2)$ are distinct vectors. See (Whitney, 1937) for more background on this material (although using slightly different terminology).

**Lemma 6.8** *If $\gamma$ is a generic closed plane curve that intersects itself $p$ number of times, then*

$$\int_{\gamma} d\theta \equiv 2(p + 1)\pi \quad \mathrm{mod}\ 4\pi.$$

*Proof*  Note that any such curve $\gamma$ is homotopy equivalent to a curve with a minimum number of self-intersections. First we assume that $\gamma$ itself has a minimum number of self-intersections (i.e., that there is no other homotopy-equivalent curve with a smaller number of self-intersections). See Figure 6.8. If $p$ is the number of self-intersections, then we have that $p = turn(\gamma) - 1$. (To see this, note that if $turn(\gamma) = 1$, then $\gamma$ is homotopic to a circle, and thus $p = 0$. Otherwise, $\gamma$ contains of a sequence of $turn(\gamma) - 1$ loops, each of which cause a distinct self-intersection.)

Therefore, we have the following expression for the total rotation of $\gamma$:

$$\int_{\gamma} d\theta = 2\pi \ turn(\gamma) = 2(p+1)\pi.$$

Now, if $\gamma$ has more self-intersections than the minimum number in its homotopy equivalence class, then these must come in pairs, and each pair will add $4\pi$ to the quantity $2(p+1)\pi$. Thus, $\int_{\gamma} d\theta \equiv 2(p+1)\pi \mod 4\pi$.  □

**Justin's Theorem**  *Let $\gamma$ be a simple, closed, vertex-avoiding curve on a flat-foldable crease pattern $G$ on a region $R$, where $R$ is allowed to be not simply connected. Let $\alpha_i$ be the signed angles, in order, between the consecutive creases that $\gamma$ crosses, for $1 \le i \le 2n$. Also let $M$ and $V$ be the number of mountain and valley creases, respectively, that $\gamma$ crosses. Then,*

$$\alpha_1 + \alpha_3 + \cdots + \alpha_{2n-1} \equiv \alpha_2 + \alpha_4 + \cdots + \alpha_{2n} \equiv \frac{M-V}{2}\pi \quad \mod 2\pi. \qquad (6.2)$$

*Proof*  Let $\mathcal{R}$ be a curve as stipulated in the statement of the theorem. Let $l_i$ be the crease line that $\mathcal{R}$ crosses between angles $\alpha_i$ and $\alpha_{i+1}$, with $l_{2n}$ between $\alpha_{2n}$ and $\alpha_1$. Since $\mathcal{R}$ divides the plane into two regions, direct the crease lines $l_i$ to be vectors oriented so that they are pointing from the interior to the exterior of $\mathcal{R}$. Let $x_i$ be the point on $l_i$ at which $\mathcal{R}$ crosses. Then $\mathcal{R}$ is homotopy equivalent to a curve that is just like $\mathcal{R}$ but passes through each point $x_i$ in a direction parallel to the directed crease $l_i$. Thus we may assume that $\mathcal{R}$ itself has this property, so that each $l_i$ is a directed tangent to $\mathcal{R}$ at $x_i$ and $x_i$ is an inflection point of $\mathcal{R}$. (See Figure 6.9.)

We insist that $\mathcal{R}$ has these properties because it ensures that when we fold $G$ the image $\mathcal{R}'$ of $\mathcal{R}$ will still be a smooth curve; at the crease line $l_i$ the curve $\mathcal{R}$ will fold at the inflection point $x_i$, bending into a smooth curve at $x_i$. Also, since $\mathcal{R}$ is tangent and in the same direction as the directed crease $l_i$, the amount of angle rotation of $\mathcal{R}$ between crease $l_i$ and $l_{i+1}$ will be $\alpha_i$. Or, more precisely, if $d\theta$ is the differential

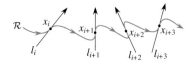

**Figure 6.9**  We make $\mathcal{R}$ cross the crease lines $l_i$ at inflection points.

element of rotation of the curve $\mathcal{R}$, we have that $\alpha_i = \int_{\mathcal{R}(x_i, x_{i+1})} d\theta$, where $\mathcal{R}(x_i, x_{i+1})$ denotes the section of $\mathcal{R}$ between $x_i$ and $x_{i+1}$.

Then, we also have that

$$\alpha_1 + \alpha_2 + \cdots + \alpha_{2n} = \int_{\mathcal{R}} d\theta = \pm 2\pi$$

and

$$\alpha_1 - \alpha_2 + \cdots - \alpha_{2n} = \int_{\mathcal{R}'} d\theta,$$

assuming that we take the face where $\mathcal{R}$ starts, with angle $\alpha_1$, to be in the positive direction. But $\mathcal{R}'$ is just another curve winding around in the plane some number of times, so if we let $p$ be the number of times $\mathcal{R}'$ intersects itself, then we have by Lemma 6.8 that $\int_{\mathcal{R}'} d\theta \equiv 2(p+1)\pi \mod 4\pi$. Combining these we have

$$\alpha_1 + \alpha_3 + \cdots + \alpha_{2n-1} \equiv \alpha_2 + \alpha_4 + \cdots + \alpha_{2n} \equiv p\pi \mod 2\pi.$$

All that's left is to show that $p \equiv (M - V)/2 \mod 2$. That is, we want to show that $2p \equiv M - V \mod 4$.

Let $\mathcal{L}$ be a closed curve on our crease pattern $G$ that is close to $\mathcal{R}$ and "parallel" to it, so that $\mathcal{L}$ crosses exactly the same creases as $\mathcal{R}$ and is tangent to those creases at points of inflection. (Think of $\mathcal{L}$ as being drawn alongside of $\mathcal{R}$ and just in the interior of $\mathcal{R}$.) Let $\mathcal{L}'$ be the image of $\mathcal{L}$ under our folding map $\sigma = \lim_{k \to \infty} \sigma_k$. Also, let $\mathcal{R}^*$ and $\mathcal{L}^*$ be the images of $\mathcal{R}$ and $\mathcal{L}$, respectively, under one the the maps $\sigma_k$ for suitably large $k$, so that $\mathcal{R}^*$ and $\mathcal{L}^*$ are simple closed curves in $\mathbb{R}^3$ on the surface of our folded paper just before it is completely folded flat. We then know that the linking number of the curves $\mathcal{R}^*$ and $\mathcal{L}^*$ is zero, since $\mathcal{R}$ and $\mathcal{L}$ are not linked on the unfolded paper. But we can compute the linking number of $\mathcal{R}^*$ and $\mathcal{L}^*$ another way, as well.

Following the standard method for computing the linking number of a link (see (Adams, 1994) or (Cromwell, 2004)), we let $I$ be any intersection point of $\mathcal{R}'$ and $\mathcal{L}'$. When looking at this intersection point on the 3-dimensional curves $\mathcal{R}^*$ and $\mathcal{L}^*$, one of these curves will be above the other at $I$. If the curve that is below can be rotated clockwise to be in the same direction as the curve above, then we assign that value $+1$ to the point $I$. Otherwise, if the curve below must be rotated counterclockwise to line up with the direction of the above curve, we assign $-1$ to $I$. The linking number is then the sum of the $\pm 1$ values of all the intersection points of $\mathcal{R}^*$ and $\mathcal{L}^*$.

However, the intersection points $I$ are of two types, as shown in Figure 6.10:

(1) Near each folded crease line there is exactly one intersection point, as seen in Figure 6.10(a). If this crease line is a mountain, then $\mathcal{R}^*$ will be above $\mathcal{L}^*$ and $I$ will be given the value $+1$. If the crease is a valley, then $\mathcal{L}^*$ will be on top, and $I$ will get $-1$. So the contribution of these intersection points to the linking number is $M - V$.

(2) Away from the crease lines, every self-intersection point of $\mathcal{R}'$ gives exactly two intersection points of $\mathcal{R}^*$ and $\mathcal{L}^*$, and both will receive the same value.

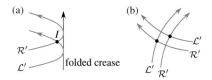

**Figure 6.10**  Examining intersection points of $\mathcal{R}'$ and $\mathcal{L}'$.

(See Figure 6.10(b).) So the contribution of these points will be of the form $2(a_1 + a_2 + \cdots + a_p)$ where each $a_i = \pm 1$ and $p$ is the number of self-intersections of the plane curve $\mathcal{R}'$. This is congruent to $2p \mod 4$.

Therefore, since the linking number of $\mathcal{R}^*$ and $\mathcal{L}^*$ is 0, we get $M - V + 2p \equiv 0 \mod 4$, which gives us Equation (6.2), and we're done.    □

This proof is interesting for a few reasons. For one, it doesn't use any of our previous flat-foldabilty results. Also, it illustrates a connection between paper folding and elementary knot theory.

---

**Diversion 6.6**    Use Justin's Theorem to show that Maekawa's Theorem is true if and only if the necessary direction of Kawasaki's Theorem is true.

---

---

**Diversion 6.7**    In the proof of Justin's Theorem, the argument for the "type (1)" intersection points $I$ might not seem very convincing. In particular, if the folded crease in Figure 6.10(a) is a mountain, then the intersection $I$ will get the value $+1$, since $\mathcal{R}^*$ is above $\mathcal{L}^*$ and $\mathcal{L}^*$ must be rotated clockwise to line up with $\mathcal{R}^*$ at that point. But if we consider the case that is the mirror image of Figure 6.10(a) and have the folded crease be a mountain, then it would seem that $I$ should be assigned $-1$, since $\mathcal{R}^*$ is still above $\mathcal{L}^*$ and $\mathcal{L}^*$ needs to rotate counterclockwise to line up with $\mathcal{R}^*$. Explain why this, in fact, is erroneous thinking.

---

---

**Diversion 6.8**    Cut out a piece of paper into the shape of the annular ring shown in Figure 6.11 with the creases $l_1, \ldots, l_4$ as shown. What values of $M - V$ does Justin's Theorem say are possible? Fold your paper model to show that any MV assignment that achieves these values is valid. (You should get eight different MV assignments for these creases.)

---

---

**Diversion 6.9**    Develop a different proof of Justin's Theorem for the case where the paper has no holes.

---

**Figure 6.11**  An example of a flat-foldable crease pattern on an annular piece of paper.

## 6.5        Global Flat-Foldability

In Section 6.1 we saw how the global flat-foldability problem (deciding whether or not a multiple-vertex crease pattern is flat-foldable) could be approached by cutting the crease pattern into faces, using a folding map $\sigma_f$ for some fixed face $f$ of the crease pattern $G$ to map all of the other faces into their proper positions, and then trying to tape all the faces back together without the paper self-intersecting. As was also mentioned in Section 6.1, this approach is inherently flawed. We will now explain why this is so.

Consider the square twist fold shown in Figure 6.12. Hopefully the reader gained experience with folding this crease pattern in Diversion 5.2. If not, then the reader should try folding one now in order to see that the faces labeled $f_1, \ldots, f_4$ do not form a linear, bottom-to-top ordering when this crease pattern is folded flat with the given MV assignment. Thus for this crease pattern, which is flat-foldable, we wouldn't be able to simply map the faces into their positions, impose an ordering on these faces, and then tape them back together.

Another way to see this is in Figure 6.13, where we've tried to illustrate a possible layer ordering of the faces (in 3-dimensional perspective) after they've been mapped into position. Notice that the crease lines $l_1$, $l_3$, and $l_4$ can be taped back together without any obstructions, but that $l_2$ cannot, since the face $f_3$ is in the way. (There are other obstructions in this picture as well.) Because this version of the square twist has faces that layer in a perfect over-under-over-under pattern, as we saw in Figure 6.12, no linear layer ordering will be able to avoid such obstructions. In other words, trying to find a valid layer ordering of the faces is impossible, which would lead one to believe that the MV assignment shown in Figure 6.12 for the square twist is impossible to fold, which is very much incorrect.

The square twist is not merely an esoteric fold in this way. Other origami models exhibit this same layering phenomenon.

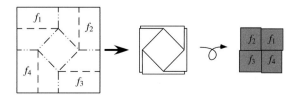

**Figure 6.12**  The square twist fold has faces that cannot be put into a linear layer ordering.

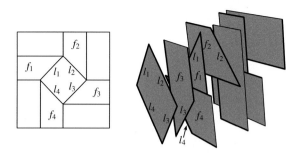

**Figure 6.13** The square twist fold with the faces cut apart and a layer ordering attempted.

---

**Diversion 6.10**    Show that it is impossible to construct simple, valid layer orderings for the crease pattern faces of the origami models shown in Figure 6.14(a), which is the traditional pinwheel model, and Figure 6.14(b), which employs the same over-under-over-under layering that is used to close the top of a sturdy cardboard box.

---

Examples like these indicate that an origami model's crease pattern alone, even with a valid MV assignment, is not enough information to fully describe how the paper layers when it folds flat. For this reason, Justin (1997) created the idea of a **superposition pattern**, or **s-net** for short, which contains the crease pattern as a subset. The intuitive idea is as follows: Imagine that your piece of paper was coated with a light ink or carbon on both sides. Then we fold the model flat using the crease pattern. With the model folded flat, we put the model on a table and rub the surface of the model with a hard tube, like the side of a pen. This will imprint lines on the paper wherever a crease line or a boundary line of the paper touches a face of the crease pattern. Unfolding the model will reveal the superposition pattern. Let us define this more formally.

**Definition 6.9**    Given a crease pattern $G = (V, E)$ on a region $R$, let $L = E \cup \partial R$ be the set of crease lines $E$ embedded in the plane together with the boundary of the region $R$. Fix a face $f$ of the crease pattern. Then the **superposition pattern**, or **s-net**, of $G$ is the set $\sigma_f^{-1}(\sigma_f(L))$.

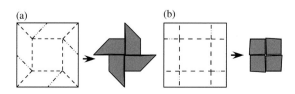

**Figure 6.14** Two more origami models that are layer-order-challenged.

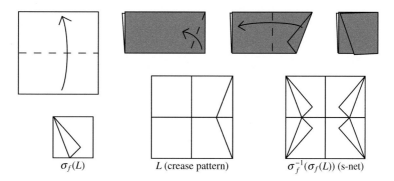

$\sigma_f(L)$           $L$ (crease pattern)           $\sigma_f^{-1}(\sigma_f(L))$ (s-net)

**Figure 6.15** An example of an origami model, its crease pattern $L$, it's folded image $\sigma_f(L)$, and its s-net. (Taken from (Justin, 1997).)

**Example 6.10** Figure 6.15 reproduces an example given in (Justin, 1997). A square piece of paper is folded in half from bottom to top, then a small triangle flap is folded from the lower right corner. Then the result is folded in half again from right to left. Unfolding this shows us the crease pattern, which together with the boundary of the original square gives us the planar graph $L$ (which essentially is the crease pattern). If we fix a face of the crease pattern $f$, then the folding map $\sigma_f(L)$ gives us the folded image of the crease pattern, viewed as if all the layers of paper were transparent and we can see all the crease lines through the paper. Taking the inverse image of this, $\sigma_f^{-1}(\sigma_f(L))$ gives us the superposition pattern, or s-net. It shows us the crease pattern together with all regions of the paper that get mapped on top of faces of the crease pattern under $\sigma_f$.

**Diversion 6.11** Prove that an s-net $\sigma_f^{-1}(\sigma_f(L))$ of a crease pattern $L$ is independent of the choice of the face $f$.

The purpose of the s-net is to give us a more refined decomposition of the paper, the faces of which we can then use to determine a layer ordering. What makes the s-net more viable in this regard is that the faces of the s-net, which we will call **s-faces**, neatly stack on top of one another under the folding map $\sigma_f$.

**Lemma 6.11** *If $s_1$ and $s_2$ are the interiors of two s-faces, then we have either $\sigma_f(s_1) = \sigma_f(s_2)$ or $\sigma_f(s_1) \cap \sigma_f(s_2) = \emptyset$.*

*Proof* Let $C = \{\, \text{int}(F) \mid F$ is a face of the s-net$\}$, where $\text{int}(F)$ denotes the interior of the set $F$. Similarly, let $X = \{\, \text{int}(F) \mid F$ is a face of $\sigma_f(L)\}$. If we view $C$ and $X$ as topological spaces with the subspace topology of the Euclidean plane, then we have

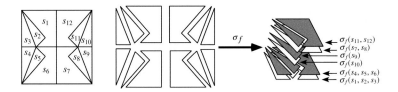

**Figure 6.16** An illustration of the covering map $\sigma_f$ for the example in Figure 6.15. The purple s-faces in the folded image are those whose orientations were flipped under the folding map.

that $\sigma_f : C \to X$ is a covering map. Indeed, since every connected component $F$ of $X$ is the interior of a face of $\sigma_f(L)$, we know that $F$ will not contain the images of any crease lines, vertices, or boundary points of $L$. So, for every $x \in F$, if we take a small enough open neighborhood $U$ so that $x \in U \subseteq F$, then $\sigma_f^{-1}(U)$ will be disjoint open sets in $L$, each of which are homeomorphic to $U$.

Thus, if $s_1$ and $s_2$ are the interiors of two s-faces, then they are connected components of $C$. Suppose that there exists an $x \in \sigma_f(s_1) \cap \sigma_f(s_2)$. Let $F$ be the largest component in $X$ containing $x$. Then $F$ lifts to $\sigma_f^{-1}(F)$, which is a union of connected components of $C$, including $s_1$ and $s_2$. Thus we have that $\sigma_f(s_1) = F = \sigma_f(s_2)$. □

Figure 6.16 shows how the different s-faces of the origami fold from Figure 6.15 get positioned under the folding map $\sigma_f$, which also illustrates how $\sigma_f$ acts as a covering map. Note that if we had tried to define such a covering map from the crease pattern faces instead of the s-net, it would not have worked. If we had defined the domain $C$ to be the interiors of the faces of the crease pattern $L$, then the map $\sigma_f$ would not map the components of $C$ to components of $X$; there would be points in faces of the crease pattern that would map onto the images of crease lines in $\sigma_f(L)$, which are outside of the space $X$.

Lemma 6.11 assures us that if we try to determine a layer ordering for a flat origami fold using the faces of the s-net, then we will not run into a circular contradiction in the ordering like we saw with the square twist in Figure 6.13 and in Diversion 6.10. We now turn our attention to the creation of such an ordering. First we will need a bit more terminology.

**Definition 6.12**   The s-net can be considered as a planar drawing of a graph with vertices, edges, and faces. The edge set can be partitioned into three disjoint sets: the **boundary edges**, the **c-creases**, which are simply the creases of the original crease pattern, and the **s-creases**, which are pre-images of crease lines in $\sigma_f(L)$ but not actually folded creases themselves. That is, if $E$ is the edge set of the graph $L$ and $E_s$ is the edge set of the s-net $\sigma_f^{-1}(\sigma_f(L))$, then the s-creases are the edges in $E_s \setminus E$.

The idea, then, for determining if a given crease pattern is flat-foldable is to (1) make sure that it is at least a phantom fold by checking the Kawasaki condition at each vertex, (2) fix a face $f$ of the crease pattern to define a folding map $\sigma_f$, (3) use $\sigma_f$ to determine the s-net of the crease pattern, (4) map each face $s_i$ in the s-net to its folded state $\sigma_f(s_i)$, and then (5) try to find a valid layer ordering on the image faces

$\sigma_f(s_i)$ that will allow us to glue adjacent s-faces back together in their folded state to preserve continuity of the map $\sigma_f$ over the entire crease pattern.

Finding such a layer ordering will not be easy in general, as we will detail later in Section 6.6. Before we can consider such layerings, however, we need to examine more closely how we can glue the mapped s-faces $\sigma_f(s_i)$ together while making sure that the resulting flat fold will not intersect itself.

Recall that, like the crease-pattern-with-boundary $L$, $\sigma_f(L)$ has the structure of a planar embedding of a graph. Let $e$ be an edge of $\sigma_f(L)$. Then the inverse image $\sigma_f^{-1}(\sigma_f(e))$ will be a collection of edges in the s-net, each of which may either be a boundary edge (which we can ignore, since they do not glue together with anything), a c-crease, or an s-crease. Whenever c- or s-creases get mapped by $\sigma_f$ onto a single crease $e$ in the folded image, we will need to check that our layer ordering doesn't cause the paper to cross itself at $e$.

Define an equivalence relation on the set of s-faces as follows: Two s-faces are equivalent, and we write $s_1 \sim s_2$, if they get mapped to the same face of the folded image, so $\sigma_f(s_1) = \sigma_f(s_2)$. We then let $>$ be any partial ordering on the set of all s-faces made by totally ordering the equivalence classes of $\sim$ (i.e., it totally orders the layers of s-faces that get mapped on top of each other). We call $>$ a **superposition order** of the s-net, and writing $s_1 > s_2$ means that $\sigma_f(s_1)$ will be "above" $\sigma_f(s_2)$ in the folded image. What this really means, of course, is that in the limiting process of our flat origami, where $\sigma_f = \lim_{n\to\infty} \sigma_n$ and $\{\sigma_n\}_{n=1}^{\infty}$ is our infinite sequence of origamis, we have that, for suitably large $n$, $\sigma_n(s_1)$ is above $\sigma_n(s_2)$ in $\mathbb{R}^3$. (We could even assume that our fixed face $f$ lies in the $xy$-plane, so that "above" means "has greater $z$-coordinates.")

To check that $>$ does not cause the paper to self-intersect, we define, for any non-boundary edge $e$ of the s-net, $F_1(e)$ and $F_2(e)$ to be the two s-faces that border $e$. Then for two s-net edges $e_1$ and $e_2$ that map onto each other, there are only three possibilities: they are both s-creases, they are both c-creases, or one is an s-crease and the other is a c-crease. Any self-intersection of the paper that $>$ causes will fall into one of these three categories between two such creases, and thus we arrive at Justin's three **non-crossing conditions** (see (Justin, 1997), though the colorful names for these conditions are borrowed from (Akitaya et al., 2016)):

(1) **Tortilla-Tortilla Condition:** If $e_1$ and $e_2$ are both s-creases and $\sigma_f(e_1) = \sigma_f(e_2)$, then we can't have $F_1(e_1) > F_1(e_2)$ and $F_2(e_2) > F_2(e_1)$.
(2) **Taco-Tortilla Condition:** If $e_1$ is a c-crease and $e_2$ is an s-crease, and $\sigma_f(e_1) = \sigma_f(e_2)$, then we can't have $F_1(e_1) > F_1(e_2) > F_2(e_1)$.
(3) **Taco-Taco Condition:** If $e_1$ and $e_2$ are both c-creases and $\sigma_f(e_1) = \sigma_f(e_2)$, then we can't have $F_1(e_1) > F_1(e_2) > F_2(e_1) > F_2(e_2)$.

The non-crossing conditions are illustrated in Figure 6.17. Notice that they encompass all ways in which two creases from the s-net can be mapped onto each other and have a superposition order that forces the paper to self-intersect. For example, if we have two s-creases that map onto each other and the superposition order on their adjacent faces satisfies the Tortilla-Tortilla Condition, then we will be able to glue the

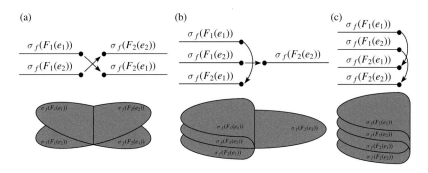

**Figure 6.17** Justin's non-crossing conditions for the folded image of an s-net together with a superposition order: (a) Tortilla-Tortilla, (b) Taco-Tortilla, and (c) Taco-Taco.

required faces together along the edges and not have those parts of the paper intersect. Thus, if we can find a superposition order that satisfies the non-crossing conditions for every pair of non-boundary edges of the s-net that map onto each other, then we'll be able to glue all the s-faces together without crossings of the paper.

Therefore, with some trepidation (which will be explained shortly), we propose the following:

**Proposition 6.13**   *A crease pattern $G = (V, E)$ on a simply connected region $R$ that is flat-foldable at each vertex (locally) will be flat-foldable (globally) if and only if there exists a superposition order on the s-net of the crease pattern that satisfies the non-crossing conditions.*

*Proof*   The necessary direction of this proposition is clear, since if the crease pattern is flat-foldable, then there exists a sequence of flat origamis $\{\sigma_n\}_{n=1}^{\infty}$ that converge to a folding map $\sigma_f$ and satisfy Definition 5.4. We then induce a superposition order on the s-net from one of the origamis $\sigma_n$ for suitably large $n$ (large enough so that the s-faces will be stacked up neatly and allow us to read off the superposition order). Since $\sigma_n$ is one-to-one, the superposition order obtained in this way will satisfy the non-crossing conditions.

For the other direction, having a folding map $\sigma_f$ and a superposition order that satisfies the non-crossing conditions means, essentially, that we could create an origami $\sigma : R \to \mathbb{R}^3$ by letting our folded image $\sigma_f(L)$ (where $L$ is the region $R$ with the crease pattern graph drawn on it) be in the $xy$-plane, and then lifting the image of the s-faces along the $z$-axis in accordance with the superposition order. We then connect the appropriate s-faces together with half-cylinders so that the resulting map $\sigma$ is one-to-one (since the superposition order given to us satisfies the non-crossing conditions), continuous, and smooth except, if needed, at the vertices or along the edges of $G$. Then $\sigma$ meets our definition of an origami on $G$, and its image, $\sigma(R)$, is homeomorphic to $R$ (i.e., homeomorphic to a disc). Therefore, $\sigma(R)$ can be continuously deformed to $R$ through a sequence of origamis $\sigma_n$, which is possible because our definition of origami allows the maps $\sigma_n$ to be elastic. (That is, even if we have to stretch the paper to deform it from $\sigma(R)$ back to $R$, that's OK.)   □

Of course, this "proof" is somewhat wanting. In real life, paper does not stretch, and so the above proof does not mesh with reality's interpretation of how paper folds. In practice, however, origami artists will bend and even crumple the paper in order to make a crease pattern reach its flat-folded state, and so the idea of contorting the paper massively during the sequence of origamis $\sigma_n$ isn't completely foreign.

For completeness, we include a proposition of Justin (1997) that covers the case where the paper has holes.

**Proposition 6.14**  *A crease pattern $G = (V, E)$ on a region $R$, not necessarily simply connected, that is flat-foldable at each vertex (locally) will be flat-foldable (globally) if and only if there exists a superposition order on the s-net of the crease pattern that satisfies the non-crossing conditions and the image of the set of simple closed curves that form $\partial R$ under the folding map is equivalent to the trivial link or knot.*

However, it is the non-crossing conditions that capture the main difficulty in determining whether or not a given crease pattern is flat-foldable.

---

**Example 6.15**  To illustrate how difficult it can be to check the non-crossing conditions, even for very simple origami models, we return to the two-vertex example of an impossible crease pattern shown in Figure 6.1(b) in Section 6.1. Figure 6.18 shows this create pattern $L$ again in (a), together with the folded image $\sigma_f(L)$ in (b) and the s-net in (c). We will argue that there exists no superposition order on the s-faces that satisfies the non-crossing conditions.

Suppose that we have a superposition order $>$ on the s-faces. Let us denote by $s(f_i)$ any one of the s-faces inside the crease pattern face $f_i$, so that if we write $s(f_i) > s(f_j)$, we mean that in the superposition order any s-face in $f_i$ is greater than any comparable s-faces in $f_j$.

We may assume without loss of generality that crease $l_7$ is a mountain. Recall that we saw from the origami line graph that the creases $l_2$ and $l_3$ cannot both have the same mountain-valley parity (otherwise the faces $f_2$ and $f_4$ would intersect each other).

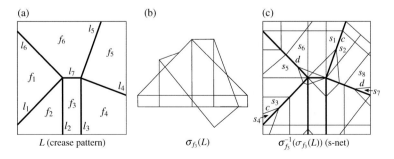

(a)                              (b)                              (c)

$L$ (crease pattern)          $\sigma_{f_3}(L)$          $\sigma_{f_3}^{-1}(\sigma_{f_3}(L))$ (s-net)

**Figure 6.18**  The folded image and s-net of the two-vertex impossible crease pattern from Figure 6.1(b)

**Case 1:** Crease $l_2$ is a mountain and $l_3$ is a valley.

Then by the Big-Little-Big Lemma (Lemma 5.25) $l_1$ is a valley, and so $l_6$ is a mountain by Maekawa's Theorem. Similarly, $l_4$ and $l_5$ are both mountains. This means that

$$s(f_6) > s(f_1) > s(f_2) > s(f_3) > s(f_4) > s(f_5)$$

is true or the exact reverse is true. That is, $f_6$ must be the face "on top" and $f_5$ "on the bottom," or vice versa, because they are the only faces whose boundary creases are all mountains. Specifically, this means that, using the labelings of the s-faces in Figure 6.18(c),

$$s_1 > s_3 \quad \text{and} \quad s_3 > s_2 \quad \text{and} \quad s_1 > s_2.$$

However, the creases marked $c$ in Figure 6.18(c) are mapped onto the same line in $\sigma_f(L)$, where the crease between $s_1$ and $s_2$ is a c-crease and the one between $s_3$ and $s_4$ is an s-crease. This violates the Taco-Tortilla Condition, since we have $s_1 > s_3 > s_2$.

**Case 2:** Crease $l_2$ is a valley and $l_3$ is a mountain.

This case is similar, but involves different s-creases. By the Big-Little-Big Lemma and Maekawa's Theorem, we have that $l_1$ and $l_6$ are both mountains, $l_4$ is a valley, and $l_5$ is a mountain. Thus $f_1$ and $f_6$ are surrounded by mountains, and so we may assume that

$$s(f_1) > s(f_2) > s(f_3) > s(f_4) > s(f_5) > s(f_6).$$

Therefore, we have

$$s_5 > s_6 \quad \text{and} \quad s_5 > s_8 \quad \text{and} \quad s_8 > s_6,$$

which violates the Taco-Tortilla Condition, again, since the creases marked $d$ in Figure 6.18(c) map to the same line in $\sigma_f(L)$.

---

**Diversion 6.12** Prove that the crease pattern in Figure 6.4(b) is impossible to fold flat by showing that any MV assignment will violate the non-crossing conditions.

---

**Remark 6.16** (Superposition orders and MV assignments)    When a crease pattern $G = (V, E)$ on a region $R$ is globally folded flat under $\sigma_f$, then there is an inherent relationship between the resulting superposition order $>$ and the corresponding mountain-valley assignment $\mu: E \to \{-1, 1\}$.

Since MV assignments are relative to one side of the paper, our fixed face $f$ under $\sigma_f$ is used to make the MV assignment $\mu$ consistent. It is also helpful to use the proper 2-coloring $c: F \to \{0, 1\}$ (where $F$ is the set of faces of $G$) of the flat folding as guaranteed by Theorem 5.8. If we assume that $c(f) = 0$, then the relationship between $>$ and $\mu$ is as follows: Let $e \in E$ be a c-crease of the crease pattern with $F_1(e)$ and $F_2(e)$ the two s-faces of the s-net that border $e$. Then there exists two faces $f_1, f_2 \in F$ such that $F_1(e) \subseteq f_1$ and $F_2(e) \subseteq f_2$. Thus, the mountain crease condition is

$$\mu(e) = -1 \Leftrightarrow \big(c(f_1) = 0 \text{ and } F_1(e) > F_2(e)\big) \text{ or } \big(c(f_1) = 1 \text{ and } F_2(e) > F_1(e)\big),$$

and the valley crease condition is

$$\mu(e) = 1 \Leftrightarrow \big(c(f_1) = 0 \text{ and } F_2(e) > F_1(e)\big) \text{ or } \big(c(f_1) = 1 \text{ and } F_1(e) > F_2(e)\big).$$

## 6.6  Flat-Foldability Is NP-Hard

If one has a physical piece of paper with the crease pattern folded on it, producing arguments like that in Example 6.15 to prove that it is impossible to fold flat is not as hard as it might seem. All one needs to do is to experiment with various MV assignments and see where in the paper a self-intersection is forced. Such a self-intersection must be where creases from the s-net line up and cause one of the non-crossing conditions to be violated. Thus, aside from actually drawing the s-net, which can be cumbersome, the only difficult part is finding a way to cover all MV assignment cases systematically.

One can easily see, however, how producing an algorithm for, say, a computer to check whether a given crease pattern is flat-foldable would be arduous. Even if an MV assignment were given, one would have to check every crease image in the folded image $\sigma_f(L)$, determine what creases in the s-net map to that crease, and then check **every pair** of such s-net creases to see if a non-crossing condition is violated. And that's only if the superposition order is forced by the MV assignment; it might not be, and if not then the various superposition orderings must be checked systematically.

Still, perhaps the concept of superposition orderings and the non-crossing conditions are merely too unwieldy. Perhaps there is a better, more efficient way to check flat-foldability.

Unfortunately, this is not the case. In fact, the main goal of this section is to prove the following theorem.

**Theorem 6.17**  *The flat-foldability question is NP-hard.*

Here, by "flat-foldability question" we mean, "Given a crease pattern $G$ on a region $R$, is it possible to (globally) fold the crease pattern flat?"

In 1996 Marshal Bern and Barry Hayes conjectured and offered a proof of this theorem (Bern and Hayes, 1996). Their proof was accepted for 20 years until a flaw was found in 2016 (Akitaya et al., 2016). Bern and Hayes' approach was sound, however. We will describe their approach, indicate its flaw, and then present a correct proof (but of a stronger result) from (Akitaya et al., 2016).

Bern and Hayes follow the approach of showing that the flat-foldability question is equivalent (i.e., can be **reduced to**) a different question that is known to be NP-hard. The reduction problem used is Not-All-Equal 3-Satisfiability, or NAE 3-SAT for short. In NAE 3-SAT, we are given $n$ boolean variables, $x_1, \ldots, x_n$ and a collection of $m$ clauses $C_1, \ldots, C_m$, where each clause is a function of the form

$$\text{NotAllEqual}(x_i, x_j, x_k) = \begin{cases} \text{FALSE} & \text{if } x_i = x_j = x_k, \\ \text{TRUE} & \text{otherwise.} \end{cases}$$

The question for NAE 3-SAT is, "Given the clauses $C_1, \ldots, C_m$, is there an assignment of boolean values to the variables $x_1, \ldots, x_n$ that makes all the $C_i$ clauses TRUE?"

In order to reduce flat-foldability to NAE 3-SAT, we will establish the following system and convention: We will be folding many (possibly hundreds of) narrow pleats in the paper, where by **pleat** we mean two crease lines $l_i$ and $l_j$ that are parallel, close together, and of sufficient distance, somewhere along the lines, from any other creases that the only way for $l_i$ and $l_j$ to fold flat would be for them to have different mountain-valley parity. (The word "pleat" used in this way is standard in origami art.) We also establish a **direction** to the pleats, as if an electrical signal were traveling along the pleat in this predetermined direction. We do this so that the pleat can be labeled TRUE if it has its left crease a mountain and its right crease a valley along this direction; it will be labeled as FALSE if the left crease is a valley and the right a mountain. (See Figure 6.19.) To complete the analogy, Bern and Hayes call these pleats **wires** and the TRUE/FALSE value the **signal** of the wire.

Here, then, is the strategy of Bern and Hayes' proof:

(1) Create origami "gadgets" whose flat-foldability requirements mimic the boolean values of the $x_i$ variables and the $C_i$ clauses.
(2) Show that these gadgets can be put together into an origami crease pattern $L$ based on a given set of clauses $\{C_i\}_{i=1}^{m}$ so that $L$ will fold flat if and only if the clauses satisfy NAE 3-SAT.

Our first gadget, called a **clause gadget**, is based on the classic triangle twist, which is the equilateral triangle version of the square twist previously seen in Diversion 5.2 and Figures 5.4 and 6.12. Triangle and square twists are standard elements seen in origami tessellations, which we will encounter in Section 9.3, and because the folding mechanisms in such twists are a bit counterintuitive, we will take some time to explain them here. (For more information on twist folds, see (Lang, 2018).) Figure 6.20 shows one of the ways to fold the classic triangle twist, where the creases of the central triangle all have the same mountain-valley parity. There are other ways to assign mountains and valleys to these creases, however. In fact, the central triangle in Figure 6.20(a) can be folded with any combination of mountains or valleys around it.

This symmetric triangle twist, where the angle labeled $\theta$ in Figure 6.20(a) is 30°, is called a **closed-back** twist because the three "arms" of the twist come together at a point on the backside (marked with circles in (a) and (c) in the figure). If we have

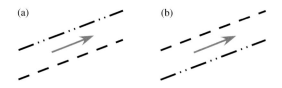

(a)                    (b)

**Figure 6.19** Wires (i.e., pleats) with (a) a TRUE signal and (b) a FALSE signal.

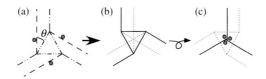

**Figure 6.20** The classic triangle twist, where the "arm angle" $\theta = 30°$. Purple dashed lines represent folded creases on the other side of the paper.

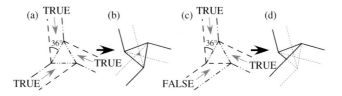

**Figure 6.21** The triangle twist with $\theta = 36°$, with an invalid MV assignment (a) and (b), and a valid MV assignment (c) and (d), forming a clause gadget.

$0° < \theta < 30°$, then it would be an **open-back** twist. In all of these examples, the central triangle creases can be any combination of mountains and valleys.

This changes, however, if $30° < \theta < 60°$. (If $\theta \geq 60°$ we get into the territory of the impossible crease pattern in Figure 6.1(a).) If, for example, $\theta = 36°$, as seen in Figure 6.21(a), the three "arms" of the twist will not come together at a point, but instead will overlap each other in the shaded region marked in Figure 6.21(b). This means that the mountain-valley assignment shown in Figure 6.21(a), where the central triangle is all mountain creases, is not valid; the "arms" will be forced to fold in back of the central triangle and will intersect each other in the shaded region. The same kind of self-intersection will happen if the central triangle is all valley creases.

If instead we make the central triangle a mixture of mountains and valleys, like the MV assignment shown in Figure 6.21(c), then some of the twist pleats will be in front of the central triangle and some will be in back, avoiding a self-intersection, as shown in Figure 6.21(d).

This triangle twist, with $\theta = 36°$, is our gadget that mimics the clauses $C_i = $ NotAllEqual($x_i, x_j, x_k$). If we have three wires $x_i$, $x_j$, and $x_k$ entering into such a twist, then notice that if the wires are all TRUE (as in Figure 6.21(a)) or all FALSE, then the central triangle will be forced to be either all mountains or all valleys, respectively. (This is because the Big-Little-Big Lemma is at play across all of the $36°$ angles.) Thus we have the following:

**Lemma 6.18** *A $\theta = 36°$ triangle twist gadget with incoming wires $x_i$, $x_j$, and $x_k$ is flat-foldable if and only if NotAllEqual($x_i, x_j, x_k$) is true.*

One can see more clearly now the scheme of the proof. We can create equal-sized pleats for the variables $x_1, \ldots, x_n$ on the side of a large piece of paper and then extend them and try to bend them around to make triples of signals enter a triangle twist

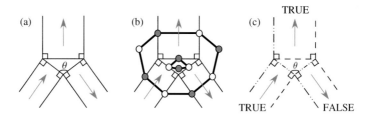

**Figure 6.22** (a) The reflector gadget with (b) its origami line graph and (c) a sample MV assignment.

gadget for each clause $C_i$. To do this, we will need to **bend** pleats around on the paper while preserving their signal as well as **split** signals so that a given variable can be used for as many clauses as needed.

The gadget shown in Figure 6.22(a) does both of these tasks. It is called a **reflector** gadget, and it is more general than the clause gadget in that its angle $\theta$ can be any value in $[90°, 180°)$, where the central triangle is isosceles. In the reflector gadget there are three wires, one incoming and two outgoing, and the central isosceles triangle at right angles to the pleats forces two of the pleats to be smaller and of equal width and the third to be wider.

The origami line graph (see Section 6.1) of this gadget is shown in Figure 6.21(b). Notice that the pleats of the gadget form an octagon cycle in the origami line graph, whereas the central triangle forms a disjoint 4-cycle. (If $\theta = 90°$ then this central triangle part of the origami line graph is slightly different, but the octagon cycle is the same.) The octagon cycle proves that the truth value of the incoming pleat wire will be the same as the wider outgoing pleat and different than the smaller outgoing one. Thus we have the next lemma:

**Lemma 6.19**    *A reflector gadget with $90° \leq \theta < 180°$ is flat-foldable if and only if the incoming wire agrees with the outgoing wide wire and disagrees with the outgoing narrow wire.*

Reflector gadgets allow us to bend and split wires so that we can have them meet at triangle twists to form our given clauses. But this will certainly result in pleat wires crossing each other, so we need a way to handle such crossings without destroying the signals of the wires.

Figure 6.23 shows **crossover gadgets** that Bern and Hayes proposed to handle such situations. Consider the gadget in (a), where the wires meet at right angles. In the center square, the creases labeled $l_1$ and $l_3$ can't have the same mountain-valley parity (else they will force the vertical pleat to self-intersect), and the same is true for creases $l_2$ and $l_4$. Combine this with the fact that Maekawa's Theorem must hold at each vertex, and we get the following:

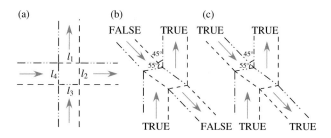

**Figure 6.23** (a) The 90° crossover gadget. Bern and Hayes' 45° crossover gadget with (b) permitted signals and (c) unpermitted signals.

**Lemma 6.20** *The 90° crossover gadget in Figure 6.22(a) is flat-foldable if and only if each pair of incoming and outgoing wires agree.*

Note that in the crossover gadget in Figure 6.23(b), the wires do not meet at right angles. Such crossover gadgets are required in Bern and Hayes' construction because the wires need to be bent at angles other than 90° in order to be made to enter the triangle twist gadgets, and therefore some wires will cross at non-right angles. Also note that Bern and Hayes were careful to have the inner parallelogram of this gadget be slightly rotated. The reason for this is because if, for example, the two labeled angles in Figure 6.23(b) are both 45°, then each vertex in this gadget will be a so-called "bird's foot" vertex, as previously seen in Figure 5.10(b). Such vertices have six different valid MV assignments, and this would give the gadget too many options for its creases to be mountains or valleys and not force the signals of the wires to be preserved. If instead the parallelogram is rotated a bit, as seen in Figure 6.23(b), then each vertex is a Big-Little-Big vertex, as previously seen in Figure 5.10(c). These vertices have only four valid MV assignments, and they provide enough constraints to force the wire signals to be preserved.

However, the problem in Bern and Hayes' proof lies with these non-90° crossover gadgets. Specifically, they do not allow the full range of signal values of the crossing wires. Figure 6.23(c) shows one example; if both of the input wires are TRUE, then the Big-Little-Big Lemma is violated at the 45° angle shown. Therefore, this crease pattern will not work as a crossover gadget, and other variations on this parallelogram-based gadget will either have the same problem or allow too many MV assignments to preserve the signals of the wires.

---

**Diversion 6.13** Compute the number of globally flat-foldable MV assignments of Bern and Hayes' 45° crossover gadget in Figure 6.23(b). Use this to give a different argument that this gadget does not allow all signal possibilities.

---

Bern and Hayes published in 1996 their proof of the NP-hardness of the flat-foldability question in 1996. It took twenty years for a group of researchers (Akitaya, Cheung, Demaine, Horiyama, Hull, Ku, Tachi, and Uehara (Akitaya et al., 2016)) to discover this error. In fact, these researchers only found this error while trying to prove a stronger result.

In (Bern and Hayes, 1996), the authors ask if placing restrictions on the crease patterns being considered would make the flat-foldability question easier. One way to explore this is to restrict ourselves to **box-pleated** crease patterns, where all vertices lie on the integer lattice of the plane and creases are only allowed to have angles with respect to the $x$-axis that are multiples of $45°$. The name "box pleating" comes from the origami literature, since such crease patterns can be used to easily make designs that have box-like features. See (Lang, 2011, Chapter 12).

However, in (Akitaya et al., 2016) the authors prove that box pleating is also NP-hard. What's more, their proof is somewhat simpler to follow. We summarize it here.

**Theorem 6.21**    *The flat-foldability question is NP-hard, even when restricted to box-pleated crease patterns.*

*Proof*    We use the same overall strategy as in Bern and Hayes' proof, in that we aim for a reduction to NAE 3-SAT by using pleat wires for TRUE/FALSE signals in the same way. Also, the $90°$ crossover gadget from Lemma 6.20 will be employed. Otherwise the gadgets we use will be different.

*Splitter gadget*: In Figure 6.24(a) the box-pleated crease pattern for a splitter gadget is shown. This is the same as the reflector gadget of Bern and Hayes with $\theta = 90°$. If we consider the wider pleat to be the input and the two narrower pleats the output (or vice versa), then all the wires have to have the same signal value (because of the octagon cycle in the origami line graph for this crease pattern, as seen in Figure 6.22(b)). If one of the narrower wires is the input, then it works as in Bern and Hayes' proof, where the signal is split into a TRUE and a FALSE signal.

*Clause gadget*: The box-pleated clause gadget that we will use is shown in Figure 6.24(b). We claim that its wires, all oriented to be pointing into the central triangle, cannot all have the same signal, and that any combination of not-all-equal signals will allow this crease pattern to fold flat.

To see this, first suppose that all the wires have the same signal, say TRUE. The points labeled $a$, $b$, and $c$ in Figure 6.24(b) are all mapped onto one another when this crease pattern is folded, and we will use them to examine the layer ordering of the three regions of paper that contain these points under the folding of the TRUE wires.

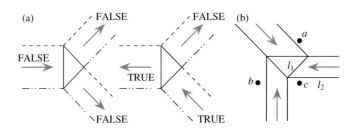

**Figure 6.24**  (a) A box-pleated splitter gadget, with two possible orientations of wires. (b) A box-pleated clause gadget.

Indeed, each wire dictates a layer order of the points: We see that $a > b$, $b > c$ and $c > a$ by following the wire orientations and the definition of a TRUE wire. But layer orders of the paper at a point when folded flat must be transitive, and so this is clearly impossible. Thus the wires cannot all have the same signal.

To show that this clause gadget is flat-foldable whenever the wires are not all the same, it suffices to show that there are exactly six globally valid MV assignments of this crease pattern. Perhaps the easiest way to see this is to realize that the MV assignment of the central triangle forces the MV assignment of the remaining creases. This is because by the Big-Little-Big Lemma applied to the bottom vertex, any MV assignment $\mu$ must have $\mu(l_1) \neq \mu(l_2)$ (see Figure 6.24(b)). Applying Maekawa's Theorem gives us that an MV assignment $\mu$ of the central triangle will determine $\mu$ for the other two creases of the bottom vertex, and then the fact that the wires must have different MV parity determines $\mu$ for the whole crease pattern. There are eight MV assignments of the central triangle, and two of them (all Ms and all Vs) make the wires all have the same signal. Those are not foldable, but the other six are, as is easily checked.

All that remains for the proof is to see that we really can put all these pieces together to make an origami crease pattern that will fold flat if and only if the given set of clauses $\{C_i\}_{i=1}^m$ satisfies NAE 3-SAT.

One way to do this is illustrated in Figure 6.25. We let our clause variables $x_i$ be horizontal pleats running from left to right along the bottom of our sheet of paper. We combine splitter gadgets as shown to either negate signals or to divert signals up toward the clauses, arranged along the top edge of the paper. Arranging four splitter gadgets into an octagon shape provides a convenient way to do this in an orderly fashion. Note that in this scheme wires will only cross at right angles and such crossings pose no problem by Lemma 6.20.

Here are some observations:

- Using splitter gadgets will create "noise" wires. Such wires need to be allowed to run off the edge of the paper and be arranged so that they do not interfere with any

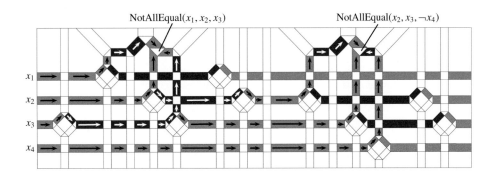

**Figure 6.25** An instance of Not-All-Equal 3-SAT using a box-pleated crease pattern. The arrows represent the flow of the TRUE/FALSE signals from the variables $x_i$ as they are diverted to the clause gadgets. Arrows changing colors indicates the signal being negated by a splitter.

gadgets along the way. Since we can make the initial wire width $\lambda$ as small as we wish and are arranging the splitters in a grid pattern, we can always handle such noise wires.

- Notice that there is a potential issue with the width of the wires. Wires start with width $\lambda$, and a splitter gadget will split a wire into two, each with width $\lambda/\sqrt{2}$. The scheme we are using makes all horizontal and vertical wires have width $\lambda$ and all wires at $\pm 45°$ angles have width $\lambda/\sqrt{2}$, except for the wires feeding directly into the clauses. This keeps the wire widths consistent throughout the crease pattern.

After creating such a crease pattern for our given set of clauses $\{C_i\}_{i=1}^m$, we know that if the crease pattern can fold flat, then such a flat folding will determine the signals of the initial wires $x_i$ along the left edge of the paper, and these signals will give us truth values for the variables $x_i$ that make the clauses $C_i$ satisfy NAE 3-SAT. In the other direction, if we know that the clauses $C_i$ satisfy NAE 3-SAT, then there exist truth values for the variables $x_i$ that satisfy all the clauses, and these values can be used to set the signals of the initial wires $x_i$ on the left edge of the paper. This sets the mountain-valley assignment for those initial wires, and this in turn forces the MV assignment for the entire rest of the crease pattern. The only place in this crease pattern that might be impossible to fold flat would be at the triangle clause gadgets, and since their corresponding clauses $C_i$ satisfy the NotAllEqual function, they will be flat-foldable.

There is one other potential issue having to do with the constructibility of such a crease pattern. From the point of view of computational complexity, one might be satisfied that a crease pattern exists whose flat-foldability is equivalent to the NEA 3-SAT question. On the other hand, one might insist that the crease pattern actually be geometrically constructible in the sense of Part I of this book, to make sure that the necessary gadget locations and angles could actually be folded starting with a large enough rectangle of paper using the basic origami operations (BOOs).

Luckily, this is not a problem. Box-pleated crease patterns have their vertices on an integer lattice, meaning that all the points and line segments of the NEA 3-SAT crease patterns are origami constructible. (Even in Bern and Hayes' original proof this was not a problem; the triangle twist clause gadgets require constructing $36°$ angles, which is possible because regular pentagons are constructible via origami; see Diversions 1.5 and 2.2). This completes the proof. ☐

The flat-foldability question that Theorem 6.21 covers is also known as **Unassigned Flat-Foldability** because the crease pattern input to the question has its MV assignments unassigned. A logical question to ask is, "if we provide an MV assignment $\mu$ along with our crease pattern $C$, can we solve the flat-foldability question in polynomial time?" This is called the **Assigned Flat-Foldability** problem.

Rather amazingly, the Assigned Flat-Foldability problem is also NP-hard, even for box-pleated crease patterns! Bern and Hayes (1996) provided a proof for general crease patterns, but their gadgets were very complicated. The box-pleat proof is easier to follow (Akitaya et al., 2016), but the gadgets are still rather complex. We will

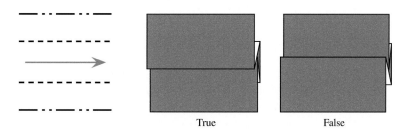

Figure 6.26 The wires used in the Assigned Flat-Foldability reduction.

only mention here that the wires for the Assigned Flat-Foldability problem are pairs of pleats with their MV assignment being MVVM, as shown in Figure 6.26. When folded, this pair of pleats will overlap, and if the left side (in the direction of the signal) is on top in this overlap, then we say the signal is TRUE; if the right pleat is on top, we say that it is FALSE. The crossover, splitter, and clause gadgets needed in the Assigned Flat-Foldability reduction to NAE-3SAT are similar to those for the Unassigned version. However, the wires have four creases instead of two and the ordering of the layers of paper needs to be carefully controlled, making the Assigned gadgets much more intricate. See (Akitaya et al., 2016) for full details.

The real point is that the Assigned Flat-Foldability problem is also NP-hard, and this means that the main difficulty of flat origami lies in finding an ordering of the layers of the folded paper so that no self-intersections occur. Globally consistent MV assignments do affect the layer ordering of the paper, but knowing the MV assignment is not enough to reduce the complexity.

Given all this work on the computational complexity of flat origami, it might be surprising to learn that there are still basic questions in this area that are open. For example, suppose that we restrict ourselves even more to a subset of box-pleated crease patterns, where the entire crease pattern is an $m \times n$ grid of squares. This is called the **stamp-folding problem** and will be discussed in detail in Section 7.3. However, it is easy to see that in the stamp-folding problem the Unassigned Flat-Foldability case is trivial—an $m \times n$ grid of creases is very easy to fold flat by first folding all vertical creases, say alternating mountains and valleys, and then folding the horizontal creases. However, the computational complexity of the Assigned Flat-Foldability problem remains very much open for $m \times n$ grid crease patterns.

## 6.7 Open Problems

The flat-foldability of crease patterns has been studied extensively. There are nuances that rise to the attention of computational origami researchers, and many of these are mentioned in the open problems sections of Chapters 5 and 10. There has definitely been an emphasis on the study of **local** flat-foldability (i.e. phantom folds) without considering global flat-foldability because it is so hard. Therefore, one general, albeit vague, area for future study is developing ways to consider global foldability that can lead to more insights. We will offer one way to frame this question.

**Open Problem 6.1**     The superposition net and the non-crossing conditions (Tortilla-Tortilla, Taco-Tortilla, and Taco-Taco) are one way to characterize the ways in which paper can self-intersect when being folded flat. Are there any other ways to characterize this? Are there other tools than can be developed to help keep track of the difficult issues in global flat-foldability, like layer ordering?

# 7    Counting Flat Folds

There are many combinatorial questions one can ask about flat origami. We have already seen some in the form of Maekawa's Theorem and enumerating the number of valid MV assignments of single-vertex folds in Section 5.5.

    In this chapter we will explore more problems of this type, starting with counting the number of valid MV assignments of multiple-vertex crease patterns. We will also examine other combinatorial questions that can be asked about flat folds that are motivated by problems in polymer physics. **Note:** For the rest of this book, in the figures we will denote a mountain crease with a bold line and a valley crease by a non-bold line. The prior notation used (dashed lines for valleys and dash-dot-dot-dash for mountains) is standard in origami instruction books, but since we will only be depicting crease patterns with MV assignments for mathematically-illustrative purposes, simply using bold and non-bold lines will look more clear.

## 7.1    Two-Colorable Crease Patterns

Recall from Section 6.1 the origami line graph $C_L$ of a flat origami crease pattern or phantom fold $C$. We saw in Theorem 6.3 that if the origami line graph $C_L$ of a phantom fold $C$ is not 2-vertex colorable, then $C$ is not flat-foldable. For some multiple-vertex flat origami crease patterns, the origami line graph will capture all of constraints that determine whether or not an MV assignment is valid. The valid MV assignments for such crease patterns are therefore easy to enumerate.

**Theorem 7.1**    *Let $C$ be a flat-foldable crease pattern whose valid MV assignments are completely determined by $C_L$ and let $n$ be the number of connected components of $C_L$. Then the number of valid MV assignments of $C$ (that locally fold flat) is $2^n$.*

---

**Example 7.2** (Square twist tessellations)    We've previously encountered the square twist in Section 5.1. In fact, Diversion 5.2 asked the reader to enumerate the number of valid MV assignments for a square twist. By now the reader might see that Maekawa's Theorem and the Big-Little-Big Lemma make this diversion a simple matter; each vertex must have three mountains and one valley or vice versa, and the creases surrounding the 45° angle at the vertex must have different MV parity. This

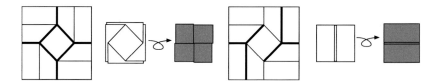

**Figure 7.1** Two different ways to fold a square twist. Bold lines are mountains and non-bold lines are valleys.

implies that there are four valid MV assignments for one vertex, which leaves only two ways to assign Ms and Vs to its two neighboring vertices, and this then forces the MV assignment for the fourth vertex. Thus we have $4 \times 2 \times 2 \times 1 = 16$ ways to fold a square twist, where we count symmetric MV assignments as different (say, MV assignments that are rotations of one another).

This multitude of ways to fold the square twist crease pattern may surprise readers who do not have direct experience with folding square twists. Examples of two different MV assignments for the square twist are shown in Figure 7.1. The variety of interesting ways that twist folds can be folded have made them popular among origami artists. See, for example, (Fujimoto and Nishikawa, 1982; Gjerde, 2008).

Let $S(m, n)$ denote the crease pattern of an $m \times n$ square twist tessellation. The $S(2, 2)$ case is shown in Figure 7.2(a), along with its origami line graph $S(2, 2)_L$. Notice how at each vertex of this crease pattern we may employ the Big-Little-Big Lemma (Lemma 5.25) at the 45° angle. Also, parallel creases that are close to each other must get different MV parity. This explains the structure of $S(2, 2)_L$ shown in Figure 7.2(a).

The general square twist origami line graph $S(m, n)_L$ will have four components at each square twist. The $2(m+n)$ components on the border of the paper will only touch one square twist; the others will touch two. This gives $(4mn - 2(m+n))/2 + 2(m+n) = 2mn + m + n$ connected components of $S(m, n)_L$.

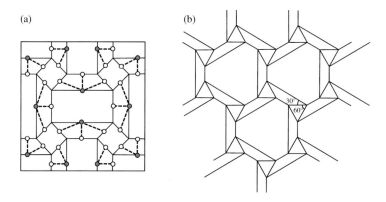

**Figure 7.2** (a) A $2 \times 2$ square twists tessellation (solid lines) with the 2-colored origami line graph superimposed (round dots and dashed lines). (b) A triangle twist tessellation.

**Theorem 7.3** *The number of valid MV assignments of $S(m, n)$ that locally fold flat is $2^{2mn+m+n}$.*

---

**Diversion 7.1** Construct a theorem similar to Theorem 7.3 for the triangle twist tessellation, shown in Figure 7.2(b).

---

We conjecture that all of these MV assignments of $S(m, n)$ are globally flat-foldable as well, but Theorem 7.3 should be viewed as an upper bound for global flat-foldability. In fact, as we saw in Section 6.1 (and in Figure 6.1(b) in particular), it is quite possible for a crease pattern to be flat-foldable at every vertex and to have no mountain-valley contradictions that the origami line graph would detect, yet still be unfoldable. We will see more unfortunate examples of this in Section 7.3.

---

Two-colorability does not capture all the ways in which mountains and valleys can be constrained, however. The next section describes a major example that is more complex.

## 7.2   Phantom Folds of the Miura-ori

A good example of an origami crease pattern for which the origami line graph does not help enumerate valid MV assignments is the **Miura-ori**, also known as the **Miura map fold**. See Figure 7.3. It was discovered by the Japanese astrophysicist Koryo Miura in the 1970s as a possible way to deploy solar panels into outer space, since it can take a large sheet and smoothly collapse it into a small package. The Miura-ori is an origami tessellation, since the crease pattern is made of congruent parallelograms, but if we fold it from the infinite plane $\mathbb{R}^2$, then it will fold into the infinite strip $\mathbb{R} \times [0, 1]$ (if the height of the parallelograms is 1).

The Miura-ori crease pattern is made of vertices that look like bird's feet, with three creases surrounding two congruent acute angles $\alpha$, which make the **toes** of the bird's foot, and the **heel** crease making two congruent obtuse angles $\pi - \alpha$ with the other

—————— = mountain
—————— = valley

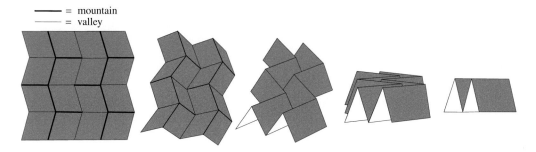

**Figure 7.3** The Miura map fold, or Miura-ori. Here, bold creases are mountains and non-bold are valleys.

**Table 7.1** Values of $M(m, n)$ for small $m$ and $n$.

| $n \backslash m$ | 2 | 3 | 4 | 5 | 6 | 7 | 8 |
|---|---|---|---|---|---|---|---|
| 2 | 6 | 18 | 54 | 162 | 486 | 1458 | 4374 |
| 3 | 18 | 82 | 374 | 1706 | 7782 | 35498 | 161926 |
| 4 | 54 | 374 | 2604 | 18150 | 126534 | 882180 | 6150510 |
| 5 | 162 | 1706 | 18150 | 193662 | 2068146 | 22091514 | 235994086 |

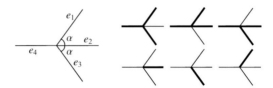

**Figure 7.4** The Miura-ori "bird's foot" vertex, whose MV restrictions are not captured by the origami line graph.

creases. As we saw in Section 5.5, there are six valid MV assignments for such a degree-4 vertex, and they are depicted in Figure 7.4. Notice how the creases labeled $e_1$, $e_2$, and $e_3$ (the toes) switch in pairs from having the same to having different MV assignments, which means that there would be no edges between these creases in the origami line graph. Nonetheless, there are MV restrictions between these creases, namely that the heel crease $e_4$ must always have the same MV parity as the majority of the creases. The origami line graph will not capture these restrictions. Other means must be used to count the number of valid MV assignments for such vertices.

Let $M(m, n)$ denote the number of phantom folds for a Miura-ori crease pattern that is made by an $m \times n$ tiling ($m$ rows and $n$ columns) of parallelograms. (We restrict ourselves to only locally flat-foldable crease patterns, i.e., phantom, because of problems that can occur with global flat-foldability, as we will see in Section 7.3.) Explicit formulas for $M(m, n)$ for $n \in \{2, 3, 4\}$ and recursions for $n \in \{3, 4, 5\}$ are derived in (Ginepro and Hull, 2014), and some of the data from these formulas can be seen in Table 7.1.

The numbers for $M(m, n)$ fit exactly with the sequence A078099 in the On-Line Encyclopedia of Integer Sequences (http://oeis.org/), which encodes $T(m, n)$, the number of ways of 3-coloring the vertices of a grid graph (with $m$ rows and $n$ columns of vertices) with one vertex pre-colored. We thus wish to prove the following:

**Theorem 7.4** (Ginepro and Hull, 2014)   $M(m, n) = T(m, n)$ *for all* $m, n \geq 1$.

We will prove this via a series of lemmas, following (Ginepro and Hull, 2014). We say that a Miura-ori vertex **points left** if the heel of the bird's foot points left, as they do in Figure 7.4. **Points right** is defined similarly.

**Lemma 7.5**   *Let C be the crease pattern for a* $2 \times n$ *Miura-ori with all the vertices pointing left, and let* $\mu$ *be an MV assignment for C. Let c be the left-most crease*

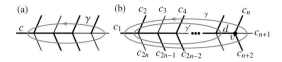

**Figure 7.5** As per Lemma 7.5, (a) a $2 \times 5$ Miura-ori with $M_\gamma = 6$, $V_\gamma = 4$, and $\mu(c) = 1$ and (b) a $2 \times n$ Miura-ori as used in the proof.

in $C$ and let $\gamma$ be a simple closed curve around all the vertices in $C$ (i.e., crossing all the non-internal creases). Let $M_\gamma$ (resp. $V_\gamma$) be the number of mountain (resp. valley) creases that $\gamma$ crosses. Then we have that if $\mu$ is locally flat-foldable then $\mu(c) = (M_\gamma - V_\gamma)/2$.

As will be seen in the proof, what Lemma 7.5 is really saying is that $M_\gamma - V_\gamma$ will always equal $\pm 2$, just as in Maekawa's Theorem, and that the MV parity of the crease $c$ will control the plus/minus parity of $M_\gamma - V_\gamma$. See Figure 7.5(a) for an example.

*Proof* We proceed by induction on $n$. When $n = 2$ we have only one vertex. If the vertex folds flat, then Maekawa's Theorem gives us that $M_\gamma - V_\gamma = \pm 2$, and consulting the six possible MV assignments in Figure 7.4 confirms the result.

Now let $C$ be a $2 \times n$ Miura-ori with MV assignment $\mu$ as in the statement of the lemma, and label the external creases $c_1, \ldots, c_{2n}$ going clockwise around the crease pattern, with $c_1$ the left-most crease. Then the right-most crease is $c_{n+1}$. Let $d$ be the remaining unlabeled crease of the right-most vertex $v$, so that $v$ has creases $d$, $c_n$, $c_{n+1}$, and $c_{n+2}$ going clockwise, as seen in Figure 7.5(b).

Suppose that $\mu$ is locally flat-foldable in $C$. Then $\mu$ is also locally flat-foldable on the crease pattern $C - v$, which is a $2 \times (n - 1)$ Miura-ori. Using the induction hypotheses, we may assume without loss of generality that $M_{\gamma'} - V_{\gamma'} = 2$ for $\gamma'$ surrounding this $2 \times (n - 1)$ Miura-ori, so that $\mu(c_1) = 1$ (a mountain). (The $M_{\gamma'} - V_{\gamma'} = -2$ case is attained by simply switching all the Ms to Vs and vice versa.) We now look at the crease $d$.

If $d$ is an M then $c_n$, $c_{n+1}$, and $c_{n+2}$ must have two Ms and one V in order for $v$ to be flat-foldable. So, if we extend $\gamma'$ to $\gamma$, which includes $v$, then we still have $M_\gamma - V_\gamma = 2$ since $\gamma$ no longer crosses $d$ but now crosses $c_n$, $c_{n+1}$, and $c_{n+2}$ (we subtract an M and add two Ms and a V). The case where $d$ is a V is similar, and in both cases, we conclude that $\mu(c_1) = (M_\gamma - V_\gamma)/2$. This completes the proof of the lemma. $\square$

We will now let $C = (V_C, E_C)$ be our crease pattern for an $m \times n$ (parallelograms) Miura-ori, where we assume that the top row of vertices all point left. Also let $G = (V_G, E_G)$ be the $m \times n$ (vertices) grid graph, which we think of as the dual graph to $C$ (ignoring the outside face). Let $\mu \colon E_C \to \{-1, 1\}$ be an MV assignment for $C$ and $c \colon V_G \to \mathbb{Z}_3$ be a 3-coloring of $G$.

We label the vertices in $G$ as follows: let $v_1, \ldots, v_n$ be the top row of vertices, starting from the left, then $v_{n+1}, \ldots, v_{2n}$ be the second row, but **starting from the right**, then $v_{2n+1}, \ldots, v_{3n}$ be the third row of vertices starting from the left, and so on

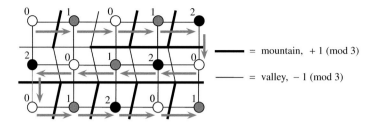

**Figure 7.6** The bijection between a locally flat-foldable Miura-ori MV assignment to a proper 3-vertex coloring of a grid graph.

in a boustrophedon pattern. This path is illustrated by the purple arrows in Figure 7.6. We will assume that $c(v_1) = 0$ because we are allowed to have one vertex pre-colored in our coloring.

Our aim is to construct a bijection between the set of all locally flat-foldable MV assignments $\mu$ of the $m \times n$ Miura-ori and the set of all proper 3-vertex colorings $c$ of the $m \times n$ grid graph with one vertex pre-colored. We will do this by showing how to convert from $\mu$ to $c$ and vice versa and then prove that these conversions form the desired bijection.

Label the creases that the boustrophedon pattern crosses, in order, $c_2, \ldots, c_{mn}$. That is, crease $c_i$ is between vertices $v_{i-1}$ and $v_i$ in the grid graph.

- **To convert from $\mu$ to $c$:** Let $c(v_1) = 0$ and then recursively define

$$c(v_i) = c(v_{i-1}) + \mu(c_i) \quad \text{(where the addition is mod 3).} \tag{7.1}$$

- **To convert from $c$ to $\mu$:** Define

$$\mu(c_i) = \begin{cases} 1 & \text{if } c(v_i) - c(v_{i-1}) \equiv 1 \ (\text{mod } 3), \\ -1 & \text{if } c(v_i) - c(v_{i-1}) \equiv 2 \ (\text{mod } 3). \end{cases} \tag{7.2}$$

For the other creases, let $d_i \in E_C$ be the crease directly below vertex $v_i$ and above $v_j$ in the superimposed grid graph. Then define

$$\mu(d_i) = \begin{cases} 1 & \text{if } c(v_i) - c(v_j) \equiv 1 \ (\text{mod } 3), \\ -1 & \text{if } c(v_i) - c(v_j) \equiv 2 \ (\text{mod } 3). \end{cases} \tag{7.3}$$

First we'll prove that a valid MV assignment $\mu$ will give us a proper coloring of the grid graph.

**Lemma 7.6** *Given a locally flat-foldable MV assignment $\mu$ for an $m \times n$ Miura-ori crease pattern and coloring function $c \colon V_G \to \mathbb{Z}_3$ defined by Equation (7.1), consider nonnegative integers $i, j$ such that vertex $v_i$ is directly above $v_j$ in the grid graph $G$, and let $d_i$ be the crease between $v_i$ and $v_j$. Then $c(v_j) \equiv c(v_i) + \mu(d_i) \ (\text{mod } 3)$.*

*Proof* Pick nonnegative integers $j > i$ where $v_i$ is directly above $v_j$ in the grid graph that is dual to our crease pattern. Telescoping the recursion for $c(v_j)$, we obtain

$$c(v_j) \equiv c(v_i) + \sum_{k=i+1}^{j} \mu(c_k) \quad (\text{mod } 3).$$

Therefore we want to show that $\mu(d_i) = \sum_{k=i+1}^{j} \mu(c_k)$. For this, we apply Lemma 7.5. Let $\gamma$ be a simple closed curve through the creases, in order, $c_{i+1}, \dots, c_j, d_i$. Then, by Lemma 7.5 we have

$$2\mu(d_i) = M_\gamma - V_\gamma = \mu(d_i) + \sum_{k=i+1}^{j} \mu(c_k),$$

where the second equality is obtained by simply adding all the values of $\mu$ along $\gamma$. Thus $\mu(d_i) = \sum_{k=i+1}^{j} \mu(c_k)$ and we are done. $\qquad\square$

Equation (7.1) guarantees that the colors given along the boustrophedon path will be different for consecutive vertices, and Lemma 7.6 tells us that we will get different colors between all the other adjacent vertices. Thus the graph coloring created by Equation (7.1) will be proper.

**Lemma 7.7** *Given a proper 3-vertex coloring c of an m × n grid graph, the m × n Miura-ori MV assignment $\mu$ created by Equations (7.2) and (7.3) is locally flat-foldable.*

*Proof* All that needs to be shown is that at each bird's foot vertex we have that the heel crease has the same MV parity as the majority of the creases around the vertex.

Take an arbitrary vertex in the crease pattern, and suppose, that it points left. Label the creases and consider the section of the superimposed grid graph around the vertex, as in Figure 7.7. From the construction of $\mu$, we have

$$c(v_i) \equiv c(v_{i-1}) + \mu(c_i) \quad (\text{mod } 3),$$
$$c(v_j) \equiv c(v_i) + \mu(d_i) \quad (\text{mod } 3),$$
$$c(v_{j+1}) \equiv c(v_j) + \mu(c_{j+1}) \quad (\text{mod } 3),$$
$$c(v_{j+1}) \equiv c(v_{i-1}) + \mu(d_{i-1}) \quad (\text{mod } 3).$$

Substituting the first three equations into one another, we obtain $c(v_{j+1}) \equiv c(v_{i-1}) + \mu(c_i) + \mu(d_i) + \mu(c_{j+1})$ (mod 3), and since each of the $\mu$ values are only $\pm 1$, we have

$$\mu(d_{i-1}) = \mu(c_i) + \mu(d_i) + \mu(c_{j+1}).$$

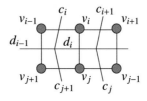

**Figure 7.7** Two Miura-ori vertices that point left with the creases labeled and the superimposed grid graph section.

Therefore, crease $d_{i-1}$, which is the bird's foot heel, must have the same MV parity as the majority of the creases at this vertex. The case where the vertex points right is similar.    □

Lemmas 7.6 and 7.7 prove that our coloring-to-MV assignment conversions are a bijection, which proves Theorem 7.4.

One immediate consequence of Theorem 7.4 is the following:

**Corollary 7.8**    $M(m, n) = M(n, m)$ *for all* $m, n \geq 1$.

This simply follows from the symmetry of the grid graph colorings, giving us $T(n, m) = T(m, n)$. Corollary 7.8 is surprising because the Miura-ori crease pattern itself does not follow this symmetry; an $m \times n$ Miura-ori does not look the same as an $n \times m$ Miura-ori.

Unfortunately, there are no known formulas for 3-coloring grid graphs. The problem of finding the chromatic polynomial of an $n \times m$ grid graph is unsolved (Jensen and Toft, 1995, Problem 14.7). However, in the OEIS listing for sequence A078099, a recursive transfer matrix for $T(m, n)$ is given. Specifically, define matrices $M(1) = [1]$,

$$ M(m + 1) = \begin{bmatrix} M(m) & M(m)^T \\ 0 & M(m) \end{bmatrix}, $$

and $W(m) = M(m) + M(m)^T$ (where $A^T$ denotes the transpose of $A$).

**Theorem 7.9**    $T(m, n)$ *equals the sum of the entries of* $W(m)^{n-1}$.

*Proof*    It turns out that $W(m)$ is the adjacency matrix for a graph whose walks of length $(n - 1)$ correspond to proper 3-colorings of the $m \times n$ grid graph with one vertex pre-colored, that is, one instance of $T(m, n)$.

First note that

$$ W(m + 1) = \begin{bmatrix} W(m) & M(m)^T \\ M(m) & W(m) \end{bmatrix}. $$

Now consider the set $S$ of all ways to properly vertex-color an $m \times 1$ grid graph with colors in $\mathbb{Z}_3$, where each coloring can be thought of as a vector in $\mathbb{Z}_3^m$. Let $C_i \subset S$ be equivalence classes where two such colorings $\vec{x}, \vec{y} \in S$ are equivalent if either $\vec{x} + (1)^m$ or $\vec{x} + (2)^m$ is equivalent to $\vec{y}$ (mod 3) (where $(i)^m$ denotes the $m$-dimensional vector with $i$ in every coordinate). Then each equivalence class $C_i$ contains three vectors from $\mathbb{Z}_3^m$, and there will be $3 \cdot 2^{m-1}/3 = 2^{m-1}$ different equivalence classes.

Let $H$ be the graph whose vertices are the classes $C_i$ where we connect $C_i$ and $C_j$ by $q$ edges, and where for a given $\vec{x} \in C_i$, we have $q$ elements $\vec{y} \in C_j$ that differ from $\vec{x}$ in every coordinate $(x(k) \neq y(k)$ for every coordinate $k \in \{1, \ldots, m\})$. By the symmetry of the elements of the equivalence classes $C_i$, we have that this number of edges $q$ will be independent of our choice of $\vec{x} \in C_i$.

---

**Diversion 7.2**  Prove that $W(m)$ is the adjacency matrix for the graph $H$.

---

Therefore a proper 3-coloring of the $m \times n$ grid graph (assuming the upper left vertex is 0) will be equivalent to a walk of length $(n - 1)$ in the graph $H$, and the number of such walks will be the sum of the entries of $W(m)^{n-1}$. Therefore, we have that $M(m, n) = T(m, n) =$ the sum of the entries of $W(m)^{n-1}$. □

---

**Diversion 7.3**  Use the transition matrix $W(m)$ to find generating functions (and closed formulas when possible) for $M(m, n)$ for $n = 2, \ldots, 7$. (Compare with (Ginepro and Hull, 2014).)

---

We are fortunate that statistical mechanics results exist for enumerating our $m \times n$ grid graph coloring problem for very large $m$ and $n$. The problem is the same as counting the number of Eulerian orientations on an $m \times n$ grid graph on a torus, which is exactly the antiferroelectric model for 2-dimensional ice lattices, called the **square ice** model (Lieb, 1967; Propp, 2001). In 1967, Lieb used a transfer matrix method to show that a grid graph with $N$ vertices (where $N$ is very large, say $10^{23}$) will have

$$(4/3)^{3N/2}$$

ways to be properly vertex colored using three colors with one vertex pre-colored. Therefore, this gives the number of locally flat-foldable MV assignments for a Miura-ori crease pattern with $N$ parallelograms, where $N$ is very large.

## 7.3  The Stamp-Folding Problem

A much more simple-looking crease pattern than the Miura-ori is the one used in the **stamp-folding problem**, which is just a grid of squares. The problem goes back to at least the 1890s (Lucas, 1891, p. 120), and it derives from the way postage stamps used to be sold, as a regular grid of squares or rectangles with perforations between the stamps, making it easy to fold along the perforations to make a one-stamp pile.

The historical treatments of this problem (Touchard, 1950; Koehler, 1968; Lunnon, 1968, 1971) consider a version of the question "in how many ways can we fold this?" that is considerably harder than the version we have been asking in this chapter. Specifically, we have been counting valid MV assignments and trying to distinguish, when possible, the different cases of phantom folds and folds without self-intersections. The stamp-folding problem, however, has traditionally also considered two foldings that layer the paper differently to be distinct, even if their MV assignments are identical. For example, a $1 \times n$ strip of stamps will have $2^{n-1}$ different valid MV assignments, since there are $(n - 1)$ creases and each can be either an M or a V. But each of these MV assignments can have many different ways in which the flaps of paper can be

**Table 7.2** Values of the various stamp-folding problems for small $m$.

| $n$ | $Sl(1,n)$ | $Sp(1,n) = Sg(1,n)$ | $Sl(2,n)$ | $Sp(2,n)$ | $Sg(2,n)$ |
|---|---|---|---|---|---|
| 1 | 1 | 1 | 2 | 2 | 2 |
| 2 | 2 | 2 | 8 | 8 | 8 |
| 3 | 6 | 4 | 60 | 32 | 32 |
| 4 | 16 | 8 | 320 | 128 | 128 |
| 5 | 50 | 16 | 1980 | 512 | 508 |
| 6 | 144 | 32 | 10512 | 2048 | 2006 |
| 7 | 462 | 64 | 60788 | 8192 | 7876 |
| 8 | 1392 | 128 | 320896 | 32768 | 30800 |
| 9 | 4536 | 256 | 1787904 | 131072 | |
| 10 | 14060 | 512 | 9381840 | 524288 | |

arranged. If we take $n = 3$, label the stamps $A$, $B$, and $C$, and make both creases valleys, we could fold $A$ on top of $C$ or $C$ on top of $A$, giving us two ways we could fold this $1 \times 3$ strip of stamps if we count layer orderings for this one MV assignment.

With this in mind, we define different variations of the stamp-folding problem (SFP) on an $m \times n$ grid of stamps as follows:

- The **layers SFP** is global flat-foldability where we count different orderings of the layers of paper to be different stamp foldings, and denote these numbers $Sl(m,n)$.

- The **phantom SFP** is where we let $Sp(m,n)$ be the number of locally flat-foldable MV assignments, ignoring layer orderings and self-intersections of the stamps.

- The **global SFP** is where we let $Sg(m,n)$ be the number of globally flat-foldable MV assignments (with no self-intersections).

Of these problems, only the phantom SFP has been solved. In fact, it is much easier than the Miura-ori version of this problem (as seen in Section 7.2).

---

**Diversion 7.4**   Prove that $Sp(m,n) = 2^{mn-1}$.

---

Koehler (1968) and Lunnon (1968) both (independently, and in the same year) published methods to compute $Sl(1,n)$ (sometimes called the "strip of stamps folding problem"). Of the two, Koehler gives the most complete and thorough treatment, and he proves that $Sl(1,n)$ also equals the number of ways to connect $n$ dots on a circle with chords where the chords are alternately colored red and blue and we are not allowed to have two chords of the same color intersect.

Unfortunately, closed formulas for $Sl(1,n)$ are too much to hope for. Koehler and Lunnon resort to writing computer programs to calculate the values (see Table 7.2; for more values see (Koehler, 1968; Lunnon, 1968)). Still, some of their analysis is worth noting.

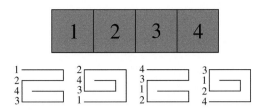

**Figure 7.8** The rotation (shift) of a $1 \times 4$ strip of stamps folding.

Both Koehler and Lunnon make use of partitioning the set of possible $1 \times n$ strips of stamps folded with layers counted into equivalence classes. Specifically, they define a **shift** or **rotation** of a folded strip to be an operation where we take the top-most stamp and move it to the bottom of the pile, changing mountains and valleys in order to make this work. See the example in Figure 7.8 to convince yourself that this is always possible. We say two such $1 \times n$ stamp foldings are equivalent if one is a rotation of another, and since these foldings can be rotated $n$ times, we have that each equivalence class will have $n$ foldings in it.

Not only does this imply that $Sl(1, n)$ will always be divisible by $n$, it means that when counting $1 \times n$ foldings where different layer permutations are considered distinct, we only need to count one from each equivalence class. For example, we may assume that the left-most stamp (labeled 1 in Figure 7.8) is always the stamp on top. This kind of equivalence class deduction (and others more subtle) are the keys that allow Koehler and Lunnon to compute $Sl(1, n)$.

Lunnon (1971) extends this work to devise algorithms to compute $Sl(m, n)$, as well as to generalize the problem to $d$-dimensional arrays of postage stamps. We will first illustrate his techniques for the 2-dimensional case.

Any $m \times n$ array of postage stamps that has been folded into a one-stamp pile can be sliced into two perpendicular cross sections, which we will call the **x-section** and the **y-section**. In this way we can reduce a 2-dimensional stamp-folding problem to a collection of 1-dimensional stamp foldings of the kind seen in Figure 7.8. The $x$-section will be a collection of cross-section curves each describing how a $1 \times n$ row of the grid folds up. Each $y$-section is a collection of curves describing how each $m \times 1$ column of the grid of stamps folds up. An example of a $2 \times 3$ stamp folding, with its sections, is shown in Figure 7.9.

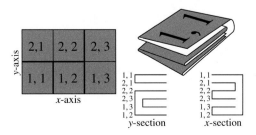

**Figure 7.9** A $2 \times 3$ stamp folding, with $x$- and $y$-sections shown.

Let us be more specific about what we are doing here. Let $M_{m,n} = [0, m] \times [0, n] \subset \mathbb{R}^2$ be our grid of postage stamps, with creases along $x = i$ for $1 \leq i \leq m - 1$ and $y = i$ for $1 \leq i \leq n - 1$ inside our rectangle. We denote each stamp by the coordinate $(i, j)$ in its upper right corner. Then a folded $m \times n$ grid of stamps will be a flat origami $\{\sigma_n\}_{n=1}^{\infty}$ where $\sigma_n \colon M_{m,n} \to \mathbb{R}^3$ for all $n$ with limit map $\sigma \colon M_{m,n} \to [0, 1] \times [0, 1]$. In order for this to be globally flat-foldable, we must have that the superposition order induced by the $\sigma_n$ satisfies Justin's non-crossing conditions as $n \to \infty$.

Lunnon (1971) claimed that such foldings of $M_{m,n}$ are completely determined by their $x$- and $y$-sections and vice versa, and that valid $x$- and $y$-sections will always (and uniquely) determine a stamp folding. Lunnon's proof is somewhat lacking in rigor, and so we provide a different (hopefully better) proof.

**Theorem 7.10**   *A folding $\sigma \colon M_{m,n} \to [0, 1]$ together with a superposition order $<$ is globally flat-foldable if and only if the $x$- and $y$-sections of this superposition order are multiple non-intersecting, 1-dimensional stamp foldings.*

*Proof*   The forward direction of this theorem is straightforward; if $\sigma$ and $<$ are globally flat-foldable, then the $x$- and $y$-sections will both be non-intersecting, and thus will be collections of valid 1-dimensional stamp foldings.

For the other direction, we are given that the $x$- and $y$-sections have no intersections. This means that the superposition order $<$ won't violate any of Justin's non-crossing conditions, for if it did, then there would be, say, a crease in the $x$-direction crossing another crease, which must also be in the $x$-direction, that would then make a crossing in the $y$-section, a contradiction. What remains to be proven is that this superposition order can actually result from a sequence of origamis $\{\sigma_n\}_{n=1}^{\infty}$. That is, can we actually unfold our pile of stamps made by these $x$- and $y$-sections? It is not obvious that some non-crossing arrangement of the stamps can't be knotted in some way that prevents it from being unfolded (or folded in the first place).

Since the individual maps $\sigma_n$ in our folding are allowed to stretch or shrink the paper, all we need to do is prove that the 3-dimensional image of the folding with its superposition order (which we can think of as some $\sigma_n$ for sufficiently large $n$) is isotopic (that is, homotopic between specific embeddings in $\mathbb{R}^3$) to a disc. This can be done by building $\sigma_n$ one stamp at a time, say, starting with stamp $(1, 1)$ and then adding all the other stamps in lexicographic order (from left to right in each row, starting at the bottom row and going up). The first stamp is isotopic to a disc, and at each step we are attaching a new stamp along at most two creases, and so the result will still be isotopic to a disc after each stamp.   $\square$

**Remark 7.11**   In the proof of Theorem 7.10, we show that any flat-folded state of an $m \times n$ array of stamps can be reached from the unfolded state by a sequence of origamis $\sigma_n$. This is actually a special case of a general theorem by Demaine, Devadoss, Mitchell, and O'Rourke (Demaine et al., 2004), which states that any folded state can be reached from the unfolded state via a sequence of folds. This result will be discussed later in Section 8.2.

Lunnon's approach, then, for calculating $Sl(m, n)$ is similar to the algorithm he uses for $Sl(1, n)$. The space of $m \times n$ stamp folds (counting layers as distinct) can be partitioned by the rotation (shift) operation just as in the $1 \times n$ case, and this is how the values of $Sl(2, n)$ in Table 7.2 were computed.

The values for $Sg(2, n)$ in Table 7.2 were calculated by Justin (2012). His strategy was to write a program that lists the MV assignments of a $2 \times n$ array of stamps that are impossible to fold flat without self-intersecting. An outline of his method follows.

First, we establish a convention to describe MV assignments for a $2 \times n$ array of stamps. Each vertex in the array will have either one mountain crease and three valleys or vice versa. Thus, at each vertex there will be one crease that is **special**, meaning it has different MV parity from the other creases at the vertex. Let us then establish the convention that in our $2 \times n$ array, the left-most crease will always be a valley. Then we can label the left-most vertex with the first letter of the word North, East, South, or West depending on the direction of the vertex's special crease. We then describe all the other vertices, from left to right, with their own directional letters, and since the left-most crease is fixed to be a valley, this will uniquely describe the MV assignment as a sequence of $(n - 1)$ letters from the set {N, E, S, W}. We call this sequence the **valuation** of the MV assignment. For example, the valuation for the $2 \times 3$ array in Figure 7.9 is SW (where we need to flip all the Ms and Vs in order to follow the convention), and the one shown in Figure 7.10 is SEWN.

Justin then takes a potential MV assignment and tests to see if it is globally flat-foldable by trying to construct a superposition order for it that does not violate the non-crossing conditions:

(1) Given a $2 \times n$ array of stamps, number the top row stamps $1, 3, \ldots$ from left to right and similarly the bottom row $2, 4, \ldots$ from left to right.
(2) We then try to create a superposition order as an $n \times n$ matrix $S$, which we call the **order matrix**, where $S_{I,J} = 1$ (and $S_{J,I} = -1$) if stamp $I$ is lower than stamp $J$ when the entire array is folded. We also let $S_{I,I} = 1$ always.
(3) First, we fill part of the matrix $S$ using the MV assignment around each vertex. For example, looking at the left-most vertex in Figure 7.10, we may assume that stamp 1 is on the bottom and set $S_{1,2} = S_{1,3} = S_{1,4} = S_{2,3} = S_{2,4} = S_{4,3} = 1$ and $S_{2,1} = S_{3,1} = S_{4,1} = S_{3,2} = S_{4,2} = S_{3,4} = -1$.
(4) If at this point the order matrix $S$ does not induce a total ordering on the stamps, then order any two stamps $I$ and $J$ for which $S_{I,J}$ is not yet defined by setting $S_{I,J} = 1$. Call these the **questionable entries** of $S$.
(6) We then check to see if the non-crossing conditions are satisfied for the $x$-section and $y$-section of the superposition order for the questionable entries of $S$ (i.e.,

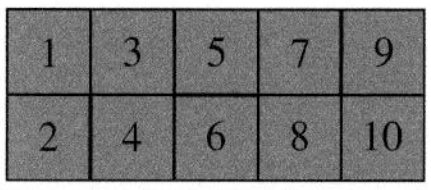

**Figure 7.10** A $2 \times 5$ array of stamps with a tricky MV assignment.

checking the Taco-Taco Condition from Section 6.5). If there is a crossing, then set $S_{I,J} = -1$ and try again. If both give a crossing the then we must go back to a previously set $S_{I,J} = 1$ and toggle it.

(7) If the tree of choices for questionable entries of $S$ runs out of options, then our MV assignment is impossible to fold. Otherwise, we end with a non-crossing superposition order and our MV assignment is globally flat-foldable.

Checking the non-crossing conditions is easier than one might think. For the $y$-section, consider two vertically adjacent stamps numbered $2i - 1$ and $2i$ and another pair $2j - 1$ and $2j$. Then consider the quantity

$$A = \text{sign}(S_{2i-1,2j-1}S_{2i,2j-1}) \cdot \text{sign}(S_{2i-1,2j}S_{2i,2j}).$$

The first term of $A$ will be negative if and only if stamp $2j - 1$ is between stamps $2i - 1$ and $2i$, and the second term will be negative if and only if stamp $2j$ is between $2i - 1$ and $2i$. Therefore, the Taco-Taco Condition from Section 6.5 will happen if and only if $A < 0$.

---

**Diversion 7.5**  Develop similar calculations that check to see if a $2 \times n$ array of stamps with a superposition order $S$ violate the Taco-Taco non-crossing condition in its $x$-section.

---

Justin also makes use of some reductions to help speed up his algorithm (although it still remains slow for large $n$). Proving these reductions makes for nice diversions.

---

**Diversion 7.6**  Prove that any $2 \times n$ array of stamps with MV assignment whose valuation is of the form $fgh$ where $f$ is a (possibly empty) sequence of Ws, $g$ contains only Ns and Ss, and $h$ is a (possibly empty) sequence of Es is globally flat-foldable.

---

**Diversion 7.7**  Prove that a valuation of the form $g$WW$h$ is globally flat-foldable if and only if $gh$ is (where $g$ and $h$ are valuation words, possibly empty).

---

**Diversion 7.8**  Given a valuation word $f$, we call the **odd skeleton** of $f$ the word we obtain by removing all the Ns and Ss from $f$, so that only Es and Ws remain. Prove that if the odd skeleton of $f$ is composed only of Ws, then $f$ is globally flat-foldable.

---

**Diversion 7.9**  What symmetries of the $2 \times n$ array of stamps can we take advantage of when counting globally flat-foldable MV assignments? Clearly one is to switch all the Ms and Vs, but what else? For example, what if we replace all the Ns with Ss and all the Ss with Ns? What else can we do? (Justin calls these **pseudo-symmetries**.)

---

**Example 7.12**   Let us demonstrate part of Justin's algorithm by proving that the $2 \times 5$ MV assignment in Figure 7.10 is impossible to fold. We begin filling in the $10 \times 10$ matrix $S$ for the superposition order by determining the layer ordering around each vertex, starting with the left-most vertex and assuming that face 1 is on the bottom. This gives us a sequence of $4 \times 4$ blocks of entries along the diagonal of $S$.

We can then determine the ordering of some non-vertex-adjacent faces by examining $2 \times 3$ parts of the MV assignment. For example, the MV assignment restricted to faces 1–6 can only be folded in one way and thus allows us to see that faces 5 and 6 must be above 1 and 2. Similarly, we can see that faces 9 and 10 must be above faces 5 and 6 by looking at the $2 \times 3$ part of the crease pattern made by faces 5–10.

For the $2 \times 3$ part made by faces 3–8, we must make a choice, which we may do without loss of generality due to the 180° symmetry of the MV assignment, that faces 7 and 8 fold on top of faces 3 and 4 when folding the two center-most vertical valley crease lines. This establishes the ordering between these faces. Finally, we may use transitivity to establish other order relations, for example, $S_{1,3} = 1$ and $S_{3,7} = 1$, so $S_{1,7} = 1$ must be true.

This determines the matrix $S$ aside from four questionable entries, indicated by question marks:

$$
S = \begin{bmatrix}
1 & 1 & 1 & 1 & 1 & 1 & \mathbf{1} & ? & \mathbf{1} & \mathbf{1} \\
-1 & 1 & 1 & 1 & 1 & 1 & \mathbf{1} & ? & \mathbf{1} & \mathbf{1} \\
-1 & -1 & 1 & -1 & -1 & -1 & 1 & -1 & \mathbf{1} & \mathbf{1} \\
-1 & -1 & 1 & 1 & 1 & 1 & 1 & -1 & \mathbf{1} & \mathbf{1} \\
-1 & -1 & 1 & -1 & 1 & -1 & 1 & -1 & \mathbf{1} & \mathbf{1} \\
-1 & -1 & 1 & -1 & 1 & 1 & 1 & -1 & \mathbf{1} & \mathbf{1} \\
-1 & -1 & -1 & -1 & -1 & -1 & 1 & -1 & \mathbf{1} & \mathbf{1} \\
? & ? & 1 & 1 & 1 & 1 & 1 & 1 & 1 & 1 \\
\mathbf{-1} & \mathbf{-1} & \mathbf{-1} & \mathbf{-1} & -1 & -1 & -1 & -1 & 1 & 1 \\
\mathbf{-1} & \mathbf{-1} & \mathbf{-1} & \mathbf{-1} & -1 & -1 & -1 & -1 & -1 & 1
\end{bmatrix}.
$$

Here we've bolded the entries established by transitivity.

We now suppose that $S_{1,8} = 1$ and check the Taco-Taco Condition between the adjacent faces 1 and 3 and the adjacent faces 8 and 10. The rule for two horizontal $1 \times 2$ strips like these is that we want

$$
\text{sign}(S_{2i-1,2j}S_{2i+1,2j}) \cdot \text{sign}(S_{2i-1,2j+2}S_{2i+1,2j+2}) > 0
$$

in order for them to be non-crossing under Taco-Taco. In our case we have

$$
\text{sign}(S_{1,8}S_{3,8}) \cdot \text{sign}(S_{1,10}S_{3,10}) = (1)(-1)(1)(1) = -1.
$$

Thus our choice for $S_{1,8}$ forced a crossing. Therefore we need $S_{1,8} = -1$. However, we then compare 1 and 3 with 6 and 8 and obtain for our condition

$$
\text{sign}(S_{1,6}S_{3,6}) \cdot \text{sign}(S_{1,8}S_{3,8}) = (1)(-1)(-1)(-1) = -1.
$$

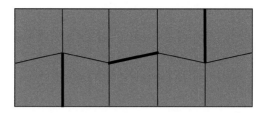

**Figure 7.11**  A $2 \times 5$ Miura-ori MV assignment that is a phantom fold but not globally flat-foldable.

Thus we have another crossing. We conclude that no choice for $S_{1,8}$ will avoid self-intersections of the paper, and therefore the MV assignment in Figure 7.10 is not globally flat-foldable.

**Remark 7.13**  The impossible-to-fold MV assignment in Figure 7.10 can be implemented in a $2 \times 5$ Miura-ori to create an example of a Miura map fold that is locally, but not globally, flat-foldable. Such an example is shown in Figure 7.11. This is why in Section 7.4 we were careful to only claim that $M(m, n)$ was counting phantom folds of the $m \times n$ Miura-ori. Counting actual globally flat-foldable MV assignments of the $m \times n$ Miura-ori is as difficult as the corresponding stamp folding problem.

There are many interesting questions still open about the computational complexity of stamp/map folding problems. In 2012 Tom Morgan (2012) created a complex reduction of the problem of determining if a $2 \times n$ MV assignment is globally flat-foldable that runs in $O(n^9)$ time. (In contrast, Justin's algorithm appears to run in exponential time.) However, as of this writing, it is not known if the corresponding problem for $3 \times n$ stamps/maps can be done in polynomial time or is NP-hard.

## 7.4     Tethered Membrane Lattice Folding

Hopefully it is becoming apparent to the reader that counting flat-folding problems contain rich combinatorial structures. One area where this has become especially evident is in the various combinatorial problems that have arisen from the physics of polymerized membranes and polymers (Francesco, 2000). Such membranes appear in nature, such as blood cell walls, and knowing how they can deform into different spatial configurations can help us understand how they function biologically.

Polymerized membranes possess a local lattice structure, where atoms form the vertices and the edges are chemical bonds. Thus, modeling how such a membrane deforms can be done by folding along the edges of such a lattice; this is called the **tethered membrane** model. The simplest lattice structure a tethered membrane can have is the triangular lattice, where each vertex has degree 6 and each face is an equilateral triangle. Physicists have used this discrete model as a way to numerically approximate

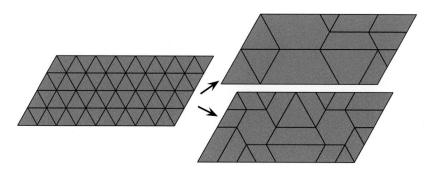

**Figure 7.12** A triangle lattice and two different phantom flat folds that can be made from subsets of the lattice.

continuous models (Kantor et al., 1986) and as a basis for transfer matrix analysis of polymerized membranes (Kantor and Jarić, 1990).

One simplified way to study the spatial configurations of such triangular tethered membranes is to classify, with an aim toward enumeration, the number of different ways such a lattice can fold flat. The physicists who study this problem frame it in a way that is different from the counting foldings problems considered thus far. Their interest is in the different final, flat-folded states, not how the material got to the state or the mountains and valleys needed to do so.

Let us state this problem more formally.

**Problem:** Given a finite section $S$ of the triangle lattice, how many different locally flat-foldable crease patterns (i.e., phantom folds) can be made using subsets of the creases in $S$, ignoring mountain and valleys? Figure 7.12 shows two examples from the same triangle lattice section.

Physicists have developed an extensive literature on analyzing this problem, and others similar to it, in terms of thermodynamics. For example, in both experimental and numerical simulation work, when an energy function is attached to each fold line (so that folding along a given crease requires a certain energy cost), researchers have noticed a geometric "crumpling transition" in the membrane between the flat, unfolded state and a flat, folded state, where temperature serves as the parameter between these extreme states (Kantor et al., 1986; Nelson and Peliti, 1987).

For the purposes of this book, we will concentrate more on the combinatorial techniques used by physicists to attack the triangle lattice folding problem, mainly following the work of Francesco and Guitter (1994). Nonetheless, the thermodynamic elements are interesting; while we will call attention to them, the details of the statistical mechanics needed to give full proofs is beyond the scope of this book.

Ignoring mountains and valleys, a vertex of the triangle lattice can fold flat in one of four different ways, up to symmetry, as shown in Figure 7.13. If we include the symmetric orientations of these vertex crease patterns, then we have 11 different types of vertices that can be used in a triangle lattice folding. It will also be useful to keep track of the relative face flip orientations in triangle lattice folds; these are indicated in Figure 7.13 by the $+$ and $-$ signs, which merely illustrate the two-colorability of

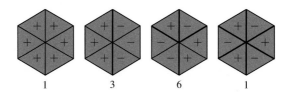

**Figure 7.13** The 11 ways a vertex of the triangle lattice can be folded flat, with symmetries indicated. Also shown are the relative spin (face flip) orientations.

flat-foldable crease patterns (as seen in Theorem 5.8). These face flips have an interesting physical interpretation for folded polymerized membranes. Whether a triangle is facing up or down after the lattice is folded flat corresponds to a spin of the 3-cycle bond between neighboring atoms, and thus the orientation of the faces gives rise to an Ising-like spin system.

Not all spin systems that we could make on a triangle lattice will correspond to a phantom fold, however. As can be seen in Figure 7.13, there is a local constraint around each vertex: The number of $+$ spins around each vertex must be a multiple of 3. Indeed, any spin configuration around a vertex with exactly 0, 3, or 6 $+$ spins will correspond to a vertex flat-fold. The number of such constrained spin vertex configurations is

$$\binom{6}{0} + \binom{6}{3} + \binom{6}{6} = 22,$$

which is twice the number of foldings of a triangle lattice vertex, as desired (since we can always flip the $+/-$ orientations and get the same crease pattern).

Like the Miura-ori counting problem from Section 7.2, the number of phantom folds from a triangle lattice (ignoring mountains and valleys) turns out to be equivalent to a graph coloring problem.

**Definition 7.14** A **Grünbaum coloring** of a triangle lattice is a 3-coloring of the edges such that every triangle face has all three colors surrounding it.

Grünbaum colorings are an important topic in topological graph theory (Albertson et al., 2010).

Let $n$ be the number of faces in our triangle lattice $L$, and let $Z_n(L)$ be the number of phantom flat-foldable crease patterns, ignoring mountains and valleys, that can be made from subsets of the lattice edges in $L$. Also let $Z_{color}(L)$ be the number of Grünbaum colorings of $L$, and let $Z_{spin}(L)$ be the number of spin systems of $L$ that meet the local constraint condition at every vertex.

**Theorem 7.15** (Francesco and Guitter, 1994)   *If a triangle lattice $L$ has n triangles, then $Z_n(L) = Z_{color}(L)/6$.*

*Proof*   Any phantom flat-foldable crease pattern from a triangle lattice $L$ will have a unique spin configuration (up to reversal of the spins), and this goes the other way as

**Figure 7.14** The spin-to-Grünbaum-coloring correspondence, where $+$ translates to counterclockwise increasing of the colors in $\mathbb{Z}_3$, and $-$ is clockwise.

well; any spin configuration that satisfies the local constraint condition (a multiple of $3$ $+$s around each vertex) will give rise to a unique crease pattern. (A lattice edge is in our crease pattern if and only if it is separating a $+$ and a $-$ face.) Thus we have that $Z_{\text{spin}}(L) = 2Z_n(L)$.

The bijection between Grünbaum colorings and phantom folds of the triangle lattice is indicated in Figure 7.14. If we use the elements of $\mathbb{Z}_3 = \{0, 1, 2\}$ as the edge colors in our Grünbaum coloring, then each lattice triangle must have these colors increasing in either clockwise or counterclockwise order. We let a counterclockwise order correspond to a $+$ face spin and clockwise correspond to a $-$ spin.

**Claim:** A flat-foldable phantom crease pattern on the triangle lattice will generate a valid Grünbaum coloring under our correspondence, and vice versa.

To see this, we need to establish that the edge coloring that our spins generate will be consistent around each vertex. Let $\sigma_1, \ldots, \sigma_6$ denote the face spins around a particular vertex in clockwise order, and let $c_1 \in \mathbb{Z}_3$ be any color we happen to assign to the lattice edge between the faces with spins $\sigma_1$ and $\sigma_6$. According to the rule of our correspondence (Figure 7.14), the colors of the other edges at our vertex, which we call $c_i$ proceeding clockwise from $c_1$, will follow the formula

$$c_{i+1} = \begin{cases} c_i + 1 & \text{if } \sigma_i \text{ is } +, \\ c_i - 1 & \text{if } \sigma_i \text{ is } -, \end{cases}$$

where the addition is done mod 3. Let $c_7$ be the color we obtain by proceeding clockwise around the vertex until we get back to our first edge. Our goal is to show that $c_1 = c_7$; see Figure 7.15.

Let $S$ be the sum of $\sigma_1$ through $\sigma_6$, where we view the positive $\sigma_i$s as "$+1$" and the negative ones as "$-1$." Then $c_1 = c_7$ if and only if $S = 0$ mod 3. But $S = 0$ mod 3 if and only if the number of positive $\sigma_i$ terms is a multiple of 3, which is exactly the local constraint condition for our spin system. Thus the 3-edge coloring around each vertex will be consistent (and a valid Grünbaum coloring) if and only if the corresponding

**Figure 7.15** Showing consistency of the constrained spin assignments $\sigma_i$ with the corresponding edge color assignments $c_i$.

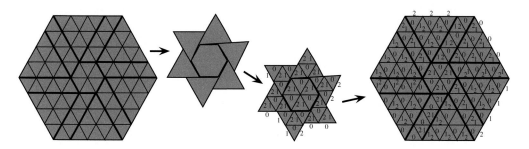

**Figure 7.16** A triangle lattice fold, where bold lines are the creases, and where we place the standard Grünbaum coloring on the folded result. We then unfold to see what Grünbaum coloring corresponds to the original crease pattern.

spin system corresponds to a flat-foldable phantom crease pattern. This establishes our claim.

Then note that we may rotate the colors of any Grünbaum coloring, by adding 1 mod 3 to all the colors, to get a different coloring with the same corresponding spin system as the original. Thus we have that $Z_{\text{color}}(L) = 3Z_{\text{spin}}(L) = 6Z_n(L)$, and the theorem is proven. ☐

**Remark 7.16** There is another way to view the bijection in Theorem 7.15. Define a **standard Grünbaum coloring** of the triangle lattice to be one where all the edges colored 0 are parallel, all the 1 edges are parallel, and all the 2 edges are parallel as well. This coloring corresponds to the null crease pattern, with no creases, since it results in all triangles having positive spin (or all negative, depending on the choice of parallel edges for the colors).

Now take any phantom flat-foldable crease pattern on a section $S$ of the triangle lattice $L$ and fold it flat. This technically gives us a mapping $f: S \to L$, since triangles will be folded on top of other triangles. We then apply the standard Grünbaum coloring to the edges of $f(S)$ and call this coloring $c$: edges of $f(S) \to \mathbb{Z}_3$. We then lift this edge coloring to a Grünbaum coloring $c'$ of $S$ as follows: For each edge $e$ in $S$, we assign $c'(e) = c(f(e))$. That is, $e$ gets the color of its image under the folding $f$. This is illustrated in the example of a hexagon twist in Figure 7.16. It is easy to see that $c'$ will also be a Grünbaum coloring (since $f$ preserves triangles), and the unfolding of $f$ will reverse the face spins of the coloring $c$, giving us that $c'$ will be the Grünbaum coloring that corresponds to our crease pattern.

**Remark 7.17** Returning to the polymerized membrane interpretation of such triangle lattice foldings, we can define the **entropy** $s$ of the folding per lattice face to be the following thermodynamic limit:

$$s = \lim_{n \to \infty} \frac{1}{n} \log Z_n = \log q.$$

(See (Francesco and Guitter, 1994).) The quantity $q$ is the geometric mean of the number of spin configurations per triangle. Prior to Francesco and Guitter's theorem, physicists had shown numerically that $q \approx 1.21$ (Kantor et al., 1986). But the correspondence with Grünbaum colorings provides an exact calculation of $q$, since in 1970 Baxter had already solved the enumeration of triangle lattice Grünbaum colorings in the thermodynamic limit (Baxter, 1970). Baxter actually solved the dual problem, enumerating proper 3-edge colorings of the hexagonal lattice (which is the planar dual graph to the triangle lattice), using the Bethe ansatz method to find the eigenvalues of the proper transfer matrix. This is the same method used by Lieb (1967) to enumerate proper 3-vertex colorings of grid graphs as related to the Miura-ori in Section 7.2. In doing so, Baxter was able to find the exact entropy $s = \log q$ as

$$q = \prod_{n-1}^{\infty} \frac{3n-1}{\sqrt{3n(3n-2)}} = \frac{\sqrt{3}}{2\pi} \Gamma(1/3)^{3/2},$$

where gamma function $\Gamma(x)$ is the continuous extension of the discrete factorial function $\Gamma(n) = (n-1)!$. This gives $q = 1.208717\ldots$, which agrees well with the prior numerical estimate in (Kantor et al., 1986).

The practical interpretation of this, for origami purposes at least, is that if we let the number of triangles $n$ in our lattice get very large, say $n \approx 10^{23}$ or so as one would have in a polymerized membrane where the vertices are atoms, then the number of phantom flat folds of the lattice will be approximately $Z_n = (1.208717\ldots)^n$.

## 7.5 Open Problems

Many aspects of flat-foldability discussed in this chapter are active areas of research. Some of the following examples of open questions are quite new and could be solved in the next few years, while others (like the stamp-folding problem) have been open for decades and have proven difficult for researchers to crack.

**Open Problem 7.1**   Is there a way to classify crease patterns whose (locally) valid MV assignments can be enumerated by a 2-colorable graph (like the origami line graph)?

**Open Problem 7.2**   Can the ideas behind Justin's algorithm for determining if an MV assignment of the $m \times n$ square grid crease pattern is globally flat-foldable be used to improve Morgan's bound on the running time of the $2 \times n$ stamp-folding problem?

**Open Problem 7.3**   Is the $3 \times n$ stamp-folding problem (the Assigned Flat-Foldability problem) NP-hard?

**Open Problem 7.4**   The bijection between locally valid MV assignments of the Miura-ori and proper 3-colorings of grid graphs makes one wonder if there are other families of graphs whose 3-colorings correspond to valid MV assignments of origami crease patterns. Such examples have been found, as well as an algorithm for constructing such graphs for crease patterns with specific (but fairly general) properties (Chiu et al., 2019). However, for a given flat-foldable crease pattern, graphs whose 3-colorings correspond to valid MV assignments in this way are not unique. Can more simple, or general, graphs than those in (Chiu et al., 2019) be found? Are there other families of flat-foldable crease patterns that correspond to simple or well-known families of 3-colorable graphs?

**Open Problem 7.5**    The examples we have seen thus far of connections between flat-foldable origami tessellations and entropy models of atomic lattices (like the Miura-ori and the square ice model, or folding triangular grids and polymer membrane folding) suggest that more such examples can be found. For instance, Michael Assis expanded on the bijection of Theorem 7.4 to perform a statistical mechanics study of several classes of flat-foldable quadrilateral origami tessellations (Assis, 2018). Are there other origami tessellations that are amenable to these kind of statistical mechanics approaches?

# 8 Other Flat-Folding Problems

We close this part of the book with a few brief sections on other problems related to flat-foldability. Some of these topics are interesting side problems of which readers may have previously heard. Others could be a chapter in their own right, but we instead refer the interested reader to other resources for details, most notably Robert J. Lang's *Origami Design Secrets* (Lang, 2011) and the graduate-level text *Geometric Folding Algorithms: Linkages, Origami, and Polyhedra* by Erik Demaine and Joseph O'Rourke (Demaine and O'Rourke, 2007). Indeed, while it is difficult to find references on the mathematics of flat origami that span breadth and depth, these two books are clear exceptions and do a much better job than the present author could of describing the areas of origami design and computational origami.

## 8.1 How Many Times Can We Fold a Sheet of Paper?

There is an old childhood challenge (which is so commonly stated that it seems difficult to historically document) stating that a person cannot fold a piece of paper in half more than seven times. The intuition is that folding paper in half doubles the number of layers each time, so that after six folds there will be $2^6 = 64$ layers of paper. With standard weight paper, making a seventh fold would be like trying to fold a block of wood, and this should be impossible. Even using a large sheet of thin newsprint verifies that six or seven folds seems to be the physical limit.

However, this line of thinking is naive for a number of reasons. One is that paper comes in many different thicknesses, and therefore using the very thinnest paper possible should allow one to increase the number of folds. Another is that the size of the paper clearly matters. One can imagine folding a sheet of paper the size of a football field more than seven times.

In 2002 Britany Gallivan, then a high-school student, physically demonstrated that more than seven folds were possible and, in fact, managed to fold a sheet of paper in half 12 times. She seems to be the first person on record to actually do this, and she also created a mathematical model by which to calculate the size and thickness of paper needed to reach as many folds as desired using her technique (Gallivan, 2002).

Specifically, Gallivan used a long strip of paper and employed **single-directional folding**, where one always folds the paper in half in the same direction, for example, making a valley fold in the middle of the folded strip each time. She used a

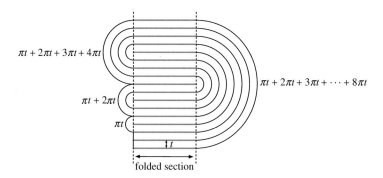

**Figure 8.1** The side view of a strip folded using single-directional folding four times, where $t$ is the thickness of the paper (exaggerated for the purposes of illustration) and the four semicircle regions indicating the folds are labeled with their respective lengths.

"jumbo roll" of tissue paper that was 4000 feet long and very thin, and performed the folding along a long expanse of an indoor shopping mall in Pomona, CA on January 27, 2002. Of mathematical interest is Gallivan's calculation of the paper strip length, $L_n$, needed to achieve $n$ folds.

**Theorem 8.1** (Gallivan, 2002)   *The minimum length of a strip of paper with thickness t needed to make n folds via single-directional folding is*

$$L_n = \frac{\pi t}{6}(2^n + 4)(2^n - 1).$$

*Proof*   We will first establish a recursive formula for $L_n$. On the $n$th fold we will be folding $2^{n-1}$ layers of paper, and the fold itself will form a semicircle disk. The $i$th layer of paper in this semicircle disk will form a semicircle of radius $it$ and thus will have length $i\pi t$. Therefore, the length of additional paper needed to make the $n$th fold will be the sum of the semicircle circumference lengths for each layer of paper (see Figure 8.1), which gives us

$$L_n = L_{n-1} + \sum_{i=1}^{2^{n-1}} i\pi t = L_{n-1} + \pi t \frac{2^{n-1}(2^{n-1}+1)}{2} = L_{n-1} + \pi t 2^{n-2}(2^{n-1}+1).$$

Telescoping this recursion results in

$$L_n = \pi t \sum_{i=1}^{n} 2^{i-2}(2^{i-1}+1) = \pi t \left( \sum_{i=1}^{n} 2^{2i-3} + \sum_{i=1}^{n} 2^{i-2} \right).$$

Using $\sum_{i=1}^{n} 2^{i-1} = 2^n - 1$ gives us the closed formula

$$L_n = \pi t \left( \frac{1}{6}(2^{2n} - 1) + \frac{1}{2}(2^n - 1) \right),$$

which simplifies to the desired result.   ☐

Gallivan refers to $L_n$ as the **loss function** because it keeps track of the amount of paper that is devoted to the bends at each fold, which can be thought of as "lost" from the rest of the folded strip (the part of the paper labeled "folded section" in Figure 8.1).

---

**Diversion 8.1**    Show that the loss function also satisfies

$$L_n = \pi t \sum_{i=1}^{n} \sum_{k=1}^{2^{i-1}} (2^{i-1} - (k-1)).$$

---

The units of $L_n$ will be the same as those used for the thickness $t$. For Gallivan's experiment, as reported in (Gallivan, 2002), she used tissue that was approximately 0.003 inches thick. Her formula then gave a minimum length of $L_{12} = 25{,}754.8$ inches, or 2146.23 feet for 12 single-direction folds. As previously mentioned, she used a 4000 foot roll of tissue for this project, 2146.23 feet of which were "lost" to the fold semicircle disks. That left 1853.77 feet of tissue for the "folded section" part of the final 12-folded strip (see Figure 8.1). Since there were $2^{12}$ layers of paper in the folded section, Gallivan's final folded section was only $1853.77/2^{12} \approx 0.45$ feet, or 5.4 inches long.

Gallivan also discusses **alternate direction folding**, where the strip is folded in half alternately via valleys and mountains.

---

**Diversion 8.2**    Derive the loss function for folding a strip of paper with thickness $t$ in half $n$ times using alternate direction folding.

---

## 8.2    Can Any Shape Be Folded and Unfolded?

In the later half of the 1990s, a new field of research in paper folding emerged, that of **computational origami**. This can be viewed as a subset of computational geometry, where questions are asked having to do with complexity or algorithms needed to solve geometric problems, only we restrict ourselves to origami-related problems. Bern and Hayes' 1996 paper, "The complexity of flat origamis" (Bern and Hayes, 1996), was probably the first published paper in this field and inspired many other researchers to explore algorithmic-geometric problems in paper folding in subsequent years.

Two surprising results emerged from this effort: the realization that, informally stated, anything can be folded and anything can be unfolded. To be more specific, any polyhedron (not necessarily convex) can be exactly folded given a large enough sheet of paper (anything can be folded), and given any folded state of a piece of paper, it can be unfolded to the flat state via continuous folding motions (anything can be unfolded). These results are surprising because it is certainly conceivable that some wildly non-convex shape, perhaps with high topological genus, might not be possible to make with a folded sheet of paper. And it is also conceivable that complex folded states of a piece of paper might exist that cannot be reached by a continuous folding process, rather than, say, requiring the paper to pass through itself or be cut and re-glued in order to reach the desired folded state.

In this section we will explain exactly how these results are formally framed, but we will outline their proofs only briefly. Full details can be found in their original sources (Demaine et al., 2000, 2004) as well as in Demaine and O'Rourke's *Geometric Folding Algorithms* text (Demaine and O'Rourke, 2007).

In (Demaine et al., 2000), Demaine, Demaine, and Mitchell not only prove that any polyhedral shape can be folded, they give a constructive algorithm to make such a folding. Unfortunately, their algorithm is extremely inefficient; they start with an extremely long rectangular strip of paper. Such a strip, say with dimensions $1 \times n$, can be made from a large $n \times n$ square of paper by dividing one side, say the left, of the square into $1/n$ths and then accordion-pleating the square with alternating mountain and valley horizontal folds at each multiple of $1/n$. Assuming mathematically thin paper, this method will turn our square into a rectangular strip as (relatively) long as we wish.

This long strip can then be folded to weave back and forth across every face of our polyhedron. In fact, with a thin enough strip, this can be done perfectly, with any excess of paper made by turning back-and-forth tucked under the paper. See Figure 8.2 for an example, but more details of how to cover the polyhedron perfectly with the strip can be found in (Demaine et al., 2000).

The meandering strip of paper can be made to cover one face of the polyhedron and then a neighboring face adjacent by an edge. To cover the whole polyhedron then requires triangulating the polyhedron's faces (or even subdividing triangle faces into smaller triangles if needed) in such a way that a Hamilton path can be drawn on the dual graph of the triangulation. (This is always possible to do; see (Arkin et al., 1996).)

The main theorem from (Demaine et al., 2000) is actually stronger than what we have described. Let us define a sheet of paper to be **bicolor** paper if the two sides of the paper have different colors.

**Theorem 8.2** (Demaine, Demaine, Mitchell (Demaine et al., 2000))   *Given any polyhedron, with each face assigned one of two colors, there is a folding of a sufficiently*

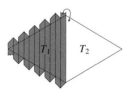

**Figure 8.2**  A triangle face $T_1$ of a polyhedron being covered by a meandering strip. The strip can be continued to cover the adjacent triangle face $T_2$. The flaps of paper where the strip makes a turn can be folded underneath along the edge of the triangle $T_1$.

*large square of bicolor paper that folds into the polyhedron with the desired colors showing on each face.*

The addition of the 2-coloring of the faces does not introduce a significant challenge. All that is needed is a way to turn our strip of paper (which can be accordion-pleated from the square to also be bicolor) over as we meander from one triangle face to another, and this is relatively easily done (see (Demaine et al., 2000; Demaine and O'Rourke, 2007)). This bicolor aspect of Theorem 8.2 was inspired by origami artists who have ingeniously used standard origami paper, which is usually white on one side and a color on the other, to create folded models of, for example, a striped zebra, a spotted cow, and an 8 × 8 chessboard, each from a single sheet of such paper.

It cannot be overemphasized that the folding construction given in the proof of Theorem 8.2 is arguably the most inefficient method possibly to make an origami model. Figure 8.3 shows a more reasonable (but still complicated!) method for folding a triangulated version of the Stanford Bunny. This crease pattern was generated by Tomohiro Tachi's Origamizer algorithm (Tachi, 2010).

Thus we have proofs that any shape, either 2-dimensional or 3-dimensional, can be folded from a square sheet of paper to any degree of approximation as we desire, and in the case of polygons and polyhedra, no approximation is needed. In simplistic and unrigorous terms, everything can be folded.

A related question, then, is, "Can everything be unfolded?" By this we mean to ask if there are any proper folded configurations (say, intrinsically isometric, non-intersecting) in which a sheet of paper can be that cannot be completely unfolded without, say, making the paper intersect itself or rip.

We addressed this question in the case of stamp/map folds in Theorem 7.10, where we argued that any non-intersecting, one-stamp pile configuration in which an $m \times n$ crease pattern grid can be put is isotopic to a disk and therefore can be achieved via a

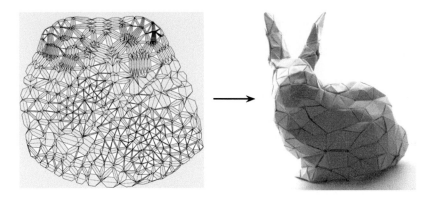

**Figure 8.3** A crease pattern, made from T. Tachi's Origamizer program (Tachi, 2010), that folds into a polyhedral version of the Stanford Bunny. This shows that complex polyhedra can be folded more efficiently than the method given in Theorem 8.2. (Images courtesy of T. Tachi.)

sequence of origamis $\sigma_n$. Recall that our definition of origami from Section 5.1 allows us to let the maps $\sigma_n$ stretch the paper, which allows us to take advantage of such an isotopy. In reality, we might need to crumple and contort the paper in order to unfold it from an especially difficult stamp fold, but the fact remains that it can be done.

In (Demaine et al., 2004) Demaine, Devadoss, Mitchell, and O'Rourke prove the general version of this result, although they use a different model of paper folding than used in this text. Specifically, they prove the following result (although this is admittedly an abridged version of their statement):

**Theorem 8.3** (Demaine, Devadoss, Mitchell, O'Rourke (Demaine et al., 2004))  *Given a folded state $f(C)$ of a simple polygonal piece of paper $C$ whose intrinsic geometry is piecewise twice-differentiable ($C^2$), there is a continuous folding motion of $C$ into $f(C)$.*

In other words, if a folded state exists, then it can be achieved by standard folding motions (see (Demaine et al., 2004; Demaine and O'Rourke, 2007) for precise definitions of this term) from the unfolded state, which also proves that any folded state can be unfolded.

The nutshell of the argument in (Demaine et al., 2004; Demaine and O'Rourke, 2007) is illustrated in Figure 8.4. Given a piece of paper $C$ with a crease pattern, which we can consider to be a triangulation of the paper by adding auxiliary creases if needed, let $T$ be a triangle in the paper that does not intersect any creases. We may then fold $C$ into the triangle $T$ using a sequence of folds that will likely have no correspondence with the crease pattern. This is proven by induction, showing that we can "roll" pieces of the paper inside the polygon until its sides touch those of $T$ and then folding the remainder into the shape of $T$. Understand, however, that we are assuming mathematically thin paper and not worrying about thickness issues, so we may have as many layers being folded as we wish.

Let the folding motions that fold $C$ into $T$ be parameterized by $P_t$ for $0 \le t \le 1$, so that $P_1(C) = T$. We then map $T$ into the folded image $f(C)$ that is given to

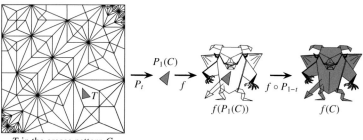

$T$ in the crease pattern $C$

**Figure 8.4** A cartoon of the argument used in (Demaine et al., 2004) to prove that any folded state can be achieved (i.e., unfolded) using standard folding motions. The crease pattern and model shown is Maekawa's Devil from (Kasahara and Maekawa, 1983).

us by the theorem. From within the geometry of this image, we then reverse our parameterization $P_{1-t}$ to unfold the triangle into the full folded image $f(C)$ using the composition $f \circ P_{1-t}$. The fact that we may do this last step relies on the hypothesis that the regions of our crease pattern are $C^2$ in the folded image, although the authors of (Demaine and O'Rourke, 2007) conjecture that the result should be provable with only a $C^1$ restriction.

## 8.3 Origami Design

One of the more direct applications of paper folding mathematics is in the design of origami models, where by "model" we mean a physical model. For example, the origami devil shown in Figure 8.4 was designed by the Japanese physicist and artist Jun Maekawa (the namesake of Maekawa's Theorem). Seeing complex origami models like this leads one to the natural question, "How do people design such models?" If we wanted to, say, fold an origami grasshopper with six legs (four shorter ones and two longer ones), antennae, wings, a head, and thorax, how would we go about finding a crease pattern that would fold (or even flat-fold) into such a thing from a single square with no cuts?

Origami artists have been solving this problem for over a century, and the history of origami in Japan seems to indicate that such creative origami explorations go back at least several centuries. For example, one approach to help the origami design process that goes back to Edo-era Japan is the use of **origami bases**, which are folded paper models that can serve as a starting point for many other origami models. The bird base is used to make the traditional Japanese crane (or **tsuru**, as it is called). Figure 8.5 shows four of the traditional origami bases along with their crease patterns.

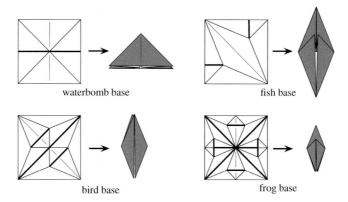

waterbomb base                    fish base

bird base                         frog base

**Figure 8.5** Four origami bases that go back to the Japanese Edo-era (1603–1867) paper folding tradition. Bold lines are mountains, non-bold lines are valleys, and the thin, shorter lines indicate auxiliary creases not used in the final folded form.

---

**Diversion 8.3**   Find a recursive pattern in the crease patterns of the fish, bird, and frog bases in Figure 8.5, and use this to develop more complex bases.

---

Note that the traditional origami bases each serve different design purposes. The waterbomb base is a very simple way to make four large 45° angle triangle flaps. (Incidentally, it is called a "waterbomb" base because it is used to make a cube-shaped, blow-up balloon, which can be filled with water and used like a water balloon.) The fish base makes two long flaps from the square using two opposite corners, but each of these flaps is more narrow (they can make a 22.5° angle when folded in half) than the waterbomb base's. The bird base makes four narrow flaps, each from a corner of the square. The frog base makes five narrow flaps, one from each corner and one from the center of the square. (The idea here is that a frog needs four legs and a head, thus five flaps.)

A more general observation can be made from these examples of bases. Notice how each of the flaps in the bases shown in Figure 8.5 are made by creases radiating from a point. The easiest example to see is the center of the frog base; the alternating mountains and valleys around this point contract the paper into a flap. In this way, we can imagine that a disc of paper is being devoted to this flap, where the disc collapses like an umbrella to form the flap.

This is illustrated for the bird base in Figure 8.6, which is shown to make a traditional flapping bird model (a simplification of the Japanese **tsuru**). Each corner of the paper makes a flap, one for the head, one for the tail, and two for the wings. Each of these four flaps will require a circle of paper to form, but since the corners of paper are the tips of the flaps, the centers of the circles will be on the corners of the paper. Thus each flap takes a quarter-circle of paper, and these quarter-circles are packed tightly together, as shown in Figure 8.6, indicating the efficiency of this model's design. This type of analysis of the crease pattern, where we draw circles indicating the amount of paper devoted to flaps, is called a **circle decomposition** of the crease pattern.

---

**Example 8.4**   The idea of circle decompositions can be used to design origami bases with specific target properties. As an easy first example, suppose that we want an

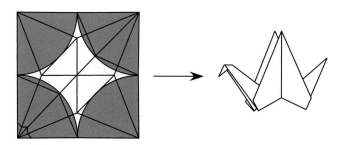

**Figure 8.6** The circle decomposition for the flapping bird crease pattern.

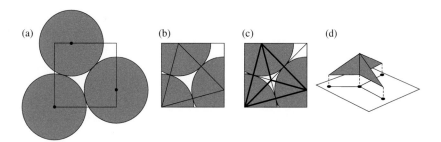

**Figure 8.7** The process of designing an origami base from a square with three equal-sized flaps that are as long as possible. Note that the result can be projected orthogonally onto a plane to make a tree representing the desired base.

origami base that gives us three long flaps of equal length. Ideally we would like a base that gives us flaps as long as possible. The bird base does this for four flaps, since its corresponding circle decomposition represents the optimal way we could place four equal-sized circles into a square. We will need to do this for three circles; that is, we need to find the best way to pack three equal-sized circles into a square, or rather so that their centers are in the square.

Figure 8.7(a) shows how to do this. Finding the optimal location for the circle centers is the same problem as trying to inscribe the biggest equilateral triangle in a square (recall Figure 1.2). The corners of this maximal equilateral triangle will be the centers of our circles (Figure 8.7(b)), and we argue that the sides of this triangle **must** be creases in the crease pattern of our base. This is because in order for the centers of the circles to be the tips of folded flaps, and in order for the radii of the circles to correspond to the length of said flaps, the centers of the three circles cannot be folded closer to each other, pairwise, in the geometry of the folded base. If there were no crease connecting two of these centers, then other creases passing between them would fold the centers closer together in the final base. Using the terminology of Lang (1996, 2011), we say that the line connecting pairwise centers of these circles form an **active path** on the crease pattern, and active paths must always be creases of the crease pattern.

The equilateral triangle by itself will not fold into our desired base, however. Following the lead given by the traditional origami bases in Figure 8.5, we can try to add angle bisectors to the equilateral triangle, as shown in Figure 8.7(c). Our three-flap base can then be folded as follows: First fold the paper outside our equilateral triangle behind to turn our square into the equilateral triangle. Then fold all layers along the angle bisectors of the triangle. This will give us a three-flap base as shown in Figure 8.7(d), where we also illustrate how the base can be projected orthogonally onto a plane to make a tree graph that forms a schematic of the desired base: just three leaf nodes connected to a central vertex by edges of equal length.

---

**Diversion 8.4**    Suppose that you want to create an origami model of a sea urchin, which we may think of structurally as a central point with lots of equal-sized spikes (or flaps) coming out of it. Design a crease pattern that would, in theory, give you a base for such an origami model. It may be easier to require the number of spikes in your sea urchin be a perfect square. Note that designing such a crease pattern does not mean it will be easy to fold! Instructions for actually folding such a model can be found in (Lang and Montroll, 1991).

---

It should be clear by now that there is a strong connection between origami model design and circle packing. This connection was developed and explored by several origami artists around the world in the 1990s, including Toshiyuki Meguro (1991–1992, 1994), Fumiaki Kawahata (1993); Kawahata (1997), and Robert J. Lang (1996, 2011). In short, non-overlapping circles may be arranged on the sheet of paper to plan where the flaps of the target origami model will be. The circles do not need to have the same radii; differently sized flaps will need circles of different radii. Arranging the circles to maximize their respective radii can lead to more efficient use of the paper and, perhaps, more elegant crease patterns. If two circles touch, then the line connecting their centers will be an active path and be included in the crease pattern.

The above process will, hopefully, result in a collection of creases that define various polygons on the paper. Folding this into the desired origami base requires collapsing the circles like a collection of umbrellas, and that means inserting creases in the various polygons formed by the active creases. There are often many ways to do this, but the following result is a simple way to deal with triangle polygons.

**Theorem 8.5** (Rabbit-Ear Theorem)    *Given a triangle ABC, let O be the incenter so that AO, BO, and CO are angle bisectors. Let XO be a perpendicular drawn from O to any one of the three sides of the triangle. Then O with the crease lines AO, BO, CO, and XO is a flat-foldable vertex.*

This Theorem gets its name from the fact that in origami parlance, this way of folding up a triangle is known as a **rabbit-ear** fold, simply because there are some classic origami rabbit models that use these folds to make the ears.

---

**Diversion 8.5**    Prove the Rabbit-Ear Theorem.

---

Thus any triangle polygon that is made from active paths may be collapsed by employing the Rabbit-Ear Theorem. Polygons of other sizes have to be handled differently, and the collection of creases that we put into such polygons to collapse them are called **bun-shi** (which is Japanese for **molecules**), a term coined by Meguro (1991–1992, 1994).

Robert J. Lang has created an algorithm for (and made more rigorous) the above origami design process for the class of models that can be folded from a **uniaxial base** (Lang, 1996). A uniaxial base is one that can project onto a connected tree graph, like the example in Figure 8.7(d), where all flaps of the origami base are on one side of the projection plane. (Note that most origami models are derived from uniaxial bases; in fact, the classic Japanese bases from Figure 8.5 can be made to be uniaxial by rearranging the flaps. However, there are bases that are not uniaxial, such as John Monrtoll's Dog Base from (Montroll, 1979).) Lang has named his algorithm **TreeMaker**, and we will now describe it in abbreviated detail. A full account of TreeMaker and its many nuances and parameters can be found in Lang's magnum opus *Origami Design Secrets* (Lang, 2011).

The input to TreeMaker is a connected tree graph $G$ with weights assigned to the edges indicating the relative lengths of the flaps. Let $G$ have $n$ terminal (leaf) nodes. Our first goal is to find coordinates $u_i = (x_i, y_i)$ for each terminal node on the square piece of paper (which we can assume is the unit square) to form our origami base. Let $m$ be the scaling factor we need to use in order for the weighted tree lengths to correspond to lengths on the square. Then if $l_{ij}$ is the distance (sum of the weights) between the terminal nodes $i$ and $j$ on the tree graph, we have $n(n-1)/2$ equations that must be satisfied,

$$\|u_i - u_j\| \geq ml_{ij}, \tag{8.1}$$

which we call the **tree conditions**. The reason for these equations is that if two terminal nodes had distance less than $ml_{ij}$ on the crease pattern, then they would never be able to have the distance equal to $ml_{ij}$ on the geometry of the folded base without ripping the paper. Finding coordinates for the $u_i$ that satisfy the Equations (8.1) while maximizing $m$ is a problem in constrained optimism and can be solved using various algorithms. (Lang has used the Augmented Lagrangian Multiplier and the Feasible Sequential Quadratic Programming algorithms, among others.) Other constraints may be added to this system to, for example, require that the nodes are symmetric about some axis of the paper.

If a solution for nodes $u_i$ and $u_j$ gives equality in Equation (8.1), then there will be an active path between these nodes, meaning that the line segment $\overline{u_i u_j}$ either is a crease line or is along the boundary of the square. It can be shown that active paths never cross, so the solutions for the points $u_i$ will give us a subset of our desired crease pattern and decompose the square into a set of polygons. Then molecules (bun-shi) are added to these polygons to give us the crease pattern for an origami base whose orthogonal projection will be proportional to our original tree graph.

An example, slightly modified from one in (Lang, 1996), is illustrated in Figure 8.8. We start with a tree graph for a critter (maybe a lizard) with a head, four legs, and a tail, where each of these flaps have the same unit length and there's a "body" in the tree of unit length as well (between non-terminal nodes $B$ and $C$ in Figure 8.8(a)). An optimal mapping of the terminal nodes of this tree to a square is shown in Figure 8.8(b) with the active paths drawn. Note that even though the tree graph is symmetric from head to tail, the corresponding mapping to the square is not. The reason for this is that

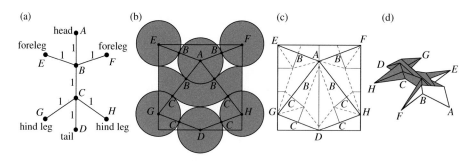

**Figure 8.8**  An example of Lang's TreeMaker algorithm, (a) starting with a weighted tree graph (perhaps for a lizard model), (b) mapping the terminal nodes optimally to the square with circles and a river, (c) adding molecules to the polygons, and (d) the folded result.

while it is optimal to place one set of the legs (say, the forelegs) at two corners of the square (i.e., nodes $E$ and $F$ in the figure), once we do this the other legs (nodes $G$ and $H$) cannot be mapped to the other two square corners, for if they were then there would not be a relative distance of length 3 between the head node $A$ and the tail node $D$. That is, in order to create as many active paths as possible, and thus optimize the radii of our circles, nodes $G$ and $H$ need to not be placed in corners of the square.

Note that the algorithm only places the terminal nodes of the tree. Internal nodes, like $B$ and $C$, can be thought of as being mapped to multiple places on the square, which then get folded together in the finished base. Also, the distance between $B$ and $C$ must still be the proportional unit length on the folded base, and this means that a ribbon-like section of the paper must be collapsed to form the $BC$ segment in the final base. This is shown in Figure 8.8(b) and is often called a **river** in the design. Thus, the process of converting a weighted tree graph to an arrangement of the terminal nodes with active paths on the square paper is called finding a **circle-river packing**.

Figure 8.8(c) shows the active paths with molecules inserted to make the full crease pattern for the base. Triangles $AEG$ and $AFH$ have rabbit-ear molecules added. (Other creases are shown, however, to indicate how the model will look in a 3-dimensional, non-flat state, such as shown in Figure 8.8(d).) Careful attention will show that the quadrilateral $AGDH$ has a different kind of molecule inserted; it uses a clever recursive generalization of the rabbit-ear molecule invented by Lang (1996) called the **universal molecule**.

The idea behind Lang's universal molecule is shown in Figure 8.9. Given a polygon whose sides are all either active paths or boundaries of the paper, we make creases at all the angle bisectors of the corners. We do not immediately extend these bisectors as far as possible, however; rather, we extend them to create an imaginary **reduced polygon** whose sides are all a perpendicular distance $h$ from the sides of the original polygon. This reduced polygon will not necessarily become part of the crease pattern. But the corners of the reduced polygon will act as proxies $p(A_i)$ of the corners of the original polygon $A_i$. Two things can happen to these proxy corners as $h$ increases:

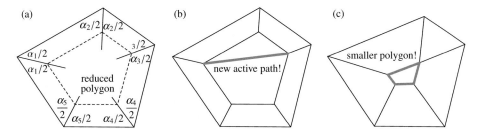

**Figure 8.9** The universal molecule's recursion in action. As we reduce the polygon (a) using angle bisectors, either (b) a new active path will be creased, giving us two smaller polygons, or (c) two bisectors will collide, giving us a polygon with fewer sides.

- Two nonadjacent proxy corners $p(A_i)$ and $p(A_j)$ will reach the minimum scaled distance $ml_{ij}$ allowed by the original corners $A_i$ and $A_j$, which means that $p(A_i)$ and $p(A_j)$ will form a new active path.
- Two adjacent proxy corners $p(A_i)$ and $p(A_{i+1})$ will collide, making the reduced polygon edge between them degenerate and giving us a reduced polygon with one fewer side.

In either case (which may happen in multiple cases simultaneously), we reduce our original polygon into a smaller polygon or polygons with fewer sides. As we do this, the only creases added to our crease pattern are the angle bisectors and any new active paths creased. As we reduce to smaller polygons, the angle bisectors generated will change. This process ends when all our reduced polygons are triangles and we employ the Rabbit-Ear Theorem or if multiple (more than three) bisectors meet at a point, which is a situation analogous to the Rabbit-Ear Theorem.

This process can be seen in the lizard example in Figure 8.8(c), where quadrilateral $AGDH$ has its angle bisectors drawn as valley creases, but then quickly the proxy corners $p(A)$ and $p(D)$ reach the minimum distance of 3 (scaled) between nodes $A$ and $D$ (since their distance on the tree graph is 3), and an active path is drawn between $p(A)$ and $p(D)$ as a mountain crease. On the left and right of this active crease, we have two reduced triangle polygons, and the angle bisectors finish the job à la the Rabbit-Ear Theorem.

---

**Diversion 8.6**   Prove that any new vertices generated by the universal molecule recursion will always be flat-foldable.

---

Diversions 8.5 and 8.6 prove that Lang's universal molecule will always be locally flat-foldable. Thus the crease patterns generated by the TreeMaker algorithm will be locally flat-foldable, and at the time of this writing, it is not known whether TreeMaker crease patterns will always be globally flat-foldable (although Robert J. Lang and Erik Demaine conjecture that they are). A more rigorous treatment of the algorithm needed to produce universal molecules form convex polygons can be found in

(Bowers and Streinu, 2015). Also, Bowers and Streinu (2016) have generalized the universal molecule for non-convex polygons.

As mentioned previously, there are many other aspects and features of Lang's TreeMaker algorithm that can aid in origami model design. There are also interesting mathematical problems, such as how to correctly assign mountains and valleys to the crease patterns this algorithm generates, on which we have not touched. Interested readers must consult *Origami Design Secrets* (Lang, 2011) for the full details.

## 8.4    The Rumpled Ruble, or Margoulis Napkin Problem

The famous Russian mathematician Vladimir Arnold was known for collecting massive lists of interesting open problems that he would use in his legendary seminars (Tabachnikov, 2007). In 1956 he posed the following problem:

"The rumpled dollar problem": is it possible to increase the perimeter of a rectangle by a sequence of foldings and unfoldings? (Arnold, 2005, Problem 1956-1)

Originally this was dubbed the "rumpled ruble problem" (Tabachnikov, 2007). This problem is also attributed to the Russian mathematician Grigory Margulis as the "napkin problem," where one asks if it is possible to fold a square napkin into a flat object with perimeter greater than the original napkin (Lang, 2011, p. 329). This problem received some attention in an email discussion initiated by Jim Propp in 1996 where several people tried to prove that it was **impossible** to gain a larger perimeter via folding. These proofs assumed that only very simple folds were allowed (say, making only one fold through the paper at a time) and not more complicated sinks, spread-squashes, or other folding moves from an origamist's repertoire.

For example, Ivan Yaschenko (1998) offers a method of folding a rectangle (say, with proportions of a dollar bill) into a 3-dimensional figure whose projection onto a plane has perimeter greater than the original rectangle. It is illustrated in Figure 8.10. The idea is that the creases create a 3-dimensional model whose base is the polygon *AEBCID*, and the perimeter of this polygon is greater than the perimeter of the original rectangle *ABCD*. The rest of the paper pops up into some tent-shaped reverse folds, but these folds can be made more numerous and thus smaller. That is, Figure 8.10

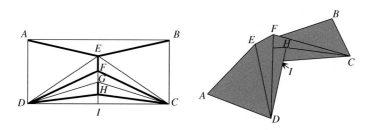

**Figure 8.10**  Yaschenko's proposed solution to the rumpled dollar problem.

shows the angles ∠*EDI* and ∠*ECI* divided into four equal angles, which determines the reverse folds where *F* and *H* are the points that poke upward. If we instead *n*-sect the angles ∠*EDI* and ∠*ECI*, then the points that poke upward will be more numerous but will be much closer to the *AEBCID* plane. As *n* increases, this model will become more and more flat.

Yaschenko's solution is, perhaps, not very satisfactory since it requires taking the limit $n \to \infty$ to make an actual flat model, and this will have an infinite number of creases.

In contrast, Lang (2011) offers a solution that shows one can, if one is a skilled enough origamist, make the perimeter of the folded model as large as one wishes. The idea is to use Diversion 8.4 as a starting point. That is, fold the base for an origami sea urchin. This will be similar to the bird and frog bases but will have many more flaps; it can have as many as you wish, in fact. We then do an origami maneuver called a **sink** to the sides of these flaps. A sink is where one inverts a point inside the model. In fact, one can sink a point in and out and in and out (called a **multiple sink**) as many times as one wishes, and the result will be a much thinner side of the flap. Sinks are **not easy** in origami! Many origami practitioners avoid models with multiple sinks, since they are difficult and can lead to ugly folds if not done with extra care.

Figure 8.11 shows the idea behind Lang's solution to the napkin folding problem. By starting with a flat sea urchin base with *n* flaps, we sink the sides of all the flaps and then splay them out so that the thin flaps form a star shape with many arms.

Recall from Section 8.3 that each of these flaps requires a disc of paper to form. If we assume for convenience that our number of points is a perfect square, say $n^2$, then the circle packing that forms this sea urchin base will be an $n \times n$ grid of packed circles, and each of these circles will have radius $r = 1/(2(n-1))$. After we make the sinks and splay the flaps, the length of our flaps will be approximately this radius $r$. We need to say "approximately" because some length will be lost in the center of the model, where the flaps are splayed. But as *n* increases, or as we make the flaps thinner and thinner with more sinks, this amount of loss will decrease relative to the length

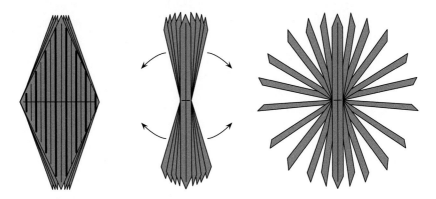

**Figure 8.11** Lang's solution: Flatten an origami sea urchin. (Illustration modified from (Lang, 2011).)

of the flaps. In any case, we will have that the perimeter of our new squished, splayed sea urchin will be twice the length of each point times the number of points, which is

$$2 \times \frac{1}{2(n-1)} \times n^2 = \frac{n^2}{n-1}.$$

This goes to infinity as $n$ increases.

Both Lang's and Yaschenko's solutions may be unsatisfying to the reader. But one thing is clear: it **is** possible to fold a piece of paper into a flat object with increased perimeter. For example, one could take a very basic case of Lang's approach, like the bird base (a packing of four circles) and with the proper sinks and splaying of flaps increase the perimeter by a small amount. Even this approach, however, is difficult for non-experts to actually fold. It remains to be seen if an easier-to-fold flat origami model can be made that increases the perimeter.

---

**Diversion 8.7**    Find another solution to the napkin folding/rumpled dollar problem. Just increasing the perimeter by a small amount is fine.

---

## 8.5    The Fold-and-Cut Problem

The famous magician Harry Houdini is known to have performed the following trick: Take a piece of paper, fold it up in a certain way, and then with one cut of a scissors reveal the unfolded paper to have been transformed into the silhouette of a bird, a duck, or any one of many shapes. Instructions for cutting a regular 5-pointed star can be found in his 1922 book *Paper Magic* (Houdini, 1922, pp. 176–177). Predating Houdini, however, is the legend of Betsy Ross and how she impressed the politicians designing the new American flag by cutting a 5-pointed star from fabric (Wilcox, 1873). Predating all of these is the 1721 Japanese book *Wakoku Chiyekurabe* (Mathematical Contests) by Kan Chu Sen, which shows how to make an outline of a Japanese crest called a **sangai bishi** by folding and then making one straight cut (Sen, 1721). An illustration of the Kan Chu Sen fold can be seen in Figure 8.12.

Can any polygonal shape be made by folding and then making one straight cut? More specifically, Given any polygon $P$ drawn on a sheet of paper $R$, we want to find a way to fold the paper into a flat object $f(R)$ such that all the sides of $P$ lie along a single line in $f(R)$ **and only** the sides of $P$ lie along this line. That way, if we cut along this line of $f(R)$, then we will be cutting $P$ from the paper and no other parts of the paper will be cut. This is what we call the **fold-and-cut problem** in the case of polygons.

Interestingly, we have already seen the solution to this problem, at least in the case of convex polygons. Lang's universal molecule from Section 8.4 will fold any convex polygon into a flat object so that the sides of the polygon all fold along an exclusive, straight line. Furthermore, Bowers and Streinu (2016) have generalized Lang's

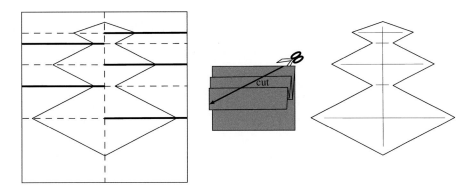

**Figure 8.12** Making a **sangai bishi** by folding and one straight cut (Sen, 1721).

universal molecule to non-convex polygons, and this generalization will solve the fold-and-cut problem in this case.

Least that appear anticlimactic, the full, universal version of the fold-and-cut problem is much stronger.

**Theorem 8.6** (Universal fold-and-cut) *Given any plane graph embedded with straight-line edges on a (suitably large) sheet of paper, we can always find a way to fold the paper that places the edges of the graph along a line of the folded paper so that the graph, and only the graph, may be cut from the paper with one straight cut.*

That is, we may start with an arbitrary plane graph drawing (with straight-line edges) $G$ instead of a polygon $P$. The full proof of this is much harder than a simple application of the universal molecule (Demaine et al., 1999; Bern et al., 2002). In fact, there are two solutions, one that uses the concept of a straight skeleton and another that uses disk (or circle) packing. We will describe only a sketch of the proof of the disk-packing version; full details of both proofs, with corrections of errors in the original papers, can be found in (Demaine and O'Rourke, 2007).

The gist of the disk-packing proof is shown in Figure 8.13. We start with our planar graph $P$ together with the boundary $R$ of our paper and consider this to be a single graph $P \cup R$. We then start forming our disk packing as follows:

(1) For each vertex $v$ of $P \cup R$, draw a circle centered at $v$ with radius equal to one-half the shortest distance between $v$ and a point of any edge $e$ of the graph that is not adjacent to $v$. (For example, this could just be half the distance between $v$ and its nearest neighboring vertex, but not if there are edge points closer to $v$.)

(2) Now consider the set of edge segments $E$ of our planar drawing $P \cup R$ that are not covered by disks. Call such an edge $e$ **crowded** if the disk $D_e$ whose diameter is $e$ (called the **diameter disk of** $e$) intersects the interior of a previous disk or the diameter disk $D_u$ of another edge $u$ of $E$.

(3) For all non-crowded edges $e$ of $E$, we add the disk whose diameter is $e$ to our packing.

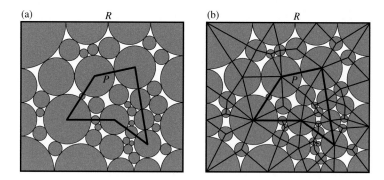

**Figure 8.13** A sketch of the disk-packing solution to the fold-and-cut problem (Bern et al., 2002).

(4) Then for all crowded edges $e$, we divide $e$ into two congruent edges by adding a vertex to its midpoint.

(5) We repeat this process of adding diameter disks until no more crowded edges exist.

(6) Once all of $P \cup R$ has been covered by disks, the regions of the paper between the disks will be filled by computing the Voronoi diagram of the centers of our current set of disks, repeatedly placing a maximal-radius disk at a Voronoi vertex and then updating the Voronoi diagram.

(7) Continue adding disks until all the gaps between the disks are surrounded by three or four disks. (These are called **3-gaps** and **4-gaps**.)

Figure 8.13(a) shows a sample result of this process. Once we have our disk packing, we connect the vertices of the disks that are tangent to each other, as shown in Figure 8.13(b); call the graph made by this set of line segments $C$. Then by our construction we have $P \cup R \subset C$, and all the interior faces of $C$ will be either triangles or quadrilaterals since our disk packing had only 3- and 4-gaps. For every triangle face of $C$ we add rabbit-ear molecule creases, and for each quadrilateral we apply Lang's universal molecule.

The resulting crease pattern will, when folded flat, collapse all the edges of $C$ onto a straight line, including those of $P$. In order to separate the edges of $P$ from the rest of $C$, we need to go back to before we started our disk packing and **thicken** the edges of $P$ by some $\varepsilon > 0$. Then, after the algorithm is run, the thickness $\varepsilon$ will offset the graph $P$ from the other edges in $C$ and create a line that only contains the edges of $P$, which will be our cut line. The details of this offset, as well as proving that the collective molecules result in a globally flat-foldable crease pattern, are omitted; again, see (Demaine and O'Rourke, 2007).

In case it is not apparent, the full proof of the Universal Fold-and-cut Theorem is very difficult. In fact, the original papers that proposed proofs of this Theorem, (Demaine et al., 1999; Bern et al., 2002), contained errors that eluded the peer-review process and many other people. These errors were subsequently repaired in (Demaine and O'Rourke, 2007).

The flat folding topics covered in this Chapter are quite varied. Some encompass challenging problems in area, perimeter, and disk-packing from geometry. Others force us to think in new ways about folding, from the constraints imposed by folding physical paper to grappling with what shapes can be folded. Exploring these different paths requires us to use a variety of mathematical tools.

The next Part of this book will expand the toolkit that origami requires by investigating how aspects of flat origami can be modeled using areas of mathematics that, at first glance, seem to have nothing to do with origami: algebra, analysis, and topology.

# Part III

# Algebra, Topology, and Analysis in Origami

Flat origami can be studied in many different ways. Taking a geometric and combinatorial approach seems the most natural way to begin such a study, and that was the topic of Part II. But in mathematics it is often amazing how many different branches can be meaningfully applied to a given topic. Origami is no exception, and flat origami in particular has been investigated independently by different people from around the world using very different tools. Sometimes these investigations have resulted in multiple discoveries of very similar aspects of flat origami, and other times very different results have been uncovered.

Part III of this book describes some of these different approaches to flat origami. Chapter 9 looks at the work of Kawasaki and Yoshida (1988), where group theory is used to relate the symmetries of a flat origami crease pattern to the symmetries of its folded image. Chapter 10 considers the work of Stewart A. Robertson (1977–1978), who considered flat foldings, which he termed **isometric foldings**, of Riemannian manifolds in arbitrary dimension. Robertson's work, in fact, includes a proof of one direction of Kawasaki's Theorem, predating Justin's and Kawasaki's independent discovery of the theorem by some five to seven years. In Chapter 11 we examine an approach to flat origami from analysis, where Dacorogna, Marcellini, and Paolini discovered that isometric foldings of $\mathbb{R}^n$ offer solutions to certain partial differential equations (Dacorogna et al., 2008).

Are there other ways in which flat origami can be modeled mathematically? Quite possibly, yes. But this part of the book covers three different approaches that have been fruitful. Furthermore, some of these approaches, especially those of Robertson and of Kawasaki and Yoshida, have not been included in more recent origami research explorations. There could easily be more work to be done in these areas.

# 9 Origami Homomorphisms

In 1988 Kawasaki and Yoshida published a groundbreaking paper (Kawasaki and Yoshida, 1988), titled "Crystallographic flat origamis," on a way to describe the symmetry of flat origami models using group theory and the folding map $\sigma_f$ introduced in Section 6.2. They used their elegant method to prove that a flat origami model folded from the infinite plane $\mathbb{R}^2$ whose crease pattern has wallpaper group symmetry must have the same symmetry in its folded form. Despite the novel ideas presented in this paper, their work has received very little attention in the origami mathematics literature.

In this chapter we present Kawasaki and Yoshida's results, following the exposition of their paper while providing more details. However, their work assumes that the region of paper $R$ is all of $\mathbb{R}^2$. We will extend their work to finite paper.

## 9.1 Symmetry Groups of Flat Origami

Let $\text{Isom}(\mathbb{R}^2)$ denote the group of isometries of the plane. We will use notation from Chapters 5 and 6, such as $R(l) \in \text{Isom}(\mathbb{R}^2)$ denoting the reflection of the plane about a line $l$. Now let $C$ be a flat origami crease pattern on a region $R$, and pick two faces, $f$ and $f'$, of $C$. Let $\gamma$ be a curve in $R$ starting from an interior point of $f$ and ending in an interior point of $f'$ that avoids the vertices of $C$, is not tangent to any crease, and crosses crease lines $l_1, \ldots, l_k$ in order. We define the transformation $[f, f'] \in \text{Isom}(\mathbb{R}^2)$ as

$$[f, f'] = R(l_1)R(l_2) \cdots R(l_k).$$

We know that $[f, f']$ is independent of the choice of $\gamma$ by Theorem 6.6. We can interpret $[f, f']$ as the isometry of the plane that would map the face $f'$ to its image when flat-folding the crease pattern while holding the face $f$ fixed. For this reason we refer to the transformation $[f, f']$ as the **folding of $f'$ toward $f$**.

---

**Diversion 9.1**  Prove that for any three faces $f, f', f''$ of $G$ we have

$$[f, f'][f', f''] = [f, f''].$$

---

Note that by our previous definition of the folding map $\sigma_f$ from Chapter 6, we have that $\sigma_f(x) = [f,f'](x)$ for $x \in f'$.

Let $\Gamma(C) \leq \text{Isom}(\mathbb{R}^2)$ be the symmetry group of our crease pattern $C$ (i.e., the subgroup of isometries that leave $C$ invariant). In general, let $\Gamma(S) \leq \text{Isom}(\mathbb{R}^2)$ be the group of symmetries of $S$.

**Lemma 9.1**  *Let $f, f'$ be faces of a flat origami crease pattern $G$ on a region $R$ and let $g \in \Gamma(C)$. Then*

$$[gf, gf'] = g[f,f']g^{-1}.$$

*Proof*  Let $l_1, \ldots, l_k$ be the creases that a vertex-avoiding curve $\gamma$ in $R$ crosses to get from $f$ to $f'$. Let $gl_i$ represent the image of the crease $l_i$ under the transformation $g \in \Gamma(C)$. Since $g$ leaves the crease pattern invariant, so does $g^{-1}$, and we have that $R(gl_i) = gR(l_i)g^{-1}$ for all crease lines $l_i$. Therefore,

$$[gf, gf'] = R(gl_1)R(gl_2) \cdots R(gl_k) = gR(l_1)g^{-1}gR(l_2)g^{-1} \cdots gR(l_k)g^{-1}$$
$$= gR(l_1)R(l_2) \cdots R(l_k)g^{-1} = g[f,f']g^{-1}. \qquad \square$$

Let us describe the meaning of Lemma 9.1. It states that if $g$ is a symmetry of our crease pattern and $f, f'$ are two faces, then the folding of the face $gf'$ toward the face $gf$ is the same transformation that we would get if we first performed the transformation $g^{-1}$, then folded $f'$ toward $f$, and then performed $g$. This makes perfect sense; since the crease pattern is invariant under $g$, the action of folding $f'$ toward $f$ should be similar to the action of folding $gf'$ toward $gf$, just offset by the symmetry $g$.

**Definition 9.2**  For a fixed face $f \in C$, define a mapping $\varphi_f : \Gamma(C) \to \text{Isom}(\mathbb{R}^2)$ by

$$\varphi_f(g) = [f, gf]g \quad \text{for all } g \in \Gamma(C).$$

It is difficult to initially see what the map $\varphi_f$ does, or why we are bothering with it, so first we prove some facts about $\varphi_f$.

**Theorem 9.3**  *The map $\varphi_f$ is a homomorphism.*

*Proof*  We need to prove that $\varphi_f(gh) = \varphi_f(g)\varphi_f(h)$ for all $g, h \in \Gamma(G)$. To that end, and using Diversion 9.1 and Lemma 9.1, we obtain

$$\varphi_f(gh) = [f, ghf]gh = [f, gf][gf, ghf]gh$$
$$= [f, gf]g[f, hf]g^{-1}gh$$
$$= [f, gf]g[f, hf]h = \varphi_f(g)\varphi_f(h). \qquad \square$$

We refer to $\varphi_f$ as the **origami homomorphism**. By the first isomorphism theorem we immediately obtain the following:

**Corollary 9.4**  *The image $\varphi_f(\Gamma(C))$ is a subgroup of $\text{Isom}(\mathbb{R}^2)$.*

---

**Diversion 9.2** Prove that if $f_1$ and $f_2$ are two faces of $C$, then $\varphi_{f_1}(\Gamma(C)) \cong \varphi_{f_2}(\Gamma(C))$. (Therefore, up to isomorphism $\varphi_f(\Gamma(C))$ is not dependent on the choice of the face $f$.)

---

The idea here is that $\varphi_f(\Gamma(C))$ describes, mostly, the symmetries of the folded model $\sigma_f(C)$. We will prove this formally and then consider examples that explain the qualifier "mostly."

**Theorem 9.5** *Let $C$ be a flat-foldable origami crease pattern and $g \in \Gamma(C)$ be a symmetry of $C$. Then the following diagram commutes:*

$$
\begin{array}{ccc}
C & \xrightarrow{\ g\ } & C \\
\sigma_f \downarrow & & \downarrow \sigma_f \\
\sigma_f(C) & \xrightarrow{\ \varphi_f(g)\ } & \sigma_f(C)
\end{array}
$$

*In other words, $\varphi_f(g) \circ \sigma_f = \sigma_f \circ g$.*

*Proof* Let $f'$ be a face of $C$ and $x \in f'$ be any point. Then

$$\varphi_f(g)(\sigma_f(x)) = [f, gf]g[f, f'](x) = [f, gf][gf, gf']g(x)$$
$$= [f, gf']g(x) = \sigma_f(g(x)),$$

where we used, from Lemma 9.1, that $g[f, f'] = [gf, gf']g$. $\qquad\square$

**Theorem 9.6** *Let $g \in \Gamma(C)$ be a symmetry of the crease pattern $C$. Then $\varphi_f(g)$ is a symmetry of the flat-folded image $\sigma_f(C)$. (So $\varphi_f(g) \in \Gamma(\sigma_f(C))$.)*

*Proof* If we let $g \in \Gamma(C)$, then $C$ is invariant under the action of $g$, so $g(C) = C$. Therefore, by Theorem 9.5 we have

$$\sigma_f(C) = \sigma_f(g(C)) = \varphi_f(g)(\sigma_f(C)).$$

Thus $\sigma_f(C)$ is invariant under $\varphi_f(g)$, as desired. $\qquad\square$

Theorem 9.6 proves that, as groups, we have $\varphi_f(\Gamma(C)) \leq \Gamma(\sigma_f(C))$. Indeed, we may think of $\varphi_f$ as a homomorphism from $\Gamma(C)$ to $\Gamma(\sigma_f(C))$. The only thing keeping $\varphi_f(\Gamma(C))$ from being isomorphic to $\Gamma(\sigma_f(C))$ is the possibility that $\sigma_f(C)$ might have symmetries that are not in the image of $\varphi_f$. Can that happen? Yes, as the next example shows.

---

**Example 9.7** Consider the crease pattern $C$ in Figure 9.1, where our piece of paper is a right isosceles triangle and there is only one crease dividing the triangle into two congruent faces $f_1$ and $f_2$. Labeling the vertices as in the figure, we see that the folding map $\sigma_{f_1}$ maps $C$ to a smaller isosceles right triangle that has a vertical line of symmetry; call this symmetry transformation $h$. Then $h \in \Gamma(\sigma_{f_1}(C))$.

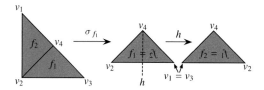

**Figure 9.1** The crease pattern $C$ from Example 9.7 on a region $R$ that is a right isosceles triangle, and its image under a folding map $\sigma_{f_1}$ then reflected under $h$, a symmetry that is not shared with $C$.

Let us compute $\varphi_{f_1}(\Gamma(C))$. We have that $\Gamma(C) = \{id, R_d\}$, where $id$ is the identity and $R_d$ is the reflection along the edge $\{v_2, v_4\}$ in the crease pattern. Using $\varphi_{f_1}(g) = [f_1, gf_1]g$, we find that $\varphi_{f_1}(id) = id$ and

$$\varphi_{f_1}(R_d) = [f_1, R_d(f_1)]R_d = R_d R_d = id.$$

Thus $\varphi_{f_1}(\Gamma(C)) = \{id\}$, which is a proper subgroup of $\Gamma(\sigma_{f_1}(C)) = \{id, h\}$. Note that this works as an example showing how $\varphi_f(\Gamma(C))$ can be a proper subgroup of $\Gamma(\sigma_f(C))$ only because the faces $f_1$ and $f_2$ of the crease pattern are congruent and have symmetries that are not shared with $C$.

## 9.2    Examples of Origami Homomorphisms

Aside from any anomalies as that seen in Example 9.7, Theorem 9.6 gives us a way to compute the symmetry group of the folded model $\sigma_f(C)$ from the symmetry group of the crease pattern $C$. In this section we will see how such a computation works for examples where the region of paper $R$ has finite area. In Section 9.3 we will examine cases where the paper is infinite.

**Example 9.8**    First, consider the crease pattern shown in Figure 9.2, where the paper is a square and the only creases are those made by the two diagonals of the square. This is a useful example to compare with Example 9.7, since it looks like the symmetries of $\sigma_f(C)$ are shared with those of $C$. Let us denote $R_\theta$ to be a rotation transformation by angle $\theta$, $R_h$ to be the reflection about a horizontal axis, $R_v$ to be the reflection about

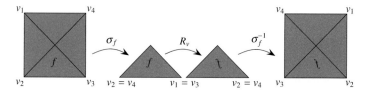

**Figure 9.2** The square-with-two-diagonals crease pattern, whose folded image has a vertical axis of reflective symmetry.

a vertical axis, and $R_{d1}$ (resp. $R_{d2}$) to be the diagonal reflection along the axis from the upper left to lower right corners (resp. upper right to lower left).

Then, letting *id* represent the identity, the symmetry group for the crease pattern is

$$\Gamma(C) = \{id, R_{90}, R_{180}, R_{270}, R_h, R_v, R_{d1}, R_{d2}\},$$

which is the dihedral group $D_4$. Theorem 9.5 tells us that to find $\Gamma(\sigma_f(C))$ we should look at $\varphi_f(\Gamma(C))$. To that end,

$$\varphi_f(id) = [f, id(f)]id = id,$$
$$\varphi_f(R_{90}) = [f, R_{90}(f)]R_{90} = R_{d1}R_{90} = R_v,$$
$$\varphi_f(R_{180}) = [f, R_{180}(f)]R_{180} = R_{d1}R_{d2}R_{180} = R_{180}R_{180} = id,$$
$$\varphi_f(R_{270}) = [f, R_{270}(f)]R_{270} = R_{d2}R_{270} = R_v,$$
$$\varphi_f(R_h) = [f, R_h(f)]R_h = R_{d2}R_{d1}R_h = R_{180}R_h = R_v,$$
$$\varphi_f(R_v) = [f, R_v(f)]R_v = R_vR_v = id,$$
$$\varphi_f(R_{d1}) = [f, R_{d1}(f)]R_{d1} = R_{d1}R_{d1} = id, \quad \text{and}$$
$$\varphi_f(R_{d2}) = [f, R_{d2}(f)]R_{d2} = R_{d2}R_{d2} = id.$$

Thus $\varphi_f(\Gamma(C)) = \{id, R_v\}$, which corresponds to the symmetry group of $\sigma_f(C)$ that we see in Figure 9.2.

**Example 9.9** (The flapping bird)  We next look at a traditional origami model, the flapping bird, which we saw at the beginning of Chapter 5. We pick a face $f$ as shown in Figure 9.3 and consider the map $\sigma_f$. Here we have $\Gamma(C) = \{id, R_{d2}\}$, using the notation in the previous example. Then we check that

$$\varphi_f(id) = id \quad \text{and} \quad \varphi_f(R_{d2}) = [f, R_{d2}(f)]R_{d2} = R_{d2}R_{d2} = id.$$

Therefore, $\varphi_f(\Gamma(C))$ is the trivial group, and indeed, the folded flapping bird $\sigma_f(C)$ has no symmetries other than the identity. (Note that there are no symmetries of $\sigma_f(C)$ because is it a flat object in the plane. If we viewed the folded flapping bird as a 3-dimensional object, as Figure 9.3 somewhat suggests, then there would be a plane of symmetry between the folded layers of the paper. But we are not considering such symmetries in flat origami.)

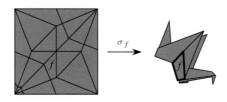

**Figure 9.3** The traditional flapping bird (crane), viewed as a folding map $\sigma_f$.

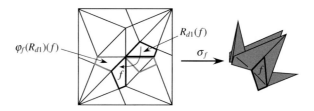

**Figure 9.4** The headless crane, with the computation of $\varphi_f(R_{d1})$ illustrated.

---

**Example 9.10** (The headless crane)   Now consider the headless crane, which is just the flapping bird from Example 9.9 without the head folded. Since the folds for the head are no longer there, the headless crane crease pattern (see Figure 9.4) has more symmetry: $\Gamma(C) = \{id, R_{180}, R_{d1}, R_{d2}\}$, which is isomorphic to the Klein four-group, $\mathbb{Z}_2 \times \mathbb{Z}_2$.

The computation of $\varphi_f(R_{d1})$ is illustrated in Figure 9.4, where the movement of the face $f$ is shown first under $R_{d1}$ and then under $[f, R_{d1}(f)]$. The product of these transformations gives us $R_{d2}$, and since these same transformations would be used to follow the movement of any other face, we have that $\varphi_f(R_{d1}) = R_{d2}$.

---

**Diversion 9.3**   Prove that in the headless crane example we have $\varphi_f(R_{180}) = R_{d2}$ and $\varphi_f(R_{d2}) = id$.

---

Thus, we have that $\varphi_f(\Gamma(C)) = \{id, R_{d2}\}$. This corresponds to the symmetries of the folded headless crane, which has an axis of reflective symmetry due to the lack of a head.

---

## 9.3  Applications to Origami Tessellations

Origami tessellations are a genre of origami models where the crease pattern is a tiling of the plane. In theory, such crease patterns can be made on the whole plane, so our region is $R = \mathbb{R}^2$. The art of origami tessellations was pioneered by Shuzo Fujimoto in the 1970s (Fujimoto and Nishikawa, 1982), but they were also independently discovered by numerous artists and designers, like Ron Resch in the 1960s and German Bauhaus students in the 1920s. A more recent instruction book and introduction to the field of origami tessellations can be found in (Gjerde, 2008).

---

**Example 9.11** (Square twist tessellation)   The classic square twist, which we saw in Section 5.1(Figure 5.4), has 4-fold rotational symmetry and thus may be tiled in a

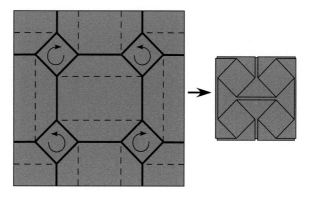

**Figure 9.5** The square twist tessellation. To make this, first crease a square sheet into an 8 × 8 grid, and then make the mountain (bold) and valley (dashed) lines as shown. All the creases need to be folded simultaneously to make the model fold flat completely.

square grid pattern to make the square twist tessellation, as we saw in Section 7.1 (Example 7.2). A 2 × 2 section of this model is shown in Figure 9.5.

The symmetry group of the infinite square twist crease pattern is the wallpaper group (or crystallographic group) *p4gm*, which has as generators two perpendicular lines of reflection, a center of 180° rotation at the intersection of the reflection lines, two centers of 90° rotation that are mirrors of each other, and two perpendicular translations that are at 45° angles to the reflection axes. These generators are illustrated in Figure 9.6(a), where we have labeled the translations $t_1, t_2$. Let us further denote the two reflections by $R_h$, $R_v$, the 90° rotations about the points labeled $a$ and $b$ by $R_a$, $R_b$, and the 180° rotation (about the rhombus in the figure) by $R_{180}$.

If we let $f$ be the face labeled in Figure 9.6, then notice that $\varphi_f(R_v) = R_v$ and $\varphi_f(R_h) = R_h$ because these symmetries leave $f$ fixed (but flipped). Similarly, $\varphi_f(R_{180}) = R_{180}$.

Figure 9.6(b) illustrates how we can compute $\varphi_f(R_a) = [f, R_a(f)]R_a$, which we see is a new 90° rotation about the point $a'$ indicated in the figure. Thus, $\varphi_f(R_a) = R_{a'}$ and similarly $\varphi_f(R_b) = R_{b'}$, where $b'$ is a new point of rotation like $a'$.

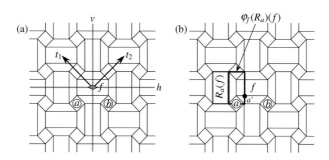

**Figure 9.6** (a) The infinite square twist tessellation crease pattern with its symmetry group generators labeled. (b) Computing $\varphi_f(R_a)$.

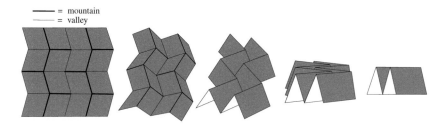

**Figure 9.7** The Miura map fold, or Miura-ori.

---

**Diversion 9.4**   Prove that $\varphi_f(t_1)$ and $\varphi_f(t_2)$ are translations parallel to, but shorter than, $t_1$ and $t_2$, respectively.

---

Therefore, we have that all the generators of $\Gamma(C)$ map under $\varphi_f$ to similar symmetries in $\Gamma(\sigma_f(C))$, and thus we must have $\Gamma(C) \cong \Gamma(\sigma_f(C))$. This can be visually confirmed by looking at the folded result of the square twist tessellation (Figure 9.5).

---

Kawasaki and Yoshida discovered that what we see happening in the square twist tessellation will always happen; if our crease pattern $C$ has wallpaper group symmetry and so does $\sigma_f(C)$, then their symmetry groups will be isomorphic.

**Theorem 9.12**   *Let $C$ be a flat origami crease pattern whose symmetry group $\Gamma(C)$ is a wallpaper group. If $\varphi_f(\Gamma(C))$ is also a wallpaper group, then $\varphi_f(\Gamma(C)) \cong \Gamma(C)$.*

*Proof*   Since $\varphi_f$ is a homomorphism, its image is isomorphic to a subgroup $G$ of $\Gamma(C)$. If $G \neq \Gamma(C)$, then $\ker(\varphi_f) \neq \emptyset$. But since $\varphi_f(\Gamma(C))$ is a wallpaper group, it must contain two linearly independent translations, as does $\Gamma(C)$. Therefore, $\ker(\varphi_f)$ does not contain any translations; it must contain only rotations and reflections. Thus $\ker(\varphi_f)$ is a finite group. But by the first isomorphism theorem, $\ker(\varphi_f)$ is a normal subgroup of $\Gamma(C)$, and wallpaper groups have no finite normal subgroups (since a finite subgroup $H$ must have no translations, and $tH$ will not equal $Ht$ for any nontrivial translation $t$). Thus we have a contradiction, and we conclude that $\varphi_f(\Gamma(C)) \cong \Gamma(C)$.
□

Not all origami tessellations meet the criteria of Theorem 9.12. One example is an infinite crease pattern $C$ that is just a grid of squares. This can, in theory, be folded into a one-square pile (of infinite thickness, but as a map from $\mathbb{R}^2$ to $\mathbb{R}^2$, we ignore this). A square does not have the same symmetry group as an infinite grid of squares in the plane, and thus $\Gamma(C)$ is not isomorphic to $\varphi_f(\Gamma(C))$. The reason this does not violate Theorem 9.12 is, of course, that the folded image $\sigma_f(C)$ does not have wallpaper group symmetry. That not example is the Miura map fold, which we also saw in Chapter 7.

**Example 9.13** (The Miura-ori)   The Miura map fold, described in Section 7.2, is an origami tessellation, since the crease pattern is made of congruent parallelograms, but if we fold it from the infinite plane $\mathbb{R}^2$, then it will fold into the infinite strip $\mathbb{R} \times [0, 1]$ (if the height of the parallelograms is 1).

**Diversion 9.5**   Prove that if $C$ is the crease pattern for the Miura-ori, then $\varphi_f(\Gamma(C))$ will be a frieze group.

## 9.4   Open Problems

As previously mentioned, there has been very little research done on origami homomorphisms aside from (Kawasaki and Yoshida, 1988) and what is presented in this book. One exception is (De las Peñas et al., 2015), which looks at how the symmetries of an origami tessellation (with wallpaper group symmetry) will be affected when its MV assignment is also considered. They use the symmetry group $G$ of the crease pattern to describe MV assignments that will have symmetries that are some (nontrivial) subgroup of $G$.

Still, this area seems ripe for more exploration. A few suggestions are offered here.

**Diversion 9.6**   Use one of the examples shown in this chapter to show that $\varphi_f(\Gamma(C))$ is not always a normal subgroup of $\Gamma(C)$.

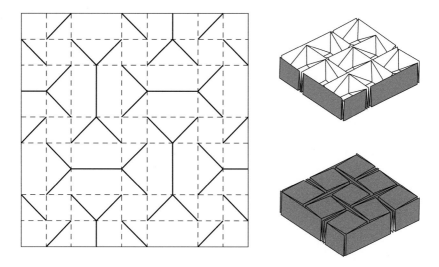

**Figure 9.8** A 3-dimensional origami tessellation by Kawasaki (1987).

**Open Problem 9.1**    Under what conditions will $\varphi_f(\Gamma(C))$ be a normal subgroup of $\Gamma(C)$?

**Open Problem 9.2**    There are many 3-dimensional origami tessellations, such as the one by Toskikazu Kawasaki shown in Figure 9.8 that folds a square into a 3-dimensional grid of cubes. Can the origami homomorphism results of this chapter be extended to such 3-dimensional origami tessellations, or to 3-dimensional origami models in general?

# 10 Folding Manifolds

It is rather surprising that many of the basic flat-folding results seen in Chapter 5 also hold when folding arbitrary 2-manifolds. Doing this requires reformulating our definition of folding, of course. In order to fold a manifold flat, we will need to fold along geodesic curves. Also surprising is that learning how to fold manifolds, in particular manifolds without boundary, gives us a natural way to obtain results on "folding" spaces like $\mathbb{R}^3$.

Most of the results in this chapter are from a groundbreaking paper by Stewart A. Robertson from 1977 (Robertson, 1977–1978) that actually proves the necessary direction of Kawasaki's Theorem nearly a decade before Justin and Kawasaki conceived of it (as far as we know). We will also include a result from Jim Lawrence and Jonathan Spingarn (Lawrence and Spingarn, 1989), called the **Angle Sum Theorem**, that provides a further generalization of the necessary direction of Kawasaki's Theorem for foldings from $\mathbb{R}^n$ to $\mathbb{R}^n$. We then extend the idea of folding manifolds to include generalized mountain-valley assignments, obtaining yet another generalization of Maekawa's Theorem as well as a necessary and sufficient generalization of Kawasaki's Theorem.

## 10.1 Isometric Foldings

We follow the treatment given by (Robertson, 1977–1978) and mostly use differential topology notation as in (Hirsch, 1976) and (Milnor, 1965). Let $\mathcal{M}$ be a smooth (that is, $C^\infty$) Riemannian manifold of dimension $m$; $\mathcal{M}$ is locally homeomorphic to a patch of $\mathbb{R}^m$, otherwise known as an $m$-**manifold**. We want to describe folding this manifold in a way that mimics the folding of paper. That is, we don't want any stretching or cutting to occur. We can do this by specifying what is preserved by such a folding; geodesics should be preserved, but only in a piecewise manner.

This motivates some definitions. Let $I = [a, b]$ be a closed interval. Then we say that a continuous curve $\gamma : I \to \mathcal{M}$ is a **zig-zag** on $\mathcal{M}$ if, for some partition $a = a_0 < a_1 < \cdots < a_k = b$ of $I$, we have that $\gamma$ restricted to $[a_i, a_{i+1}]$ is a geodesic on $\mathcal{M}$, parameterized with respect to arc-length, for each $i = 0, \ldots, k - 1$. We want $\gamma$ to be parameterized by arc-length so that the length of the zig-zag will be $L(\gamma) = b - a$.

Now let $\mathcal{M}$ and $\mathcal{N}$ be an $m$-manifold and an $n$-manifold, respectively.

**Definition 10.1**   A map $f: \mathcal{M} \to \mathcal{N}$ is called an **isometric folding** of $\mathcal{M}$ onto $\mathcal{N}$ if for every zig-zag $\gamma: I \to \mathcal{M}$, the path $\gamma^* = f \circ \gamma: I \to \mathcal{N}$ is also a zig-zag on $\mathcal{N}$.

We aim to argue in the remainder of this section that this definition of isometric folding fits what we would like "folding a manifold" to mean. Note that because of the way $\gamma$ is parameterized, we have $L(\gamma) = L(\gamma^*) = b - a$.

Now let $d_\mathcal{M}$ and $d_\mathcal{N}$ be the metrics on the manifolds $\mathcal{M}$ and $\mathcal{N}$, respectively.

**Lemma 10.2**   *Let $f: \mathcal{M} \to \mathcal{N}$ be an isometric folding. Then for all $x, y \in \mathcal{M}$, we have*

$$d_\mathcal{M}(x, y) \geq d_\mathcal{N}(f(x), f(y)).$$

*Proof*   Let $x, y \in \mathcal{M}$ and let $\gamma$ be a geodesic curve on $\mathcal{M}$ from $x$ to $y$ parameterized with respect to arc-length. Then $\gamma$ is a zig-zag on $\mathcal{M}$ and we have

$$d_\mathcal{M}(x, y) \geq L(\gamma) = L(\gamma^*) \geq d_\mathcal{N}(f(x), f(y)),$$

where we used the fact that since $\gamma^*$ is a zig-zag on $\mathcal{N}$, its total length $L(\gamma^*)$ cannot be less than $d_\mathcal{N}(f(x), f(y))$.  □

This lemma justifies the non-stretching nature of folding our manifolds. (In the literature, maps $f: \mathcal{M} \to \mathcal{N}$ with $d_\mathcal{M}(x, y) \geq d_\mathcal{N}(f(x), f(y))$ are called **nonexpansive maps**; see (Lawrence and Spingarn, 1989).) Lemma 10.2 also immediately proves another desired property of isometric foldings.

**Corollary 10.3**   *Any isometric folding is continuous.*

---

**Example 10.4**   The map $f: \mathbb{R} \to \mathbb{R}$ given by $f(x) = |x|$ is an isometric folding. It literally folds the real line in half.

Note that this example demonstrates that an isometric folding need not be differentiable.

---

**Definition 10.5**   Let $f: \mathcal{M} \to \mathcal{N}$ be an isometric folding. Then we let $\Sigma(f) \subset \mathcal{M}$ denote the set of all singularities (non-differentiable points) in the domain of $f$, and we call this the **crease pattern** of $f$.

In Example 10.4, we have $\Sigma(f) = \{0\}$.

---

**Example 10.6**   Let $f: \mathbb{R}^3 \to \mathbb{R}^3$ be given by $f(x, y, z) = (|x|, |y|, |z|)$. This isometric folding "folds" 3-space along the three coordinate planes into the first octant: $f(\mathbb{R}^3) = \{(x, y, z) : x, y, z \geq 0\}$. Here we have that the crease pattern is the union of the three coordinate planes, $\Sigma(f) = \{(x, y, z) : xyz = 0\}$.

---

**Example 10.7**   Let $M$ be the unit circle in $\mathbb{R}^2$ and $f: M \rightarrow M$ be given by $f(\cos\theta, \sin\theta) = (\cos\theta, |\sin\theta|)$ for all $0 \le \theta < 2\pi$. This "folds" the unit circle in half, and we have $\Sigma(f) = \{(-1,0),(1,0)\}$. This example also captures what happens when we fold a **unit disk** in half. However, in what follows we will want $M$ to be a compact manifold without boundary, and thus we do not want to let $M$ be a disk or any piece of paper with boundary. Instead, letting $M$ be the boundary of the disk will achieve a suitable model for folding a disk of paper. We will make this relationship formal in Section 10.2. In fact, the special case where $M$ is a $k$-dimensional sphere $S^k$ will be a very useful tool for determining the structure of $\Sigma(f)$.

Understanding the structure of the crease pattern $\Sigma(f)$ will be very important. We've already seen in the previous chapters in this book that crease patterns for isometric foldings of the form $f: \mathbb{R}^2 \rightarrow \mathbb{R}^2$ are plane graphs, begin composed of 1-dimensional edges and 0-dimensional vertices. We'll see how this generalizes to manifolds of arbitrary dimension.

**Definition 10.8**   We let $\mathcal{F}(M, \mathcal{N})$ denote the set of all isometric foldings of $M$ into $\mathcal{N}$.

## 10.2   The Local Structure of the Singular Set

From now on let us assume that the manifolds $M$ and $\mathcal{N}$ are complete. Using standard differential topology notation, we let $T(M)$ denote the set of tangent vectors of $M$ (i.e., the tangent bundle of $M$) and $T_x(M)$ the tangent space at a point $x \in M$. Similarly, let $S(M) = \{v \in T(M) : \|v\| = 1\}$ be the set of unit vectors tangent to $M$ and $S_x(M)$ the unit tangent vectors at the point $x \in M$. Then $S_x(M)$ is diffeomorphic to the unit sphere $S^{m-1}$ (where $m = \dim M$).

For each $v \in S_x(M)$ we can find a geodesic curve $\gamma_v : \mathbb{R} \rightarrow M$ with $\gamma_v(0) = x$ and $\gamma_v'(0) = v$; this is given by the exponential map $\gamma_v(t) = e^{tv}$, but we will not need to use that level of detail. Under an isometric folding $f$ this geodesic $\gamma_v$ (which we assume is parameterized with respect to arc-length) maps to a piecewise geodesic $\gamma_v^* = f \circ \gamma_v$ on $\mathcal{N}$. We let

$$\sigma_v = \{t \in \mathbb{R} : \gamma_v^* \text{ is not differentiable at } t\}.$$

Since $\gamma_v^*$ is a zig-zag, we know that $\sigma_v$ is a countable discrete subset of $\mathbb{R}$. Indeed, $\gamma_v^*$ itself is an isometric folding of $\mathbb{R}$ into $\mathcal{N}$ with $\Sigma(\gamma_v^*) = \sigma_v$.

We now establish a result showing that distances are not preserved under $f$ along a geodesic in the presence of a singularity.

**Lemma 10.9**   *Let $f \in \mathcal{F}(M, \mathcal{N})$ and let $U$ be a simple convex neighborhood of $x \in M$ such that $f(U)$ is contained in a simple convex neighborhood of $f(x)$ in $\mathcal{N}$. Let $v \in S_x(M)$ and $t \in \mathbb{R}$ be such that $y = \gamma_v(t) \in U$.*

*Then $d_{\mathcal{N}}(f(x),f(y)) < d_{\mathcal{M}}(x,y) = t$ if and only if there is a singularity of $\gamma_v^*$ on the open interval $(0,t)$.*

*Proof*  For one direction, by contrapositive suppose that there are no singularities of $\gamma_v^*$ in $(0,t)$. Then $\gamma_v^*$ is a geodesic curve from $f(x)$ to $f(y)$ in $f(U)$, and thus $d_{\mathcal{M}}(x,y) = L(\gamma_v) = L(\gamma_v^*) = d_{\mathcal{N}}(f(x),f(y))$.

For the other direction, suppose that there is a singularity $a \in \sigma_v$ of $\gamma_v^*$ on $(0,t)$. Then $d_{\mathcal{N}}(f(x),f(y)) < d_{\mathcal{N}}(f(x),\gamma_v^*(a)) + d_{\mathcal{N}}(\gamma_v^*(a),f(y)) \leq L(\gamma_v^*) = L(\gamma_v) = d_{\mathcal{M}}(x,y)$.    $\square$

We now define a tool, called the **radial function** of an isometric function, which helps us determine local regions of $\mathcal{M}$ where an isometric folding will be especially well-behaved. This will be particularly helpful when our domain manifold $\mathcal{M}$ is a sphere.

First, we note that for each $x \in \mathcal{M}$ there exists a largest number $\varepsilon_x > 0$ (possibly $\infty$) such that the open ball $B(x,\varepsilon_x) = \{z \in \mathcal{M} : d_{\mathcal{M}}(x,z) < \varepsilon_x\}$ in $\mathcal{M}$ and $B(f(x),\varepsilon_x)$ in $\mathcal{N}$ are simple convex neighborhoods of $x$ and $f(x)$, respectively. For example, if $\mathcal{M} = \mathcal{N} = \mathbb{R}$, then $\varepsilon_x = \infty$ for all $x \in \mathcal{M}$. Similarly if $\mathcal{M}$ and $\mathcal{N}$ are spheres, then $\varepsilon_x$ can be large enough to contain the whole sphere. However, if $\mathcal{M}$ is a torus, then $B(x,\varepsilon_x)$ won't be simple or (possibly) convex if $\varepsilon_x$ is too big. Thus, $\varepsilon_x$ has to do with the topology and homotopy of the manifolds and not with the isometric folding.

**Definition 10.10**    The **radial function** $r\colon S(\mathcal{M}) \to \mathbb{R}$ of $f$ is $r(v) = \min\{\varepsilon_x, 1, t\}$, where $v \in S_x(\mathcal{M})$ and $t$ is the smallest positive element in $\sigma_v$ (with $t = \infty$ if $\sigma_v = \emptyset$).

In other words, if $v \in S(\mathcal{M})$ is a unit vector tangent to $\mathcal{M}$ at the point $x$, then $r(v)$ tells us how far we can go along the geodesic $\gamma_v$ starting at $x$ before we either encounter a singularity of $f$, reach a distance of 1, or reach a point where $B(x,\varepsilon_x)$ or $B(f(x),\varepsilon_x)$ is no longer a simple convex neighborhood of $x$ or $f(x)$, respectively.

The radial function will be used to find a radius $\rho$ about points $x \in \mathcal{M}$ such that the boundary of the $(m-1)$-dimensional ball $B(x,\rho)$ in $\mathcal{M}$ will map under $f$ into the boundary of $B(f(x),\rho)$ in $\mathcal{N}$. For this to work, the important thing is for $B(x,\rho)$ to not contain any other singularities of $f$ except, perhaps, $x$ and any geodesics starting at $x$ that happen to be in $\Sigma(f)$.

However, it is not clear that our definition of isometric folding prevents anomalies to occur that would result in $r(v) = 0$ for some $v \in S_x(\mathcal{M})$ (which is not a good radius to have). For example, there could be a sequence of vertices $x_i \in \Sigma(f)$ with $\lim_{i \to \infty} x_i = x$, or we could have a set of crease lines that approach a limit.

---

**Example 10.11**    Let $\mathcal{M} = \mathcal{N} = \mathbb{R}^2$, and let $f\colon \mathcal{M} \to \mathcal{N}$ be the origami whose crease pattern is made of the lines $y = (1 - 1/2^n)x$ for $n \in \mathcal{N}$. Figure 10.1 shows the intersection of this crease pattern and the unit square $[0,1] \times [0,1]$.

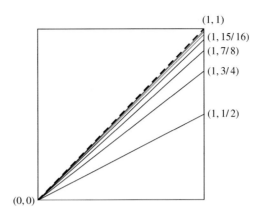

**Figure 10.1** A crease pattern whose creases, $y = (1 - 1/2^n)x$, approach a limit. Is this an isometric folding?

---

**Diversion 10.1** Prove that the function $f$ in Example 10.11 is not an isometric folding according to Definition 10.1. Then calculate what the radial function $r(v)$ is for any point $p$ on the line $y = x$ where we have the vector $v = \langle 0, -1 \rangle$.

---

Example 10.11 illustrates the need to argue why under an isometric folding we will have $r(x) > 0$ for all $x \in \mathcal{M}$, which we do in the next theorem.

**Theorem 10.12** *If $f \in \mathcal{F}(\mathcal{M}, \mathcal{N})$ and $x \in \mathcal{M}$, then $r(v) > 0$ for all $v \in S_x(\mathcal{M})$.*

*Proof* Suppose that there is an $x \in \mathcal{M}$ such that $r(v) = 0$ for some $v \in S_x(\mathcal{M})$. This means that the geodesic $\gamma_v \colon \mathbb{R} \to \mathcal{M}$ gives rise to a singular set $\sigma_v$ that has 0 as a limit point (since $\gamma_v(0) = x$).

The folding $f$ is isometric, thus for any zig-zag $\gamma \colon [a, b] \to \mathcal{M}$ we have that $\gamma^* = f \circ \gamma \colon [a, b] \to \mathcal{N}$ is also a zig-zag, which all means that $\gamma$ and $\gamma^*$ are piecewise geodesics with a finite number of singularities.

Looking at our $\gamma_v$, we can restrict its domain, say, to $\gamma_v \colon [-a, a] \to \mathcal{M}$ for some $a \in \mathbb{R}$. Our restricted $\gamma_v$ is a trivial zig-zag in that it's a geodesic curve on $\mathcal{M}$ through $x$, and $\gamma_v^*$ has an infinite number of singularities in $[-a, a]$ (since 0 is a limit point of $\sigma_v$); therefore, $\gamma_v^*$ is not a zig-zag, which contradicts $f$ being an isometric folding. $\square$

**Theorem 10.13** *The radial function of an isometric folding is continuous.*

*Proof* Let $f \in \mathcal{F}(\mathcal{M}, \mathcal{N})$ and $v \in S_x(\mathcal{M})$. Suppose $r(v) = t < 1$ where $t < \varepsilon_x$. Then $\gamma_v$ is not differentiable at $t$, and there exists a $\delta > 0$ such that $t < t + \delta < \min\{\varepsilon_x, 1\}$ and $t$ is the only singularity of $f$ in $(0, t + \delta)$. Let $y = \gamma_v(t)$ and $z = \gamma_v(t + \delta)$. Then by Lemma 10.9 we have $d_{\mathcal{M}}(x, z) > d_{\mathcal{N}}(f(x), f(z))$.

Getting back to $r$, let $W$ be an open neighborhood of $v \in S(\mathcal{M})$ and let $u \in W$, say $u \in S_p(\mathcal{M})$. Since $d_\mathcal{M}$ and $d_\mathcal{N}$ are continuous, we have that if $W$ is sufficiently small then $d_\mathcal{M}(p, q) > d_\mathcal{N}(f(p), f(q))$, where $q = \gamma_u(t + \delta)$. Lemma 10.9 then tells us that there is a singularity of $\gamma_u$ on $(0, t + \delta)$. That is, for all $u \in W$ we have $|r(v) - r(u)| \leq |t - (t + \delta)| = \delta$, and so $r$ is continuous.

The cases where $r(v) = 1$ and $r(v) = \varepsilon_x$ are easy because there will exist open neighborhoods $W$ of $v \in S(\mathcal{M})$ with $r(u) = 1$ (resp. $r(u) = \varepsilon_x$) for all $u \in W$, giving us $|r(v) - r(u)| = 0$. This completes the proof. □

Now then, since $r$ is continuous, it must attain a positive minimum value $\rho(x)$ on the closed sphere $S_x(\mathcal{M})$ for each $x \in \mathcal{M}$. We call this function $\rho : \mathcal{M} \to \mathbb{R}$ the **minimum radial function**. Further, since $r$ is continuous, $\rho$ is as well, and since $r(v) > 0$ for all $v \in S_x(\mathcal{M})$, we have that $\rho(x) > 0$.

If $x \in \mathcal{M}$ is not a singularity of $f$, then there will be no singular points of $f$ in $B(x, \rho(x))$. The following theorem will give us insight on what happens when we pick $x \in \Sigma(f)$. Recall that the notation $f|_U$ indicates the function $f$ restricted to the set $U$, which is hopefully a subset of the domain of $f$. We also let, for $x \in \mathcal{M}$, $S(x, r) = \{z \in \mathcal{M} : d_\mathcal{M}(x, z) = r\}$ denote the sphere centered at $x$ of radius $r$ in $\mathcal{M}$ (i.e., the boundary of $B(x, r)$).

**Theorem 10.14**   *Let $f \in \mathcal{F}(\mathcal{M}, \mathcal{N})$ and let $\rho : \mathcal{M} \to \mathbb{R}$ be the minimum radial function of $f$. Then for all $x \in \mathcal{M}$, $f|_{S(x, \rho(x))}$ is an isometric folding into $S(f(x), \rho(x))$.*

*Proof*   We first claim that $f$ maps $S(x, \rho(x))$ into $S(f(x), \rho(x))$: We know that $S(x, \rho(x))$ and $S(f(x), \rho(x))$ are smooth submanifolds of $\mathcal{M}$ and $\mathcal{N}$, diffeomorphic to $S^{m-1}$ and $S^{n-1}$, respectively. Let $v \in S_x(\mathcal{M})$. By definition of $\rho(x)$, we have that $\gamma_v$ is differentiable on $[0, \rho(x)]$. Thus, by Lemma 10.9, for all $v \in S_x(\mathcal{M})$ we have $d_\mathcal{M}(\gamma_v(0), \gamma_v(\rho(x))) = d_\mathcal{N}(\gamma_v^*(0), \gamma_v^*(\rho(x)))$, and thus $f$ maps $S(x, \rho(x))$ into $S(f(x), \rho(x))$. (Note that we are careful to say "into" not "onto," since $f$ still might be folding the sphere $S(x, \rho(x))$ if $x \in \Sigma(f)$.)

We will now prove that $f|_{S(x, \rho(x))}$ is an isometric folding. Let $y = f(x)$ and $r = \rho(x)$, and let $\gamma : J = [a, b] \to S(x, r)$ be a geodesic segment on $S(x, r)$ parameterized by arc-length. We want to show that $\gamma^* = f \circ \gamma$ is a zig-zag on $S(y, r)$ with $L(\gamma) = L(\gamma^*)$. Note that this is not obvious! Since $f$ is an isometric folding from $\mathcal{M}$ to $\mathcal{N}$, we have that $\gamma^*$ is a zig-zag on $\mathcal{N}$, but this does not immediately give that $\gamma^*$ is a zig-zag on $S(y, r)$. In particular, we need to prove that the differentiable arcs of $\gamma^*$ are geodesics on the submanifold $S(y, r)$ of $\mathcal{N}$.

First, we will prove that $L(\gamma) = L(\gamma^*)$. Choose any subdivision $a = a_0 < a_1 < \cdots < a_s = b$ of $J$ such that for each $i = 1, \ldots, s$ there is a unique geodesic segment $\delta_i : [a_{i-1}, a_i] \to \mathcal{M}$ joining $\gamma(a_{i-1})$ to $\gamma(a_i)$ and $L(\delta_i) = d_\mathcal{M}(\gamma(a_{i-1}), \gamma(a_i))$. (Note that the curves $\delta_i$ might not lie on $\gamma$, and the $\delta_i$ might not be parameterized with respect to arc-length, so we can't expect to have $L(\delta_i) = L(\gamma_i)$, where $\gamma_i$ is the corresponding segment of $\gamma$ between $\gamma(a_{i-1})$ and $\gamma(a_i)$.) The piecewise geodesic path $\delta : [a, b] \to \mathcal{M}$ with $\delta|_{[a_{i-1}, a_i]} = \delta_i$ then gives us a piecewise geodesic path $\delta^* = f \circ \delta$ on $\mathcal{N}$ with

$L(\delta) = L(\delta^*)$. By choosing the subdivision of $J$ to be sufficiently fine, we can make $\delta$ approximate $\gamma$ and $\delta^*$ approximate $\gamma^*$ as closely as we please. In other words, for any $\varepsilon > 0$ there exists a subdivision of $J$ such that $|L(\gamma) - L(\delta)| < \varepsilon$ and $|L(\gamma^*) - L(\delta^*)| < \varepsilon$. Thus $L(\gamma) = L(\gamma^*)$.

We now prove that $\gamma^*$ is a zig-zag on $S(y, r)$; that is, we need to argue that $\gamma^*$ is made of geodesic arcs. To this end, we note that, by definition of $S(x, r)$, we have that the geodesic curves normal to $S(x, r)$ are geodesic rays emanating from $x$, and since $f$ is an isometric folding, these rays are mapped by $f$ to geodesic rays emanating from $y$. Now, since $\gamma$ is a geodesic arc on $S(x, r)$, its principle normal vectors will be normal to $S(x, r)$. That is, $\gamma''(t)$ is normal to $S(x, r)$ for all $t \in J$, and this property is preserved by $f$; also, $(\gamma^*)''(t)$ is normal to $S(y, r)$ for all $t \in J$, which implies that as we travel along $\gamma^*$ we are traveling along geodesic arcs. Thus $\gamma^*$ is a zig-zag and the proof is complete. □

**Definition 10.15** We denote $f_x = f|_{S(x,\rho(x))} \colon S(x, \rho(x)) \to S(f(x), \rho(x))$.

Theorem 10.14 tells us that $f_x$ is an isometric folding of the submanifold $S(x, \rho(x))$ of $\mathcal{M}$ into the submanifold $S(f(x), \rho(x))$ of $\mathcal{N}$, and therefore it makes sense to talk about the set of singularities $\Sigma(f_x)$ of this folding.

**Corollary 10.16** *Let $f \in \mathcal{F}(\mathcal{M}, \mathcal{N})$ and let $x \in \mathcal{M}$. Then the set of singularities of $f$ in $B(x, \rho(x))$ is the geodesic cone $(\Sigma(f_x) \times [0, \rho(x)])/(\Sigma(f_x) \times \{0\})$ in $\mathcal{M}$ with base $\Sigma(f_x)$ and apex $x$.*

*Proof* Recall that if $\dim \mathcal{M} > 1$ then $\Sigma(f_x) = \emptyset$ if and only if $x$ is not singular. If $\dim \mathcal{M} = 1$ then we can have that $x$ is singular and $\Sigma(f_x) = \emptyset$. In either case, if $x \notin \Sigma(f)$ then the corollary is vacuously true.

Now suppose $x \in \Sigma(f)$. For every $v \in S_x(\mathcal{M})$, let $\overline{\gamma}_v = \gamma_v \cap B(x, \rho(x))$. Then we can only have $\overline{\gamma}_v \cap \Sigma(f) = \{x\}$ or $\overline{\gamma}_v \cap \Sigma(f) = \overline{\gamma}_v$; any other possibility would contradict $\rho(x)$ being the minimal radial function at $x$. Thus, every point in $\Sigma(f_x)$, which are all on the boundary of $B(x, \rho(x))$, is connected to $x$ by a geodesic segment $\overline{\gamma}_v$ that is entirely in $\Sigma(f)$. Therefore, $\Sigma(f) \cap B(x, \rho(x))$ is the geodesic cone in $\mathcal{M}$ with base $\Sigma(f_x)$ and apex $x$. □

We may now determine the structure of $\Sigma(f)$.

**Theorem 10.17** *Let $f \in \mathcal{F}(\mathcal{M}, \mathcal{N})$. Then there is a decomposition of $\mathcal{M}$ into mutually disjoint, connected, totally geodesic submanifolds, which we call **strata**, with the following properties:*

(i) *Let $\Sigma_k(f)$ denote the union of all strata of dimension $k$. Then $\Sigma(f) = \bigcup_{k=0}^{m-1} \Sigma_k(f)$.*

(ii) *For each strata $s$, $f|_s$ is a locally isometric immersion into $\mathcal{N}$.*

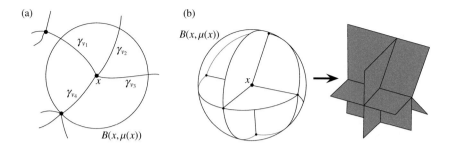

**Figure 10.2** Two local crease patterns around $x \in \Sigma(f)$ when $x$ is a 0-strata with (a) dim $\mathcal{M} = 2$ and (b) dim $\mathcal{M} = 3$.

(iii) *The boundary of each strata is a union of strata of lower dimension and, in the case where $\mathcal{M}$ is compact, of finitely many such strata.*
(iv) *The boundary of $\Sigma_k(f)$ in $\mathcal{M}$ is $\bigcup_{l=0}^{k-1} \Sigma_l(f)$.*

*Proof*   The theorem follows by induction once the lower-dimensional cases have been established. Suppose $f \in \mathcal{F}(\mathcal{M}, \mathcal{N})$ where dim $\mathcal{M} = 1$. We have already seen that if $\mathcal{M} = \mathbb{R}$ then the image under $f$ is a piecewise geodesic curve in $\mathcal{N}$ and thus $\Sigma(f)$ is a discrete countable subset of $\mathcal{M}$. Similarly, if $\mathcal{M}$ is diffeomorphic to $S^1$, then $\Sigma(f)$ is a finite subset of $\mathcal{M}$. In these cases it is either simple or vacuous to check that properties (i)–(iv) hold.

Now suppose $f \in \mathcal{F}(\mathcal{M}, \mathcal{N})$ where dim $\mathcal{M} = 2$ and dim $\mathcal{N} = n$. Then by Theorem 10.14, for all $x \in \mathcal{M}, f_x \in \mathcal{F}(P, Q)$ where $P$ and $Q$ are diffeomorphic to $S^1$ and $S^{n-1}$, respectively. So if $x \in \Sigma(f)$, we have by the previous case that $\Sigma(f_x)$ is a finite set, and then by Corollary 10.16 we have that $\Sigma(f) \cap B(x, \rho(x))$ is made of finitely many geodesic segments emanating from $x$. See Figure 10.2(a) where we have chosen $x \in \Sigma(f)$ to be a strata of dimension 0 (a 0-strata). Since this must represent what $\Sigma(f)$ looks like around every 0-strata, we conclude that when dim $\mathcal{M} = 2$, $\Sigma(f)$ is an embedded graph on $\mathcal{M}$ made of 0-strata connected by 1-strata. By this construction we immediately obtain properties (i) and (ii) of the theorem, and properties (iii) and (iv) follow from the fact that $\Sigma(f)$ is a graph.

If dim $\mathcal{M} = 3$ and dim $\mathcal{N} = n$, then for all $x \in \mathcal{M}$ we have $f_x \in \mathcal{F}(P, Q)$ where $P$ is diffeomorphic to $S^2$. By the previous case, $\Sigma(f_x)$ is a graph embedded on $P$ made of 0-strata and 1-strata. By Corollary 10.16, we have that $\Sigma(f) \cap B(x, \rho(x))$ consists of geodesic 2-strata emanating from $x$ making a cone with $\Sigma(f_x)$. An example is shown in Figure 10.2(b), together with what $\Sigma(f)$ would look like around $x$ without the minimal radial sphere $B(x, \rho(x))$. Again, since this describes the structure of $\Sigma(f)$ about every 0-strata, we find that $\mathcal{M}$ will be decomposed into 3-strata and $\Sigma(f)$, which is made of 0-, 1-, and 2-strata. This gives us properties (i) and (ii), whereby properties (iii) and (iv) are obtained inductively from the dim $\mathcal{M} = 2$ case and the fact that around every 0-strata $\Sigma(f)$ is a cone.

Induction for dim $\mathcal{M} = m$ proceeds exactly the same way as we saw in the dim $\mathcal{M} = 3$ case, and this completes the proof.    $\square$

## 10.3    Robertson's Theorem

The material in this chapter, thus far, is fairly abstract. We are developing a way to talk about and model the flat folding of manifolds, which are, in general, $n$-dimensional; when $n = 2$ this means that we are folding a surface that could be, for example, a piece of paper or part of a sphere or an infinite plane or the surface $z = x^2 + y$. While this is a generalization of normal origami, all of these 2-dimensional manifold cases are still possible to physically fold, with the proper materials (e.g., spherical paper is hard to find). And our notion of a crease pattern $\Sigma(f)$ in these cases generalizes nicely from the crease patterns of normal origami models; the creases only need to be geodesics on the 2-manifold. Even the generalizations of the crease pattern into higher dimensions make sense in that instead of merely a network of geodesic segments and points, we have a larger structure of geodesic surfaces, or 2-strata, and higher-dimensional strata. Viewing diagrams like that shown in Figure 10.2(b) allows one to actually visualize what it might be like to, say, fold a 3-dimensional cube along "crease planes" that intersect at a vertex inside the cube.

It is challenging, however, to understand why we do what comes next: Assume that our isometric foldings $f: \mathcal{M} \to \mathcal{N}$ have dim $\mathcal{M} =$ dim $\mathcal{N}$, that $\mathcal{M}$ and $\mathcal{N}$ are both oriented, that $\mathcal{N}$ is connected, and that $\mathcal{M}$ is compact and **without boundary**. In other words, we will only be considering cases such as $\mathcal{M}$ being a sphere or a torus, which are compact manifolds without boundary. This assumption seems to exclude normal origami and could be viewed as veering off into impractical abstraction. But what is amazing is that this assumption actually does capture normal origami! Rather, it will give us a tool by which we can prove theorems about folding manifolds, analogous to Kawasaki's Theorem. We will name this new tool **Robertson's Theorem**.

Thus assume that dim $\mathcal{M} =$ dim $\mathcal{N} = q$, the manifolds are oriented, $\mathcal{N}$ is connected, and $\mathcal{M}$ is compact and without boundary. Then the degree of our isometric folding map $f: \mathcal{M} \to \mathcal{N}$, denoted deg $f$, is well-defined. There are several equivalent ways to define the degree of a map between manifolds, but we'll adopt the one presented by Milnor (1965). Recall that $T_x(\mathcal{M})$ denotes the vector space of tangent vectors at $x \in \mathcal{M}$. Let $x \in \mathcal{M}$ be a regular point of $f$, meaning that the derivative $df_x: T_x(\mathcal{M}) \to T_{f(x)}(\mathcal{N})$ is a linear isomorphism between oriented vector spaces. Then define the **sign of** $df_x$ to be $+1$ or $-1$ according to whether or not $df_x$ preserves orientation. Letting $y = f(x)$, we then define

$$\deg(f; y) = \sum_{x \in f^{-1}(y)} \operatorname{sign} df_x.$$

It can then be proven that this integer $\deg(f; y)$ does not depend on $y$, that is, it will be the same no matter which $x \in \mathcal{M}$ we initially chose (see (Milnor, 1965)). Therefore, we call this value the **degree of** $f$ (denoted deg $f$). Intuitively, deg $f$ counts how many times $\mathcal{M}$ "wraps around" $f(\mathcal{M}) \subset \mathcal{N}$.

Getting back to our isometric folding $f: \mathcal{M} \to \mathcal{N}$, since each of the finitely many strata of $\Sigma(f)$ is isometrically immersed in $\mathcal{N}$ and each of these strata have positive codimension, it follows that for almost all $y \in f(\mathcal{M})$ we have that $f^{-1}(y) = \{x_1, \ldots, x_v\}$ where each $x_i \in \mathcal{M} \setminus \Sigma(f)$. If $\lambda$ points $x_i \in f^{-1}(y)$ have positive sign (that is, positive sign of $df_{x_i}$) and $\mu$ have negative sign, then $\deg f = \lambda - \mu$.

Let us also call each $q$-dimensional stratum ($q$-stratum) $S$ of $f$ **positive** or **negative** according to whether $f|_S$ is orientation-preserving or -reversing. Then let $V$, $V_+$, $V_-$, and $V_f$ denote the $q$-dimensional volumes of $\mathcal{M}$, the positive $q$-strata, the negative $q$-strata, and $f(\mathcal{M})$, respectively. Therefore $V = V_+ + V_-$.

**Robertson's Theorem**  *Let $f \in \mathcal{F}(M, N)$ where $M$ and $N$ are oriented and of the same dimension, $N$ is connected, and $M$ is compact and without boundary. Let $\deg f = k$. Then $V_+ = V_- + kV_f$.*

*Proof*  Volume $V_f$ is equal to the volume in $N$ of $f(\mathcal{M}) \setminus f(\Sigma(f))$. But each point in $f(\mathcal{M}) \setminus f(\Sigma(f))$ is counted when adding the volumes of the various oriented $q$-strata, a net total of $k$ times, where $k = \lambda - \mu$ in the above terminology. Thus $kV_f = V_+ - V_-$, which establishes the theorem.  □

For the remainder of this section, we will assume that all isometric foldings $f$ satisfy the conditions of Robertson's Theorem.

**Corollary 10.18**  *If $f \in \mathcal{F}(M, N)$ and $f$ is not surjective, then $\deg f = 0$.*

*Proof*  If $f$ is not surjective, then $f(\mathcal{M})$ has boundary in $N$. For any point $y \in f(\mathcal{M})$ near such a boundary, the pre-image $f^{-1}(y)$ will have an even number of points, half with positive orientation and half with negative orientation. Thus $\deg f = \lambda - \mu = 0$.  □

---

**Example 10.19**  Robertson's Theorem is used in (Robertson, 1977–1978) to prove the necessary direction of Kawasaki's Theorem. Take a flat-foldable vertex with consecutive angles $\alpha_i$ between the creases. Draw a copy of $S^1$ (which has radius 1) centered at the vertex. Since the vertex is flat-foldable, it induces an isometric folding $f: S^1 \to S^1$ that, assuming there are some creases, will not be surjective. (See Figure 10.3.) Thus $\deg f = 0$. Also, the 1-strata in the image will have volume equal to the lengths of arcs on $S^1$, which will equal the angles $\alpha_i$. Assuming that the odd-indexed angles give positive orientation and the even-indexed ones are negatively oriented, Robertson's Theorem gives

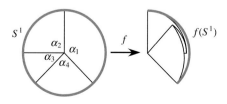

Figure 10.3 Robertson's Theorem applied to a flat vertex fold.

$$\alpha_1 + \alpha_3 + \cdots + \alpha_{2n-1} = \alpha_2 + \alpha_4 + \cdots + \alpha_{2n} + 0 = \pi,$$

which is one direction of Kawasaki's Theorem.

In fact, a more general version of this direction of Kawasaki's Theorem follows immediately from the formula in Robertson's Theorem and the fact that $V = V_+ + V_-$:

**Corollary 10.20** *If $f \in \mathcal{F}(\mathcal{M}, \mathcal{N})$ and $\deg f = 0$, then $V_+ = V_- = V/2 \geq V_f$.*

If $\dim \mathcal{M} = \dim \mathcal{N}$ and $\mathcal{M}$ is compact without boundary, then the $\deg f \neq 0$ case for isometric foldings turns out to be trivial:

**Corollary 10.21** *Let $f \in \mathcal{F}(\mathcal{M}, \mathcal{N})$. Then $\Sigma(f) \neq \emptyset$ if and only if $\deg f = 0$. Also, $\Sigma(f) = \emptyset$ if and only if $\deg f = \pm 1$.*

*Proof* Recall that we are assuming that $\mathcal{M}$ is compact and without boundary. Let us assume without loss of generality that $V = 1$. Then $V_+ = \alpha$ for some $0 \leq \alpha \leq 1$ and $V_- = 1 - \alpha$. Let $\deg f = k$. Then we have

$$\Sigma(f) \neq \emptyset \Leftrightarrow 0 < \alpha < 1 \Leftrightarrow -1 < kV_f < 1 \Leftrightarrow -1 < k < 1 \Leftrightarrow k = 0.$$

On the other hand,

$$\Sigma(f) = \emptyset \Leftrightarrow \alpha = 0 \text{ or } \alpha = 1 \quad \text{and} \quad V_f = 1 \Leftrightarrow k = 2\alpha - 1 = \pm 1. \qquad \square$$

**Example 10.22** To illustrate Corollary 10.20, consider an isometric folding $f: \mathcal{M} \to \mathcal{N}$ where $\dim \mathcal{M} = \dim \mathcal{N} = 3$. Surround a vertex $x$ (0-stratum) of $\Sigma(f)$ with a sphere of radius $\rho(x)$, and let us assume that this sphere is a copy of $S^2$. (The whole manifold and folding could be scaled to make this true if we really wanted.) Then $f$ maps this copy of $S^2$ centered at $x$ into a copy of $S^2$ in $\mathcal{N}$ centered at $f(x)$, and thus we have induced an isometric folding $f: S^2 \to S^2$, which, again, will not be surjective. Thus $\deg f|_{S^2} = 0$. The 2-strata in the domain $S^2$ will be spherical polygons, and the volume of these 2-strata will be their spherical area. (See Figure 10.4.)

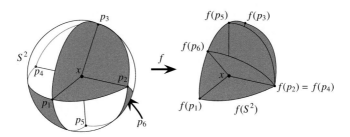

**Figure 10.4** An example of a flat-foldable vertex $x$ in a 3-manifold isometric folding, with points $p_1, \ldots, p_6$ made by the intersections of the 1-strata at $x$ and the bounding sphere $S^2$ shown, along with their images under $f$.

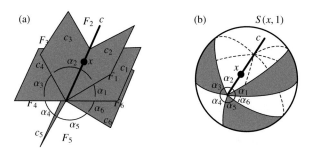

**Figure 10.5** (a) The crease pattern $\Sigma(f)$ of an isometric folding $f: \mathbb{R}^3 \to \mathbb{R}^3$ that is made of a line $c$ and half-planes $c_1, \ldots, c_6$. (b) The sphere $S(x, 1)$ with the half-planes of $\Sigma(f)$ intersecting along great circles to create lunes with angles $\alpha_i$.

Therefore, Corollary 10.20 gives us a Kawasaki-like theorem for flat-foldable vertices in a 3-manifold isometric folding that can only be stated by considering a sphere $S^2$ centered at the vertex $x$ and by viewing how the 2-strata of $\Sigma(f)$ emanating from $x$ cut spherical polygons in $S^2$: We obtain that the spherical polygons can be partitioned into those that are positively and negatively oriented, which we may think of as a proper 2-coloring, say with purple and white (see Figure 10.4). If we let $B_1, \ldots, B_k$ be the areas of the purple spherical polygons and $W_1, \ldots, W_k$ be the areas of the white spherical polygons, then Corollary 10.20 becomes

$$\sum_{i=1}^{k} B_i = \sum_{i=1}^{k} W_i = 2\pi.$$

**Example 10.23**    Consider the isometric folding $f: \mathbb{R}^3 \to \mathbb{R}^3$ whose crease pattern $\Sigma(f)$ is shown in Figure 10.5(a). This crease pattern consists of a line $c$ connected to half-planes $c_1, \ldots, c_6$, defining wedge-shaped regions $F_1, \ldots, F_6$ of $\mathbb{R}^3$ between the half-planes. If we let $x$ be a point in $c$ and consider the sphere $S(x, 1)$, then $f_x: S(x, 1) \to S(x, 1)$ (since here $\rho(x) = 1$) is an isometric folding. In this case, the 2-strata of $\Sigma(f_x)$ are lunes on a sphere of radius 1 (see Figure 10.5(b)) and thus have spherical area $2\alpha_i$ where $\alpha_i$ is the angle between the planes. Here again, $f_x$ is not surjective and so $\deg f_x = 0$. Therefore, Robertson's Theorem implies that the sum of the areas of the positively oriented lunes under $f_x$ equals the sum of the areas of the negatively oriented lunes. That is,

$$\alpha_1 + \alpha_3 + \alpha_5 = \alpha_2 + \alpha_4 + \alpha_6.$$

## 10.4 The Angle Sum and Recovery Theorems

Examples 10.19, 10.22, and 10.23 suggest that a version of the necessary direction of Kawasaki's Theorem, which we will call the **Angle Sum Theorem**, holds for isometric foldings $f\colon \mathcal{M} \to \mathcal{N}$ (where dim $\mathcal{M} =$ dim $\mathcal{N} = q$) around any strata of $\Sigma(f)$ of dimension less than $q$. While this result is hinted at in (Robertson, 1977–1978), it is first explicitly stated in (Lawrence and Spingarn, 1989). We provide a proof different from that of Lawrence and Spingarn that employs Robertson's Theorem.

First, we introduce a definition to formalize the kinds of volumes measured in the previous examples. Namely, we borrow the concept of the **inner angle** in a polytope from (Grünbaum, 2003).

**Definition 10.24** Let $P$ be a $q$-dimensional polytope in $\mathbb{R}^q$, let $F$ be a $k$-face of $P$ (for $k < q$), and let $x$ be a point in the relative interior of $F$. Let $C(x, F, P)$ denote the cone with apex $x$ of $F$ spanned by $P$, where we view $P$ as a $q$-dimensional face of itself. Then the **inner angle of $P$ at $F$**, denoted $a(F, P)$, is the $(q-1)$-volume of $S(x, r) \cap C(x, F, P)$, where $r > 0$ is chosen small enough so that the $(q-1)$-sphere $S(x, r)$ intersects only cells of $P$ that are adjacent to $x$ and we have normalized the $(q-1)$-volume measure so that the volume of $S(x, r)$ equals that of $S(x, 1)$.

In an isometric folding $f\colon \mathcal{M} \to \mathcal{N}$ with dim $\mathcal{M} =$ dim $\mathcal{N} = q$, we can take the polytope $P$ to be any $q$-strata of $\Sigma(f)$ and $F$ to be any $k$-strata for $k < q$, and if $x$ is in the relative interior of $F$, then the radius of the inner angle definition can be $r = \rho(x)$.

**Theorem 10.25** (Angle Sum Theorem (Lawrence and Spingarn, 1989)) *Let $f \in \mathcal{F}(\mathcal{M}, \mathcal{N})$, where dim $\mathcal{M} =$ dim $\mathcal{N} = q$ and $\Sigma(f) \neq \emptyset$. Let $\mathcal{A}$ (resp. $\mathcal{B}$) be the set of $q$-strata of $\Sigma_q(f)$ whose images under $f$ are positively (resp. negatively) oriented. Let $F \in \Sigma_k(f)$ for some $k < q$. Then*

$$\sum_{\substack{A \in \mathcal{A} \\ \overline{A} \supseteq F}} a(F, A) = \sum_{\substack{B \in \mathcal{B} \\ \overline{B} \supseteq F}} a(F, B), \tag{10.1}$$

*where $\overline{X}$ denotes the closure of the set X. Furthermore, the quantity in Equation (10.1) will equal one-half the $(q-1)$-volume of $S^{q-1}$.*

*Proof* Let $x$ be a point in the relative interior of $F$. Since $\Sigma(f) \neq \emptyset$, we have deg $f = 0$. Then $f_x\colon S(x, \rho(x)) \to S(f(x), \rho(x))$ is an isometric folding between $(q-1)$-dimensional manifolds. Applying Corollary 10.20 of Robertson's Theorem to $f_x$ gives us that the sum of the $(q-1)$-dimensional volumes of the positive $(q-1)$-strata of $\Sigma(f_x)$ equals the sum of volumes of the negative $(q-1)$-strata. Normalizing this gives exactly Equation (10.1). $\qquad \square$

---

**Diversion 10.2**   Lawrence and Spingarn offer a different proof of the Angle Sum (Theorem 10.25): Fix an $x$ in the relative interior of $F$, and for each $A \in \Sigma_q f$, define the characteristic function $C_A : S(f(x), \rho(x)) \to \mathbb{R}$:

$$C_A(y) = \begin{cases} 0 & \text{if } y \notin f_x(A \cap S(x, \rho(x))), \\ 1 & \text{if } y \in f_x(A \cap S(x, \rho(x))). \end{cases}$$

Explain why the function

$$\sum_{\substack{A \in \mathcal{A} \\ A \supseteq F}} C_A - \sum_{\substack{B \in \mathcal{B} \\ B \supseteq F}} C_B$$

equals the degree of $f_x$ (which is zero) almost everywhere. Then integrate to prove the Angle Sum Theorem.

---

**Remark 10.26**   The fact that Theorem 10.25 holds for strata $F$ of arbitrary dimension $k$ has a profound implication on the structure of $\Sigma(f)$. It means that versions of Equation (10.1) hold wherever strata intersect in $\Sigma(f)$. For example, suppose that we let $x$ be a vertex (0-stratum) of $\Sigma(f)$ and, as in Theorem 10.25, consider the $(q-1)$-dimensional sphere $S(x, \rho(x))$ centered at $x$, which isometrically folds under $f_x$. Then we can let $y \in \Sigma(f_x)$ be a 0-stratum (i.e., a vertex of the crease pattern on $S(x, \rho(x))$, which is the intersection of a 1-stratum of $\Sigma(f)$ and $S(x, \rho(x))$), and apply Theorem 10.25 to $f_x$ and the point $y$. That is, consider a $(q - 2)$-dimensional sphere $S(y, \rho(y))$ and the isometric folding $f_y : S(y, \rho(y)) \to S(f(y), \rho(y))$. (Note that here the minimal radial function $\rho(y)$ should be based on $f_x$, which could be different than using $f$. But since $f_x$ is a restriction of $f$, we have that $f_x(y) = f(y)$, so describing the codomain of $f_y$ as $S(f(y), \rho(y))$ makes sense.) This use of Theorem 10.25 may continue down all dimensions.

This pattern can be seen in the 3-manifold isometric folding shown in Figure 10.4. The vertex $x$ satisfies the dimension $q = 3$ version of Equation (10.1) as described in Example 10.22. But the vertices $p_1, \ldots, p_6$ on the sphere also satisfy the dimension $q = 2$ version of Equation (10.1), which is just Kawasaki's Theorem. This can actually be seen to hold at vertices $p_1$ (which is surrounded by right angles on the sphere) and $p_2$ (whose angles are, in order, $\pi/2, \pi/4, \pi/2$, and $3\pi/4$ on the surface of the sphere) in Figure 10.4.

While much of Lawrence and Spingarn's study of isometric foldings is similar to that of Robertson, they do provide one result that was new at the time: a **recovery theorem** that characterizes when a cell complex in $\mathbb{R}^n$ could be a crease pattern for an isometric folding.

**Theorem 10.27** (Recovery Theorem (Lawrence and Spingarn, 1989))   *Suppose that $K$ is a polyhedral cell complex subdividing $\mathbb{R}^q$ whose q-cells can be properly 2-colored (meaning that any two q-cells whose intersected closure has dimension $q - 1$ must* *receive different colors) with color classes $\mathcal{A}$ and $\mathcal{B}$. Further suppose that for each $(q-2)$-face $F$ of $K$ we have*

$$\sum_{\substack{A \in \mathcal{A} \\ A \supseteq F}} a(F, A) = \sum_{\substack{B \in \mathcal{B} \\ B \supseteq F}} a(F, B),$$

*then there is an isometric folding $f \colon \mathbb{R}^q \to \mathbb{R}^q$ such that $\Sigma(f) = K$.*

We will hold off on the proof of the Recovery Theorem until Chapter 8. (See Theorem 11.9.)

**Remark 10.28** Lawrence and Spingarn point out a surprising subtlety that emerges when combining the Recovery Theorem with the Angle Sum Theorem. The Recovery Theorem is a sufficient condition for an isometric folding $f \colon \mathbb{R}^q \to \mathbb{R}^q$ to exist, but its hypotheses are weaker than the necessary condition of the Angle Sum Theorem. Namely, the Recovery Theorem needs the angle sums to be equal only around the $(q-2)$-dimensional faces of the polyhedral cell complex $K$, which then implies the existence of an isometric folding $f$ with $\Sigma(f) = K$. But then, since $f$ is an isometric folding, the Angle Sum Theorem implies that the angle sum holds at **every** face of dimension less than $q$. In the $q = 2$ case (which is normal origami on 2-dimensional pieces of paper), this statement is vacuous, but for higher-dimensional folding it is quite remarkable.

For example, in $\mathbb{R}^3$, suppose that we divide the surface of the 2-sphere $S^2$ into a finite number of spherical polygons where (a) the polygons can be properly 2-colored with color classes $\mathcal{A}$ and $\mathcal{B}$ and (b) at each vertex the sum of the spherical polygon angles colored $\mathcal{A}$ equals the sum of the angles colored $\mathcal{B}$. Then our combination of the Recovery Theorem and the Angle Sum Theorem implies that the sum of the spherical polygon areas in color class $\mathcal{A}$ equals the areas in color class $\mathcal{B}$. This result in spherical geometry is not new, but the fact that it is an immediate consequence of isometric foldings is quite unexpected.

---

**Diversion 10.3** Give a stand-alone proof of this spherical geometry result, without using isometric foldings. That is, prove that if a division $K$ of the surface of a sphere into spherical polygons has the angle sum property at each vertex, then the "angle sum" property must also hold for the areas of the spherical polygons.

---

## 10.5 Maekawa and Kawasaki for Isometric Foldings

We have seen how the necessary direction of Kawasaki's Theorem generalizes to $(q-2)$-faces of isometric foldings of compact manifolds without boundary of dimension $q$. Note that "isometric foldings" has replaced our phrase "flat-foldable." It is therefore natural to ask whether Maekawa's Theorem and the converse direction of Kawasaki's Theorem also generalize to such manifolds.

Robertson and Lawrence and Spingarn do not consider these questions, probably because it is not obvious what the higher-dimensional analog of mountain and valley creases should be, and isometric foldings themselves carry no information about which direction the $q$-dimensional manifold is bending.

In this section we will see how generalizations of Mawkawa's Theorem and Kawasaki's converse are possible in higher dimensions. To do this we need to extend our definition of a mountain-valley assignment.

**Definition 10.29**  Let $f \in \mathcal{F}(\mathcal{M}, \mathcal{N})$ with dim $\mathcal{M}$ = dim $\mathcal{N}$ = $q$. Let $\Sigma_k(f)$ denote the union of all strata of dimension $k$ in $\Sigma(f)$. Then a **mountain-valley (MV) assignment** for $f$ is a function $\mu_f \colon \Sigma_{q-1}(f) \to \{-1, 1\}$.

When contemplating MV assignments in the 2-dimensional case, we are greatly aided by the intuition that folding actual paper gives us; MV assignments determine a layer ordering of the paper, and applying the geometry of the crease pattern to this layer ordering is what allows us to determine when a given MV assignment will be valid. This was the approach used in much of Chapter 5.

In higher-dimensional isometric foldings this intuition abandons us. It is no longer clear when an MV assignment of, say, an isometric folding $f \colon \mathbb{R}^3 \to \mathbb{R}^3$ will be valid. However, the basic principle should be the same: an MV assignment will determine a partial ordering of the $q$-strata of an isometric folding $f \colon \mathcal{M} \to \mathcal{N}$ where dim $\mathcal{M}$ = dim $\mathcal{N}$ = $q$. However, as we saw in Section 6.5, a partial ordering on the $q$-strata will not always suffice. Therefore, we need to extend our definitions of the superposition order to higher dimensions.

In the following assume that $f \colon \mathcal{M} \to \mathcal{N}$ is an isometric folding with dim $\mathcal{M}$ = dim $\mathcal{N}$ = $q$.

**Definition 10.30**  The **superposition pattern**, or **s-net**, of $f$ is the set $f^{-1}(f(\Sigma(f) \cup \partial \mathcal{M}))$.

---

**Diversion 10.4**  Prove that the s-net of an isometric folding $f \colon \mathcal{M} \to \mathcal{N}$ is a cellular decomposition of $\mathcal{M}$ that contains $\Sigma(f)$, satisfying all the properties of $\Sigma(f)$ in Theorem 10.17.

---

**Diversion 10.5**  Prove that if $s_1$ and $s_2$ are two $q$-strata of the s-net of an isometric folding $f \colon \mathcal{M} \to \mathcal{N}$, where dim $\mathcal{M}$ = dim $\mathcal{N}$ = $q$, then either $f(s_1) = f(s_2)$ or $f(s_1) \cap f(s_2) = \emptyset$. (See Lemma 6.11.) Is this also true for s-net strata of dimension less than $q$?

---

**Definition 10.31**  The $(q-1)$-strata of the s-net of an isometric folding $f$ can be partitioned into three disjoint sets: the **boundary** $(q-1)$-**strata**, the **c-creases**, which are also in $\Sigma(f)$, and the **s-creases**, which are the pre-images $y$ of $f(c)$ where $c$ is a c-crease but $y$ is not a c-crease itself. (Compare with Definition 6.12.)

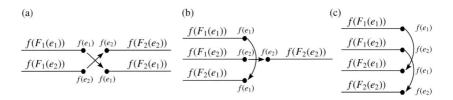

**Figure 10.6** The non-crossing conditions for isometric foldings. Note the similarity with Figure 6.17

Note that we are still using the term "crease" to refer to the $(q-1)$-strata, borrowing it from the 2-dimensional case where these strata are actually straight lines.

**Definition 10.32** Define an equivalence relation $\sim$ on the set of $q$-strata in the s-net of $f$ by $s_1 \sim s_2$ if and only if $f(s_1) = f(s_2)$. A **superposition order** $>$ of $f$ is a partial ordering on the $q$-strata of the s-net made by totally ordering the equivalence classes of $\sim$.

As in the 2-dimensional, flat paper case, the superposition order describes which $q$-strata are above or below others in the isometric folding image. This is necessary for determining if any self-intersections occur.

**Definition 10.33** If $e$ is a c-crease or an s-crease in the s-net of an isometric folding $f$, then let the two $q$-strata of the s-net that border $e$ be denoted $F_1(e)$ and $F_2(e)$. If $e_1$ and $e_2$ are c- or s-creases of the s-net with $f(e_1) = f(e_2)$, then the **non-crossing conditions** for $f$ under a superposition order $>$ are (see Figure 10.6):

(i) **Tortilla-Tortilla Condition:** If $e_1$ and $e_2$ are both s-creases and $f(e_1) = f(e_2)$, then we can't have $F_1(e_1) > F_1(e_2)$ and $F_2(e_2) > F_2(e_1)$.
(ii) **Taco-Tortilla Condition:** If $e_1$ is a c-crease, $e_2$ is an s-crease, and $f(e_1) = f(e_2)$, then we can't have $F_1(e_1) > F_1(e_2) > F_2(e_1)$.
(iii) **Taco-Taco Condition:** If $e_1$ and $e_2$ are both c-creases and $f(e_1) = f(e_2)$, then we can't have $F_1(e_1) > F_1(e_2) > F_2(e_1) > F_2(e_2)$.

**Remark 10.34** Notice that the illustration of our arbitrary-dimension non-crossing conditions in Figure 10.6 is the same as the illustration for the 2-dimensional paper case in Figure 6.17 from Chapter 6. Readers might be skeptical of this; couldn't more types of intersections be possible in higher dimensions? But no matter what dimension $q$ we are in, the superposition order only adds one additional dimension to account for the layer ordering of the $q$-strata, and the possibilities of how intersections can occur will still depend on what $(q-1)$-strata s-net creases map onto each other: either s-crease onto s-crease, s-crease onto c-crease, or c-crease onto c-crease. Given that those are the only possibilities (boundary creases do not create self-intersections), the only possible intersections along s-net creases will be those illustrated in Figure 10.6.

In line with Remark 6.16 of Chapter 6, an isometric folding $f$ with superposition order $>$ will induce an MV assignment $\mu\colon \Sigma_{q-1}(f) \to \{-1, 1\}$ that is based on the 2-colorability of the $q$-strata of $\Sigma(f)$.

**Definition 10.35**   Given an isometric folding $f$ together with a superposition order $>$ on the s-net of $f$, we say that $f$ **self-intersects with** $>$ if the non-crossing conditions are violated under $>$. If this is the case, we say that $f$ under $>$ is **not flat-foldable**. Otherwise, we say that $f$ is **flat-foldable** under $>$. We say that an MV assignment $\mu$ for $f$ is **valid** if it is induced by a superposition order $>$ that makes $f$ flat-foldable.

Before we can consider analogs of Maekawa's Theorem and the converse of Kawasaki's Theorem in the $q$-dimensional case, we need to assume a basic principle of how MV assignments should behave when interacting between objects of different dimensions under an isometric folding.

**MV Dimension Principle**   *Let $f\colon \mathcal{M} \to \mathcal{N}$ be an isometric folding, and let $\mathcal{P} \subset \mathcal{M}$ be a submanifold with dim $\mathcal{P} = k < q$ such that $f$ induces an isometric folding $f|_\mathcal{P}\colon \mathcal{P} \to \mathcal{N}$.*

*Then any MV assignment $\mu$ induced from a superposition order $>$ of $f$ can be applied to $f|_\mathcal{P}$ as well in the following way:*

- *If $e$ is a $(k-1)$-strata of $\Sigma(f|_\mathcal{P})$ bordered by $k$-strata $F_1(e), F_2(e) \in \Sigma_k(f|_\mathcal{P})$ and $F_1(e) \subset G_1 \in \Sigma_q(f)$ and $F_2(e) \subset G_2 \in \Sigma_q(f)$ where $G_1 > G_2$, then we must have $F_1(e) > F_2(e)$ if $>$ is to carry through to a superposition order on $f|_\mathcal{P}$.*
- *If any $e \in \Sigma_{k-1}(f|_\mathcal{P})$ is the intersection of $\mathcal{P}$ and a $(q-1)$-strata $g$ of $\Sigma(f)$, then $\mu(e) = \mu(g)$ must be true if $\mu$ is to carry through to an MV assignment of $f|_\mathcal{P}$.*

This principle is clearly true in standard, 2-dimensional origami: if we take a flat-foldable crease pattern $C$ under folding map $\sigma_f$ and MV assignment $\mu$, and we draw a line $L$ on $C$ so that $\sigma_f$ maps $L$ to a line (and thus $\sigma_f|_L$ can be viewed as a 1-dimensional flat folding on its own), then the folded line $\sigma_f(L)$ will have mountain and valley points that must correspond to the intersection of $L$ to mountain and valley creases of $C$ under $\mu$.

In practice we will make use of this principle by taking $x \in \Sigma(f)$ and considering the isometric folding $f_x\colon S(x, \rho(x)) \to S(f(x), \rho(x))$. Any valid MV assignment $\mu_f\colon \Sigma_{q-1}(f) \to \{-1, 1\}$ of $f$ must match with a valid MV assignment $\mu_{f_x}\colon \Sigma_{q-2}(f_x) \to \{-1, 1\}$.

**Theorem 10.36** (Generalized Maekawa's Theorem for Isometric Foldings)   *Let $f\colon \mathcal{M} \to \mathcal{N}$ be an isometric folding with dim $\mathcal{M} = $ dim $\mathcal{N} = q$, and let $\mu_f$ be a valid MV assignment for $f$. Let $F \in \Sigma_{q-2}(f)$. Then*

$$\sum_{\substack{A \in \Sigma_{q-1}(f) \\ \overline{A} \supseteq F}} \mu_f(A) = \pm 2. \tag{10.2}$$

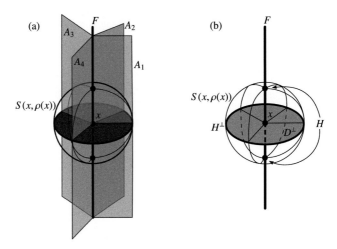

**Figure 10.7** The sketch of a possible picture, with notation, of the proofs of the Generalized Maekawa's and Kawasaki's Theorems for Isometric Foldings. (a) The point $x$ taken in the relative interior of the $(q-2)$-stratum $F$ and $S(x, \rho(x))$ intersecting the $(q-1)$-strata $A_i$, when $q = 3$. (b) Illustrating $H = F \cap S(x, \rho(x))$ and $H^\perp$ and $D^\perp$.

*Proof* Let $x$ be a point in the relative interior of $F$ and consider the $(q-1)$-sphere $S(x, \rho(x))$. We know from Theorem 10.14 that $f_x : S(x, \rho(x)) \to S(f(x), \rho(x))$ is an isometric folding, and so any valid MV assignment $\mu_{f_x}$ of $f_x$ must match $\mu_f$. That is, every $(q-2)$-strata $e$ of $\Sigma(f_x)$ is the intersection of $S(x, \rho(x))$ and a $(q-1)$-strata $g$ of $\Sigma(f)$, and we must have $\mu_f(g) = \mu_{f_x}(e)$.

Since $\dim F = q - 2$ and $\dim S(x, \rho(x)) = q - 1$, we have that $H = F \cap S(x, \rho(x))$ is a $(q-3)$-sphere. Let $H^\perp = \{v \in S(x, \rho(x)) : \langle v, u \rangle = 0 \ \forall u \in H\}$ be the orthogonal complement of $H$ in $S(x, \rho(x))$ (where $\langle \cdot, \cdot \rangle$ is the standard inner product). Then $\dim H^\perp = 2$, and thus $H^\perp$ is a circle. Let $f_{H^\perp}$ denote the isometric folding $f$ restricted to $H^\perp$. See Figure 10.7 for one possible illustration of these terms when $q = 3$.

Our goal is to prove the following claims:

**Claim 1:** Every $A \in \Sigma_{q-1}(f)$ that borders $F$ intersects $H^\perp$ at one point.
**Claim 2:** $f(H^\perp)$ is contained in a circle in $\mathcal{N}$. That is, we can consider $f$ restricted to $H^\perp$ as an isometric folding $f_{H^\perp} : S^1 \to S^1$.

Once these claims are established, the proof of the theorem becomes almost immediate: $f$ restricted to $H^\perp$ is like folding a 2-dimensional disc (shown as $D^\perp$ in Figure 10.7(b)), and if an MV assignment $\mu_f$ is to be valid for $f$, then the induced MV assignment $\mu_{f_{H^\perp}}$ must also be valid for $f_{H^\perp}$. By the original, 2-dimensional Maekawa's Theorem, if we let $a_1, \ldots, a_k$ be the intersection points of each $A_i \in \Sigma_{q-1}(f)$ with $H^\perp$, then $\sum \mu_{f_{H^\perp}}(a_i) = \pm 2$. By the MV Dimension Principle, we have $\mu_f(a_i) = \mu_{f_{H^\perp}}(a_i)$ for all $i = 1, \ldots k$, proving Equation (10.2) and the theorem.

Intuitively Claims 1 and 2 make sense: The circle $H^\perp$ goes "around" the $(q-2)$-face $F$, and thus cuts through each $(q-1)$-strata $A_i$ bordering $F$. And the orthogonality of

$H^\perp$ to $F$ means that this circle intersects each $A_i$ with a tangent vector that is orthogonal to $A_i$, and thus reflecting about $A_i$ will map the arcs of $H^\perp$ between the $A_i$ strata onto themselves, giving that $f(H^\perp)$ will be contained in a circle.

But we cannot rely on our intuition, and therefore we will prove all this with care to the details. (See (Hirsch, 1976) for definitions of the differential topology terms that we use in what follows.)

Let $\{\varphi_i, U_i\}_{i\in\Lambda}$ be an atlas for $\mathcal{M}$ where $\{U_i\}_{i\in\Lambda}$ is an open cover of $\mathcal{M}$ and $\varphi_i: U_i \to \mathbb{R}^q$. Then $x \in U_i$ for some $i \in \Lambda$, and we pick a radius $r \le \rho(x)$ small enough so that $S(x, r) \subset U_i$. Next, let $A_1 \in \Sigma_{q-1}(f)$ with $F \subset \bar{A}_1$ and pick a suitable coordinate system for $\varphi_i(U_i)$ so that $\varphi_i(A_1)$ lies in the subspace $\{(x_1, \cdots, x_{q-1}, 0) : x_i \in \mathbb{R}\}$ and $\varphi_i(F)$ lies in $\{(x_1, \ldots, x_{q-2}, 0, 0) : x_i \in \mathbb{R}\}$. Then $S(x, r)$ is the $(q-1)$-sphere with equation, in terms of the chart $\{\varphi_i, U_i\}$, $\sum_{i=1}^q x_i^2 = r^2$. Then

$$\varphi_i(H) = \varphi_i(F \cap S(x, \rho(x))) = \left\{ (x_1, \ldots, x_{q-2}, 0, 0) : \sum_{i=1}^{q-2} x_i^2 = r^2 \right\},$$

which is a $(q-3)$-sphere. Then we have

$$\varphi_i(H^\perp) = \{(0, \ldots, 0, x_{q-1}, x_q) : x_{q-1}^2 + x_q^2 = r^2\},$$

which is a circle. Parameterize $\varphi_i(H^\perp)$ with $\gamma(t) = (0, \ldots, 0, r\cos t, r\sin t)$ for $0 \le t \le 2\pi$. Then $\gamma'(t) = (0, \ldots, 0, -r\sin t, r\cos t)$.

Notice that the points $\gamma(0)$ and $\gamma(\pi)$ are $(0, \ldots, 0, \pm r, 0) \in \varphi_i(H^\perp)$. Furthermore, both $\gamma(0)$ and $\gamma(\pi)$ lie in the subspace $K = \{(x_1, \ldots, x_{q-1}, 0) : x_i \in \mathbb{R}\}$ that contains $\varphi(A_1)$. Now, $\varphi_i(F) \subset K$ as well, and $\varphi_i(A_1)$ borders $\varphi_i(F)$ in $K$, whereas $\gamma(0)$ and $\gamma(\pi)$ straddle $\varphi_i(F)$ in $K$. Therefore, exactly one of $\gamma(0)$ and $\gamma(\pi)$ will be in $\varphi_i(A_1)$; assume without loss of generality that $\gamma(0) \in \varphi_i(A_1)$ and let $a = \varphi_i^{-1}(\gamma(0))$. Since $\gamma(0)$ and $\gamma(\pi)$ are the only possible elements of $\varphi_i(H^\perp)$ that could also be in $\varphi_i(A_1)$, we have that $a = H^\perp \cap A_1$, proving Claim 1.

Now let $A_1, \ldots A_k$ be the $(q-1)$-strata of $\Sigma(f)$ that have $F$ on their boundary and let $a_j = H^\perp \cap A_j$ for $j = 1, \ldots, k$. Then the points $a_j$ divide the circle $H^\perp$ into $k$ arcs $\widehat{a_j a_{j+1}}$. The strata $\varphi_i(A_j)$ will lie in hyperplanes $\sum_{n=1}^q b_{j,n} x_n = 0$, but they all border $\varphi_i(F)$, so if $(x_1, \ldots, x_q) \in \varphi_i(A_j)$ then by projecting to $\varphi_i(F)$ we have that $(x_1, \ldots, x_{q-2}, 0, 0)$ also lies in this hyperplane. Subtracting the hyperplane equations for these two points gives $b_{j,q-1} x_{q-1} + b_{j,q} x_q = 0$, and so $x_q = -(b_{j,q-1}/b_{j,q}) x_{q-1}$. Setting $c = \arctan(-b_{j,q-1}/b_{j,q})$, we have coordinates $\varphi_i(a_j) = (0, \ldots, 0, r\cos c, r\sin c)$, and so the tangent vector of $\gamma$ at $\varphi_i(a_j)$ is $(0, \ldots, 0, -r\sin c, r\cos c)$. The inner product

$$\langle (0, \ldots, 0, -r\sin c, r\cos c), (x_1, \ldots, x_{q-1}, -(b_{j,q-1}/b_{j,q}) x_{q-1}) \rangle$$

$$= \frac{b_{j,q-1}r}{\sqrt{b_{j,q-1}^2 + b_{j,q}^2}} x_{q-1} + \frac{b_{j,q}r}{\sqrt{b_{j,q-1}^2 + b_{j,q}^2}} \left( \frac{-b_{j,q-1}}{b_{j,q}} \right) x_{q-1} = 0$$

shows that the tangent vector of $\gamma$ at $\varphi_i(a_j)$ is orthogonal to any point in $\varphi_i(A_j)$. Thus, reflecting about $\varphi_i(A_j)$ will map the arcs $\varphi_i(\widehat{a_{j-1}a_j})$ and $\varphi_i(\widehat{a_j a_{j+1}})$ onto each other.

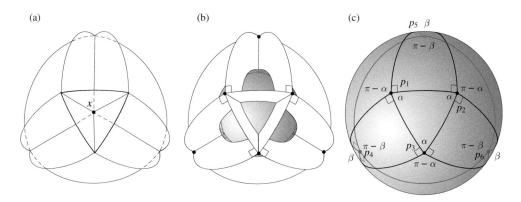

**Figure 10.8** (a) Example 10.38's proposed single-vertex crease pattern $\Sigma(f)$ in $\mathbb{R}^3$. (b) The planes of the crease pattern being projected onto a sphere centered at $x$. (c) The spherical projection, showing $\Sigma(f_x)$.

Since $\varphi_i$ is a homeomorphism, this means that reflecting about the $(q-1)$-strata $A_j$ in $\mathcal{M}$ will also map the arcs of $H^{\perp}$ onto each other, proving Claim 2 and the theorem. □

**Remark 10.37** Viewed through the lens of the Generalized Maekawa's Theorem for Isometric Foldings, we see that in the dim $\mathcal{M} = 2$ case of folding paper, we take $F$ to be a vertex in the crease pattern (a 0-strata) and the 1-strata $A_i$ will be crease lines adjacent to $F$ to arrive at the original Maekawa's Theorem. Increasing everything up by one dimension, we may consider the isometric folding $f: \mathbb{R}^3 \to \mathbb{R}^3$ in Example 10.23 (illustrated in Figure 10.5), where $F$ is the 1-strata $c$ and the $A_i$ are the 2-strata $c_1, \ldots, c_6$. The Generalized Maekawa's Theorem tells us that any valid MV assignment $\mu_f$ for this isometric folding must have $\sum_{i=1}^{6} \mu_f(c_i) = \pm 2$.

Thus, the aspect that makes Maekawa's Theorem work is the jump in dimension from dim $\mathcal{M} = q$ to dim $F = q - 2$. This specific shift in strata dimension allowed us to consider how $f$ acts on a 2-dimensional disc around $F$, which must obey the original Maekawa's Theorem.

Attempts to find other ways to generalize Maekawa's Theorem to isometric foldings do not work as well. We will see this in the next example.

**Example 10.38** We shall describe an example of an isometric folding $f: \mathbb{R}^3 \to \mathbb{R}^3$ whose crease pattern $\Sigma(f)$ at a 0-stratum $x$ satisfies the conclusion of Theorem 10.25 (the Angle Sum Theorem) but for which other versions of Maekawa's Theorem can't work and there cannot be a superposition order $>$ that satisfies the non-crossing conditions. Figure 10.8(a) shows this crease pattern, made of planes emanating from a vertex $x$, the geometry of which is made clear when we see how these planes project onto a sphere of radius 1, depicted in Figure 10.8(b) and (c).

On the sphere $S(x, 1)$, the crease planes cut out an equilateral triangle with angles $\alpha$ and corners $p_1, p_2, p_3$. From these corners we have geodesics extending at right angles to the sides of the triangle, meeting pairwise at vertices $p_4$, $p_5$, and $p_6$. Recalling that geodesics on a sphere are great circles, we have that the point $p_5$ is the polar point of the geodesic line containing $p_1p_2$. (That is, if we think of $p_1p_2$ as an equator of our sphere, then $p_5$ can be thought of as the "north pole" of that equator.) Similarly, $p_4$ is the polar point of the segment $p_1p_3$. Therefore, if we connect $p_4$ and $p_5$ with a geodesic, we will have that the point $p_1$ is the polar point to $p_4p_5$, which implies that the geodesics $p_1p_4$ and $p_1p_5$ intersect $p_4p_5$ at right angles. This justifies the angles labeled in Figure 10.8(c), and it also applies to the other angles at points $p_4$, $p_5$, and $p_6$. Thus this example demonstrates the Angle Sum Theorem in the case where $\dim \mathcal{M} = q = 3$ and $k = 1$. For example, if we take $F$ to be the 1-strata in $\Sigma(f)$ connecting points $p_1$ and $p_6$ (through the center of the sphere in Figure 10.8(c)), then the inner angles $a(A_i, F)$ for 3-strata $A_i$ adjacent to $F$ will be proportional to the angles $\alpha, \pi/2, \pi - \alpha$, and $\pi/2$ shown around the vertex $p_1$ of $\Sigma(f_x)$ in Figure 10.8(c). One can also demonstrate the Angle Sum Theorem in this example when $F$ is the 0-strata $x$:

---

**Diversion 10.6**   Prove that the spherical triangles of the crease pattern in Figure 10.8(c) have areas that satisfy the conclusion of the Angle Sum Theorem, Theorem 10.25.

---

However, this example also shows how other versions of Maekawa's Theorem generalization will fail. Suppose that we could find a valid MV assignment $\mu_f \colon \Sigma_2(f) \to \{-1, 1\}$. We know from Theorem 10.36 that the sum of $\mu_f$ around any 1-strata will be $\pm 2$. Is it possible that the sum of $\mu_f$ for all 2-strata adjacent to the 0-stratum $x$ could be constant? Looking at $\Sigma(f_x)$ in Figure 10.8(c), we see that the sum of $\mu_{f_x}$ around each vertex $p_i$ will equal $\pm 2$. That is, on the surface of a sphere, Maekawa's Theorem will hold at each vertex using the same proof we used in the flat paper case in Section 5.2. Given a vertex $p \in \Sigma_0(f_x)$, let $M_p$ (resp. $V_p$) equal the number of 1-strata $s$ adjacent to $p$ with $\mu_{f_x}(s) = -1$ (resp. $\mu_{f_x}(s) = 1$). We then say that $p$ **folds up** if $M_p - V_p = 2$ and **folds down** if $M_p - V_p = -2$. Let $U_x$ be the number of 0-strata in $f_x$ on $S(x, 1)$ that fold up and $D_x$ the number that fold down. Since $\mu_f = \mu_{f_x}$ by the MV Dimension Principle, we obtain that at our 3-manifold isometric folding vertex $x$,

$$\sum_{\substack{A \in \Sigma_2(f) \\ x \in \bar{A}}} \mu_f(A) = \frac{1}{2} \sum_{p \in \Sigma_0(f_x)} (M_p - V_p) = (2U_x - 2D_x)/2 = U_x - D_x.$$

Note that we needed to divide our sum by two because each 1-stratum is adjacent to two 0-strata in $\Sigma(f_x)$ and thus is counted twice in the sum. Clearly $U_x - D_x$ can equal many different values, in the same way that $M - V$ for multiple-vertex flat paper crease patterns can attain many different values. (Compare with Theorem 6.7.)

Next, we will argue that the situation in this example is much worse than a matter of the difference between mountains and valleys not being a constant around a 0-stratum; no valid MV assignments actually exist for this isometric folding! Consider that the equilateral triangle on the sphere has the same configuration as the equilateral triangle in the 2-dimensional impossible-to-fold-flat crease pattern in Figure 6.1(a). That is, if the crease pattern $\Sigma(f_x)$ drawn on the sphere in Figure 10.8(c) were flat-foldable, then the folding would induce a valid MV assignment $\mu_{f_x}$ on the creases. As long as we have $\alpha < \pi/2$, then we must have $\mu_{f_x}(p_1p_2) \neq \mu_{f_x}(p_2p_3)$ by Lemma 5.25 (the Big-Little-Big Lemma), and similarly we have $\mu_{f_x}(p_2p_3) \neq \mu_{f_x}(p_3p_1)$ and $\mu_{f_x}(p_3p_1) \neq \mu_{f_x}(p_1p_2)$. This chain of inequalities is impossible to satisfy with the two values in the range of $\mu_{f_x}$, $\{-1, 1\}$, and therefore the spherical crease pattern $\Sigma(f_x)$ is impossible to fold flat with a valid MV assignment, even though $f_x \colon S(x, 1) \to S(x, 1)$ is an isometric folding. By the MV Dimension Principle, this implies that the isometric folding $f \colon \mathbb{R}^3 \to \mathbb{R}^3$ with crease pattern shown in Figure 10.8(a) is not flat-foldable. That is, both $f$ and $f_x$ are not flat-foldable despite satisfying the conclusion of the Angle Sum Theorem. Stated another way, both $f$ and $f_x$ "fold flat" as phantom folds, but cannot do so without forcing a self-intersection of the 3-strata of $\Sigma(f)$ or the 2-strata of $\Sigma(f_x)$, respectively.

---

From the above example we see that trying to prove the converse direction of the Angle Sum Theorem where $F$ is a 0-stratum (vertex) only works when $q = 2$. Indeed, for crease patterns on 2-manifolds, proving the converse of the Angle Sum Theorem is the same as proving the converse direction of Kawasaki's Theorem (from Section 5.3).

Instead, extending the converse of Kawasaki's Theorem to higher-dimensional foldings needs to proceed similarly to our extension of Maekawa's Theorem. That is, if $f \colon \mathcal{M} \to \mathcal{N}$ is an isometric folding with $\dim \mathcal{M} = \dim \mathcal{N} = q$ and $F$ is a $(q-2)$-stratum, then, using the same techniques as in the proof of the Generalized Maekawa's Theorem for Isometric Foldings (Theorem 10.36), we take $x$ to be a point in the relative interior of $F$ and let $H = F \cap S(x, \rho(x))$. Then $H^\perp$ is 1-dimensional (a circle) and isometrically folds into a 1-dimensional submanifold in $\mathcal{N}$ under $f$. In fact, we can think of $H^\perp$ as the boundary of a disc $D^\perp$ of 2-dimensional paper, where $f_{D^\perp} = f$ restricted to $D^\perp$ is an isometric folding (see Figure 10.7(b)). If the angles $\alpha_i$ between the 1-strata of $\Sigma(f_{D^\perp})$ meeting at $x$ satisfy $\sum(-1)^i \alpha_i = 0$, then by (the 2-dimensional) Kawasaki's Theorem there exists a superposition order $>_{D^\perp}$ on the s-faces of the s-net of $f_{D^\perp}$ and an associated MV assignment $\mu_{D^\perp}$ that make $f_{D^\perp}$ flat-foldable.

Using the MV Dimension Principle, we can naturally extend $>_{D^\perp}$ and $\mu_{D^\perp}$ to a superposition order $>$ on the s-net of $f$ around the $(q-2)$-face $F$ and an MV assignment $\mu_f$ on the $(q-1)$-faces of $\Sigma(f)$ adjacent to $F$.

**Claim:** The folding $f$ will be flat-foldable under $>$, and $\mu_f$ will be valid around the face $F$.

To see this, suppose that $>$ violates the Taco-Tortilla non-crossing condition. Then there is a c-crease $e_1$ and an s-crease $e_2$ of the s-net of $f$ (thus, $e_1$ and $e_2$ are $(q-1)$-strata of the s-net of $f$ adjacent to $F$) such that $f(e_1) = f(e_2)$ and $F_1(e_1) > F_1(e_2) > F_2(e_1)$, using the notation of Definition 10.33. Since $x \in \overline{e_1} \cap \overline{e_2}$, we have that $e_1$ and $e_2$ both intersect $H^\perp$ and $D^\perp$. Let $d_1 = e_1 \cap D^\perp$ and $d_2 = e_2 \cap D^\perp$. Then the 2-strata of the s-net of $f_{D^\perp}$ that border $d_1$ will be $F_1(d_1) = F_1(e_1) \cap D^\perp$ and $F_2(d_1) = F_2(e_1) \cap D^\perp$. Define $F_1(d_2)$ and $F_2(d_2)$ similarly. Then $f(d_1) = f(d_2)$ and $F_1(d_1) >_{D^\perp} F_1(d_2) >_{D^\perp} F_2(d_1)$, which means that $>_{D^\perp}$ violates the Taco-Tortilla Condition. This contradicts our assumption that $f_{D^\perp}$ is flat-foldable under $>_{D^\perp}$.

Similar arguments can be made in the cases where $>$ violates the Tortilla-Tortilla and Taco-Tortilla Conditions. We have proven most of the following:

**Theorem 10.39** (Generalized Kawasaki's Theorem for Isometric Foldings)    *Let $f: \mathcal{M} \to \mathcal{N}$ be an isometric folding with dim $\mathcal{M} = $ dim $\mathcal{N} = q$, and let $F$ be a $(q-2)$-strata of $\Sigma(f)$. Let $\mathcal{A}$ (resp. $\mathcal{B}$) be the set of q-strata of $\Sigma(f)$ whose images under $f$ are positively (resp. negatively) oriented. Then $f$ is flat-foldable around $F$ if and only if*

$$\sum_{\substack{A \in \mathcal{A} \\ \overline{A} \supseteq F}} a(F, A) = \sum_{\substack{B \in \mathcal{B} \\ \overline{B} \supseteq F}} a(F, B). \tag{10.3}$$

*Proof*    All that remains to be shown is that the angle sum condition $\sum(-1)^i \alpha_i = 0$ around $x$ in $D^\perp$ is equivalent to the angle sum condition stated in the theorem. First, since each 1-strata of $\Sigma(f_{H^\perp})$ is contained in a $q$-strata adjacent to $F$ of $\Sigma(f)$, we can naturally match the 2-color classes $(\mathcal{A}, \mathcal{B})$ of the $q$-strata to 2-color classes $(\mathcal{A}_{H^\perp}, \mathcal{B}_{H^\perp})$ of the 1-strata of $\Sigma(f_{H^\perp})$. Then $\sum(-1)^i \alpha_i = 0$ is equivalent to

$$\sum_{\substack{A \in \mathcal{A}_{H^\perp} \\ x \in \overline{A}}} a(x, A) = \sum_{\substack{B \in \mathcal{B}_{H^\perp} \\ x \in \overline{B}}} a(x, B), \tag{10.4}$$

since each angle $\alpha_i$ equals $a(x, A)$ where $A$ is the 1-strata of $\Sigma(f_{H^\perp})$ subtended by the angle $\alpha_i$.

Let $A_{H^\perp}$ be the 1-strata of $\Sigma(f_{H^\perp})$ that is contained in the $q$-strata $A$ of $\Sigma(f)$ adjacent to $F$. That is, $A_{H^\perp} = A \cap H^\perp$. Then $a(x, A_{H^\perp})$ is the fraction of $S^1$ filled by $S^1 \cap C(x, A_{H^\perp})$, where $C(x, A_{H^\perp})$ is the 2-dimensional cone with apex $x$ spanned by $A_{H^\perp}$. Since $A_{H^\perp} = A \cap H^\perp$, this fraction is the same as the fraction of $S^{q-1}$ filled by $S^{q-1} \cap C(x, F, A)$, which equals $a(F, A)$ as given in Definition 10.24. Thus the angle sum equations (10.3) and (10.4) are equivalent.    □

Other implications of the necessary direction of Kawasaki's Theorem for 2-dimensional flat origamis can carry through to higher-dimensional analogs. For example, the Generalized Kawasaki's Theorem (Theorem 6.4) extends naturally to higher dimensions with a proof very similar to the 2-dimensional case. (This is stated, without proof, in (Kawasaki, 1989a).)

**Theorem 10.40**    *Let $f: \mathcal{M} \to \mathcal{N}$ be an isometric folding between two manifolds, with dim $\mathcal{M} = q$ and $\mathcal{M}$ simply connected. Let $\gamma$ be a closed curve on $\mathcal{M}$ whose image crosses a finite number of $(q-1)$-strata and does not cross any k-strata of $\Sigma(f)$ for*

$k < (q-1)$. *Let $s_i$ be the $(q-1)$-strata that $\gamma$ crosses, in order, for $i = 1, \ldots, n$, and let $R(s_i)$ denote the reflection isometry in $\mathcal{M}$ along $s_i$. Then we have*

$$R(s_1)R(s_2) \cdots R(s_n) = I.$$

---

**Diversion 10.7**  Prove Theorem 10.40.

---

## 10.6  Open Problems

As in the previous chapter, the results presented here seem ripe for further work. Stewart A. Robertson and some of his PhD students did continue to work on isometric foldings, often in directions of increased abstraction, such as (Robertson and El-Kholy, 1986), and sometimes trying to classify the kinds of isometric foldings of the form $f: S^2 \to S^2$, such as (Azevédo Breda, 1992).

Robertson suggests an especially interesting problem in (Robertson, 1977–1978), which he stated was still open (and interesting) in 1997 (CIM, 1997):

**Open Problem 10.1**  Let $f \in \mathcal{F}(S^m, S^m)$ not be the identity map. Is it true that any non-trivial isometric folding $g$ of $S^m$ into itself is homotopic, through isometric foldings, to $f$ or an isometry of $f$?

In other words, can one isometric folding of the sphere $S^m$ be "rolled" into any other (nontrivial) one? Robertson (1977–1978) established that this is true when $m = 1$, but that even the $m = 2$ case is open.

Another avenue one can pursue is to combine this chapter with Part I and explore higher-dimensional origami geometric constructions. For example, when folding $\mathbb{R}^3$ our "crease lines" become planes, and we construct new points, lines, and planes by reflecting about the crease planes. First steps in this kind of 3-dimensional origami geometric constructions can be found in (Alperin, 2018; Lucero, 2018).

**Open Problem 10.2**  What is the progression of algebraic power generated by single-fold origami geometric constructions when folding $\mathbb{R}^n$ along $(n-1)$-dimensional hyperplanes? (We know when $n = 2$ that we can solve up to quartic equations. The problem is unexplored for $n > 2$.)

## 10.7  Historical Remarks

Stewart A. Robertson seems to have been the first person to fully investigate, and publish, the generalization of paper folding to folding manifolds. Independently, and with a different motivation, Jim Lawrence and Jonathan Spingarn undertook a study in the mid-1980s on foldings from $\mathbb{R}^q$ to $\mathbb{R}^q$. Robertson states that his work was inspired by supposing that "a plane sheet of paper is crumpled gently in the hands, and then is crushed flat against a desk top" (Robertson, 1977–1978). In contrast, Lawrence and Spingarn state that their interest in isometric foldings grew from their studies of non-expansive mappings $f\colon \mathbb{R}^q \to \mathbb{R}^q$, in particular their observation that the sequence of **average iterates of** $f$, defined by $x_{n+1} = (f(x_n) + x_n)/2$, behave as if $f$ were a global isometry except for a finite number of initial terms (Lawrence and Spingarn, 1987). While our treatment of this topic follows Robertson's fairly closely, the proofs of many results of Robertson and of Lawrence and Spingarn are very similar.

Interestingly, Toshikazu Kawasaki himself explored extending flat origami mappings $f\colon \mathbb{R}^2 \to \mathbb{R}^2$ to higher-dimensional flat origami on $\mathbb{R}^n$ and on the sphere $S^n$ (Kawasaki, 1989a). He conjectured the cellular decomposition structure of $\Sigma(f)$, but offered no proofs.

In the 2000s, Dacorogna, Marcellini, and Paolini undertook a similar study, unaware of the prior work of Robertson, Lawrence and Spingarn, and Kawasaki (Dacorogna et al., 2008). However, this more recent work took a very different approach, based on analysis instead of topology, and came from yet another area of motivation. We will examine their work in Chapter 11.

# 11 An Analytic Approach to Isometric Foldings

One major theme of this book is that origami can be modeled mathematically in many different ways. Sometimes the approach is dictated by the specific aspect of folding that captures one's interest, such as geometric constructions requiring the methods discussed in Part I of this book. Other times an application or one's own background will lead to a certain approach.

In (Dacorogna et al., 2008), Bernard Dacorogna, Paolo Marcellini, and Emanuele Paolini introduce an analytic approach to origami. In fact, their work is done in arbitrary dimension, and thus can be compared to the isometric foldings from Chapter 10 of the form $f\colon M \subset \mathbb{R}^n \to \mathbb{R}^m$. Their approach is to consider such maps $f$ that are **Lipschitz continuous** and whose Jacobian is an orthogonal matrix almost everywhere (that is, except for a set of measure zero in $M$, which would be the singular set $\Sigma(f)$). Dacorogna, Marcellini, and Paolini's work is motivated by their search for explicit solutions to some Dirichlet problems in partial differential equations, which explains their analytical approach.

Rather than develop this analytical approach completely from scratch (as done in (Dacorogna et al., 2008)), we will develop the theory enough to prove its equivalence to Robertson's and Lawrence and Spingarn's work from Chapter 10, explore some of its aspects, and then briefly see how it gives solutions to certain Dirichlet problems.

## 11.1 Lipschitz Continuous and Rigid Maps

In what follows we let $\Omega \subset \mathbb{R}^n$ be a bounded, open set and consider functions $f\colon \Omega \subset \mathbb{R}^n \to \mathbb{R}^m$ for $n \leq m$. Since $f$ is a function of $n$ variables, let us write

$$f(x) = f(x_1, \ldots, x_n) = (f_1(x_1, \ldots, x_n), \ldots, f_m(x_1, \ldots, x_n)).$$

Then the **Jacobian** of $f$ at a point $p$ is the $m \times n$ matrix

$$J_p(f) = \begin{pmatrix} \frac{\partial f_1}{\partial x_1}(p) & \cdots & \frac{\partial f_1}{\partial x_n}(p) \\ \vdots & \ddots & \vdots \\ \frac{\partial f_m}{\partial x_1}(p) & \cdots & \frac{\partial f_m}{\partial x_n}(p) \end{pmatrix}.$$

Recall that an $m \times n$ matrix $A$ is **orthogonal** if $A^T A = I$, where $A^T$ denotes the transpose of $A$. In fact, let $O(m, n)$ denote the set of all $m \times n$ orthogonal matrices. Also recall that as a linear transformation, orthogonal matrices are isometries.

**Definition 11.1**    A function $f \colon \mathbb{R}^n \to \mathbb{R}^m$ is **Lipschitz continuous** if there exists a real constant $k \geq 0$ such that $\|f(x) - f(y)\| \leq k\|x - y\|$ for all $x, y \in \mathbb{R}^m$.

**Definition 11.2**    Let $f \colon \Omega \subset \mathbb{R}^n \to \mathbb{R}^m$. We say that $f$ is a **rigid map** if $f$ is Lipschitz continuous and $J_p(f) \in O(m, n)$ for almost every $p \in \Omega$.

As in Chapter 10, we let $\Sigma(f)$ denote the set of points in the domain $\Omega$ where $f$ is not differentiable.

**Definition 11.3**    We say that a rigid map $f$ is **piecewise** $C^1$ if the following conditions hold:

(i)  $\Sigma(f)$ is closed in $\Omega$.
(ii)  $f$ is $C^1$ on every connected component of $\Omega \setminus \Sigma(f)$.
(iii)  For every compact set $K \subset \Omega$, the number of connected components of $\Omega \setminus \Sigma(f)$ that intersect $K$ is finite.

This treatment, where we will model origami foldings by piecewise $C^1$ rigid maps, gives us what we need to emulate Robertson's treatment in Chapter 10 in the case where the domain and co-domain have the same dimension. The proofs given in (Dacorogna et al., 2008) are considerably different from Robertson's (and Lawrence and Spingarn's) and worth examining on their own. Therefore, we will reproduce some of them here.

**Lemma 11.4**    *Let $f \colon \Omega \subset \mathbb{R}^n \to \mathbb{R}^n$ be a piecewise $C^1$ rigid map. Then $J_p(f)$ is constant (or equivalently, $f$ is affine) on every connected component of $\Omega \setminus \Sigma(f)$.*

*Proof*    Let $C$ be a connected component of $\Omega \setminus \Sigma(f)$. Then $\Sigma(f|_C) = \emptyset$ and thus $J_p(f|_C) \in O(n, n)$ for all $p \in C$. Thus $f$ is an isometry on $C$ (i.e., $f$ is affine). □

**Lemma 11.5**    *Let $f$ be a rigid map where the domain $\Omega$ is convex. Then $\|f(x) - f(y)\| \leq \|x - y\|$ for all $x, y \in \Omega$.*

This lemma, demonstrating the non-stretching nature of paper, is equivalent to Lemma 10.2, but because of the different setup here, we need an analytical proof.

*Proof*    Let $x, y \in \Omega$. By the convexity of $\Omega$, we can, for every $\varepsilon > 0$, find a curve $\gamma \colon [0, 1] \to \Omega$ such that $\gamma(0) = x$, $\gamma(1) = y$, the length of $\gamma$ $\ell(\gamma) \leq \|x - y\| + \varepsilon$, and $f$ is differentiable on the points $\gamma(t)$ for almost every $t \in [0, 1]$. Now, since $f$ is a rigid map, $J_p(f)$ is an orthogonal matrix, which means that $\|J_p(f)x\| = \|x\|$ for all $p \in \Omega$ and almost every $x \in \Omega$, because the Jacobian is an isometry. Thus we have

$$\|f(x) - f(y)\| = \|f(\gamma(1)) - f(\gamma(0))\| \le \int_0^1 \left\| \frac{d}{dt} f(\gamma(t)) \right\| dt$$

$$= \int_0^1 \left\| J_{\gamma(t)}(f)\gamma'(t) \right\| dt = \int_0^1 \left\| \gamma'(t) \right\| dt \le \|x - y\| + \varepsilon.$$

Letting $\varepsilon \to 0$, we obtain the desired result. $\qquad\square$

**Definition 11.6** A set $F \subset \mathbb{R}^n$ is a $k$-dimensional **convex polyhedral facet** if $F$ is bounded, nonempty, and closed and there exists an affine $k$-dimensional plane $\Pi$ and a finite number of open affine half-spaces $H_1, \ldots, H_N$ such that

$$F = \overline{\Pi \cap H_1 \cap \cdots \cap H_N}$$

(where $\overline{A}$ denotes the closure of $A$). The plane $\Pi$ is called the **supporting plane** of $F$.

We say a set $F \subset \mathbb{R}^n$ is a $k$-dimensional **polyhedral set** if $F$ is the union of a finite number of $k$-dimensional convex polyhedral facets.

A set $F$ is a **locally finite $k$-dimensional polyhedral set** in $\Omega$ if given any point $x \in F$ there exists a neighborhood $U \subset \Omega$ of $x$ such that $F \cap \overline{U}$ is a polyhedral set.

We are now ready for our analytical equivalent to Theorem 10.17.

**Theorem 11.7** (Dacorogna et al., 2008) *Let $f : \Omega \subset \mathbb{R}^n \to \mathbb{R}^n$ be a piecewise $C^1$ rigid map (where $\Omega$ is open). Then $\Sigma(f)$ is a locally finite $(n-1)$-dimensional polyhedral set. Furthermore, if $\Omega$ is convex then every connected component of $\Omega \setminus \Sigma(f)$ is a convex set.*

*Proof* We will prove the "furthermore" first. Assume that $\Omega$ is convex and let $A$ be a connected component of $\Omega \setminus \Sigma(f)$. Then on $A$ our rigid map has the form $f(x) = Jx + q$ for some $J \in O(n, n)$ and $q \in \mathbb{R}^n$. Let $x_1, x_2 \in A$ and $t \in [0, 1]$. Consider the point $x = tx_1 + (1 - t)x_2$; we would like to show that $x \in A$. Since $J$ is an isometry and by Lemma 11.5,

$$\|x_1 - x_2\| = \|f(x_1) - f(x_2)\| \le \|f(x_1) - f(x)\| + \|f(x) - f(x_2)\|$$

$$\le \|x - x_1\| + \|x - x_2\| = \|x_1 - x_2\|.$$

Thus the inequalities in the previous expression are all equalities, and also (by the triangle inequality) $\|f(x) - f(x_1)\| = \|x - x_1\|$ and $\|f(x) - f(x_2)\| = \|x - x_2\|$. In other words, $f$ preserves the distance between $x$ and $x_1$ and $x_2$ for all $x$ in the convex hull of $A$, conv($A$). This means that for all $x \in$ conv($A$) we have $f(x) = Jx + q$, which implies that $f$ is differentiable for all $x \in$ conv($A$). That is, $\Sigma(f)$ is outside the convex hull of $A$, implying that $A$ is convex.

We now consider the special case where $\Omega$ is an open cube in $\mathbb{R}^n$ and $\Omega \setminus \Sigma(f)$ has only a finite number of ($n$-dimensional) connected components. We want to show that $\overline{\Sigma(f)}$ (which denotes the closure in $\mathbb{R}^n$, not just $\Omega$) is a polyhedral set. Since $\Omega$ is convex, we know that each connected component of $\Omega \setminus \Sigma(f)$ is convex.

Let $A$ and $A'$ be two such convex, connected components. Then there exists an $(n-1)$-dimensional affine plane $\Pi$ that separates $A$ from $A'$. Let $\Pi_1, \ldots, \Pi_N$ be the $(n-1)$-dimensional planes that separate $A$ from all other connected components, and let $\Pi_{N+1}, \ldots, \Pi_M$ be all the planes containing $(n-1)$-facets of the cube $\Omega$ (thus $M = N + 2n$). Then let $H_1, \ldots, H_M$ be the half-spaces such that $\partial H_i = \Pi_i$ and such that $A \subset H_i$. Now let $K = \bigcap_{i=1}^{M} H_i$, which by definition is an $n$-dimensional polyhedral set whose boundary $\partial K$ is an $(n-1)$-dimensional polyhedral set.

**Claim:** $K = \overline{A}$.

*Proof of claim:* The fact that $A \subset K$ follows from the definitions of $\Pi_i$ and $H_i$, and thus we have that $\overline{A} \subseteq K$. On the other hand, let $x \notin A$. If $x \notin \Omega$ then clearly $x \notin \overline{K}$ since $K \subset \overline{\Omega}$. Otherwise, if $x \in \Omega \setminus A$ then there exists another connected component $A'$ of $\Omega \setminus \Sigma(f)$ such that $x \in \overline{A'}$. Now, for some $i$ we have that $A' \subset \mathbb{R}^n \setminus H_i$, and thus $x \in \overline{A'} \subset \mathbb{R}^n \setminus K$. Therefore, $\mathbb{R}^n \setminus A \subset \mathbb{R}^n \setminus K$, which together with $A \subset K$ gives us $\overline{A} = K$. Our claim is proven.

We have established that if $\Omega$ is a cube then every connected component $A$ of $\Omega \setminus \Sigma(f)$ is an $n$-dimensional polyhedral set. Hence the boundary $\partial A$ is an $(n-1)$-dimensional polyhedral set, as is $\overline{\Sigma(f)}$ since $\Sigma(f) \cup \partial\Omega$ is the union of all the boundaries of the connected components of $\Omega \setminus \Sigma(f)$ and $\partial\Omega$ is itself a polyhedral set.

In general, when $\Omega$ is any open set and $f$ is a piecewise $C^1$ rigid map, we take any point $x \in \Sigma(f)$ and consider a cubic neighborhood $U$ of $x$ that intersects a finite number of connected components of $\Omega \setminus \Sigma(f)$. We know that $\overline{\Sigma(f) \cap U} = \Sigma(f) \cap \overline{U}$ is polyhedral and hence $\Sigma(f)$ itself is a locally finite polyhedral set. $\qquad\square$

The following shows the connection between piecewise $C^1$ rigid maps and isometric foldings (from Chapter 10).

**Theorem 11.8**   *A function $f: \Omega \subset \mathbb{R}^n \to \mathbb{R}^m$ is an isometric folding if and only if it is a piecewise $C^1$ rigid map.*

*Proof*   Suppose $f$ as stated is an isometric folding. Then by Lemma 10.2 we know that $f$ is Lipschitz continuous with $k = 1$. Further, we know that $\Sigma(f)$ has $m$-dimensional volume equal to zero, and so let $x \in \Omega$ be an element of an $m$-strata $A$ of $f$ (thus anything we say about $x$ will qualify as an "almost everywhere" statement). Then $f|_A$ is an isometry, since $A$ contains no singularities of $f$ and thus maps geodesics to geodesics. Therefore, for all $p \in A$, $J_p(f)$ is an orthogonal matrix.

For the converse, let $f$ be a piecewise $C^1$ rigid map and let $\gamma: [a,b] \to \Omega$ be a zig-zag on $\Omega$. Since $\gamma$ is continuous, we have that $\gamma([a,b]) \subset \Omega$ is compact and thus intersects a finite number of components of $\Omega \setminus \Sigma(f)$; call these components $C_1, \ldots C_k$, each of which is an open set since $\Sigma(f)$ is closed in $\Omega$.

If $f|_{C_i}$ denotes $f$ restricted to the component $C_i$, then we have that $\Sigma(f|_{C_i}) = \emptyset$, thus $J_p(f|_{C_i}) \in O(m,n)$ for all $p \in C_i$, and thus $f$ is an isometry on $C_i$. Therefore, $f(\gamma([a,b]) \cap C_i)$ is a zig-zag for all $i = 1, \ldots, k$.

Let $[a_i, b_i]$ be the longest section of $[a, b]$ such that $\gamma((a_i, b_i)) \subset C_i$. (That is, we can't expand the endpoints of $(a_i, b_i)$ and still have $\gamma$ of it being in $C_i$.)

Then if $b_i = a_{i+1}$ for all $i$, we have that $f(\gamma([a, b]))$ is a zig-zag and we're done (i.e., $f$ is an isometric folding).

If not, then there exists an $i$ such that $b_i < a_{i+1}$ and thus $\gamma([b_i, a_{i+1}]) \subset \Sigma(f)$. Since $\Sigma(f)$ is an $(n-1)$-dimensional polyhedral set, we have that $\gamma([b_i, a_{i+1}])$ is contained in the union of a finite number of $(n-1)$-dimensional convex polyhedral facets.

Take $x \in [b_i, a_{i+1})$. Then since $\Sigma(f)$ has zero $n$-dimensional volume in $\Omega$ (else $J_p(f)$ would not be an orthogonal matrix almost everywhere in $\Omega$), we have that for any open neighborhood $U \subset \Omega$ containing $\gamma(x)$, $U$ also intersects at least two connected components, say $U_1$ and $U_2$ of $\Omega \setminus \Sigma(f)$. We get at least two because if, say, $U \setminus \Sigma(f) \subset U_1$ then on $U_1$ we have that $f$ is almost everywhere an affine map, which by continuity means that $f$ is differentiable everywhere on $U$, contradicting that $\gamma(x) \in \Sigma(f)$.

Now fix a neighborhood $U$ of $\gamma(x)$ with $U_1, \ldots U_l$ the connected components of $\Omega \setminus \Sigma(f)$ that intersect $U$. Since $f$ is affine on each $U_j$, we can write $f(x) = L_j(x) = f(p_j) + J_{p_j}(f)(x - p_j)$ on $U_j$ for $j = 1, \ldots, l$, where $p_j \in U_j$.

Pick a real number $c$ as big as possible with $x < c \le a_{i+1}$ such that $\gamma([x, c)) \subset U$ and $\gamma([x, c])$ is contained in a single $(n-1)$-dimensional convex polyhedral facet. Then

$$\gamma([x, c]) \subset U \cap \Sigma(f) \subset \overline{U_1} \cap \cdots \cap \overline{U_l}.$$

By continuity of $f$, we conclude that the affine maps $L_1, \ldots, L_l$ coincide on $U \cap \Sigma(f)$. This means that $f$ acts as an isometry on the curve $\gamma([x, c])$, and thus $f$ preserves its length.

We may now do the above process multiple times, starting with $x = b_i$ to get an interval $[b_i, c_1]$ where $f$ preserves the length of $\gamma([b_i, c_1])$, then letting $x = c_1$ to get an interval $[c_1, c_2]$ where $f$ preserves the length of $\gamma([c_1, c_2])$, and so on.

**Claim:** This process must end at some point with some $c_q = a_{i+1}$.

*Proof of claim:* Suppose not. Then we have an infinite partition of $[b_i, a_{i+1}]$, that is, $b_i < c_1 < c_2 < \cdots < a_{i+1}$, where the lengths of $[c_j, c_{j+1}]$ must get arbitrarily small. In other words, there is a limit point $c$ of the $c_j$ in $[b_i, a_{i+1}]$. Since each set $\gamma([c_j, c_{j+1}])$ is in an $(n-1)$-dimensional convex polyhedral facet of $\Sigma(f)$, $\gamma([c_{j+1}, c_{j+2}])$ must be in a different $(n-1)$-dimensional facet (by construction of the $c_j$), and $c$ is a limit point of all the $c_j$s, we conclude that the compact set $\gamma([b_i, a_{i+1}])$ must intersect an infinite number of $(n-1)$-dimensional convex polyhedral facets of $\Sigma(f)$. That means that if we let $K$ be a compact subset of $\Omega$ with $\gamma([b_i, a_{i+1}]) \subset K$ (which must exist since $\Omega$ is open), then $K$ intersects an infinite number of components of $\Omega \setminus \Sigma(f)$, contradicting the fact that $f$ is piecewise $C^1$. This proves our claim.

We conclude that the curve $f(\gamma([b_i, a_{i+1}]))$ is made of a finite number of geodesic curves and is thus a zig-zag. Since this can be done to any other sections of $\gamma([a, b])$ that are contained in $\Sigma(f)$, we have that the whole curve $f(\gamma([a, b]))$ is a zig-zag, and thus $f$ is an isometric folding. $\qquad\square$

The development given in (Dacorogna et al., 2008) proves that a piecewise $C^1$ rigid map $f$ satisfies Kawasaki's Theorem about every $(n-2)$-dimensional facet $P$ of $\Sigma(f)$, where the angles are taken to be the dihedral angles between the $(n-1)$-facets that meet at $P$. However, since we have that piecewise $C^1$ rigid maps are isometric foldings, we obtain this result immediately from Theorem 10.17 and the Angle Sum Theorem 10.25 (where we would have to replace $\Omega$ with $\overline{\Omega}$ to achieve compactness of the domain of $f$), as well as the other Kawasaki-like results for lower-dimensional facets of $\Sigma(f)$.

The authors of (Dacorogna et al., 2008) also independently derived a version of the Recovery Theorem along the lines of that given by Lawrence and Spingarn (Theorem 10.27), describing how a rigid map can be uniquely created from a given crease pattern. Since we did not prove this result in Section 10.4, we will prove it here using Dacorogna, Marcellini, and Paolini's approach.

**Theorem 11.9** (Recovery Theorem (Dacorogna et al., 2008))   *Let $\Omega$ be a simply connected subset of $\mathbb{R}^n$. Let $\Sigma \subset \Omega$ be a locally finite polyhedral set that satisfies the Kawasaki condition ($\sum \alpha_{2i} = \sum \alpha_{2i+1} = \pi$) around every $(n-2)$-dimensional facet. Then there exists a rigid map $f$ such that $\Sigma = \Sigma(f)$. Moreover, $f$ is uniquely determined once we fix a value $y_0 = f(x_0)$ and the Jacobian $J_{x_0}(f)$ for a point $x_0 \in \Omega \setminus \Sigma$.*

*Proof*   The basic idea of the proof is to use the idea of a folding map $\sigma_f$ and vertex-avoiding curves $\gamma$ from Chapter 6, except generalized to higher dimensions, to construct the rigid map $f$ from the singular set $\Sigma$. We follow the proof from (Dacorogna et al., 2008) but with our notation.

Let $\Gamma$ be the set of all continuous curves $\gamma : [0,1] \to \Omega$ whose image

(1) intersects a finite number of $(n-1)$-dimensional polyhedral facets of $\Sigma$ (i.e., $(n-1)$-strata),
(2) Crosses no $k$-dimensional polyhedral facets of $\Sigma$ for $k < (n-1)$,
(3) is not tangent to any of the $(n-1)$-dimensional polyhedral facets.

Given a $\gamma \in \Gamma$, let $0 < t_1 < t_2 < \cdots < t_N < 1$ be the points where $\gamma(t)$ passes through an $(n-1)$-dimensional facet $F_j \ni \gamma(t_j)$ of $\Sigma$. Let $R(F_j)$ denote the reflection in $\mathbb{R}^n$ about the $(n-1)$-dimensional plane containing $F_j$. Let

$$A_\gamma = R(F_1)R(F_2) \cdots R(F_N).$$

We claim that $A_\gamma$ depends only on the endpoints $\gamma(0)$ and $\gamma(1)$ and not on the curve itself. (Compare to Theorem 6.6.) To see this, consider another curve $\mu \in \Gamma$ with $\mu(0) = \gamma(0)$ and $\mu(1) = \gamma(1)$. Then the combined curve $\gamma\mu^{-1}$ made by traveling along $\gamma$ and then backward along $\mu$ is a closed curve, and by the proof of Theorem 10.40 (which uses the Kawasaki condition given in our hypothesis), we have that $A_\gamma A_{\mu^{-1}} = I$, which implies that $A_\gamma = A_\mu$.

We will now construct our rigid map $f : \Omega \to \mathbb{R}^n$. Pick points $x_0 \in \Omega$ and $y_0 \in \mathbb{R}^n$ and set $f(x_0) = y_0$. Also pick a matrix $J_0 \in O(n,n)$, which will be the Jacobian $J_{x_0}(f)$.

Take an $x \in \Omega \setminus \Sigma$ and define $f(x) = L_0 A_\gamma x$, where $\gamma$ is any curve in $\Gamma$ starting at $x_0$ and ending at $x$ and $L_0$ is defined as $L_0 x = y_0 + J_0 x$. Then by continuity we may extend $f$ to all of $\Omega$. Indeed, for every $(n-1)$-dimensional facet $P$ of $\Sigma$, the affine functions defining $f$ on either side of $P$ differ only by a reflection that leaves $P$ fixed. This is also true on the lower-dimensional facets of $\Sigma$ since they all live in the intersections of $(n-1)$-dimensional facets.

Therefore, $f \colon \Omega \to \mathbb{R}^n$ is a rigid map with $\Sigma(f) = \Sigma$, $J_{x_0}(f) = J_0$, and $f(x_0) = y_0$. By construction we also have that $f$ is the unique rigid map with these properties. $\quad\square$

## 11.2 Applications to Dirichlet Problems

In (Dacorogna et al., 2008), Dacorogna, Marcellini, and Paolini state that their work on rigid piecewise $C^1$ rigid maps was motivated by the search for explicit solutions for certain types of Dirichlet problems. A **Dirichlet problem** is a partial differential equation of the following type: For an open set $\Omega \subset \mathbb{R}^n$ we are given a Lipschitz continuous function $\varphi \colon \overline{\Omega} \to \mathbb{R}^m$ (called the **boundary data** of the problem) and a family $E$ of $m \times n$ matrices. The problem is to find a Lipschitz continuous function $f \colon \overline{\Omega} \to \mathbb{R}^m$ such that

$$\begin{cases} J_p(f) \in E & \text{for almost every } p \in \Omega, \\ f(x) = \varphi(x) & \text{for all } x \in \partial\Omega. \end{cases}$$

Dirichlet problems have their roots in mathematical physics and harmonic functions. We limit ourselves to the case where $m = n \geq 2$ and $\varphi$ is an affine map where the eigenvalues of $J_p(\varphi)$ are less than 1. We also let our set $E$ be the set of orthogonal $n \times n$ matrices. In other words, we want $f$ to be an isometry almost everywhere in $\Omega$ but an affine contraction on $\partial\Omega$. The existence of a solution $f$ to the Dirichlet problem under these conditions has been proven (Dacorogna and Marcellini, 1999), but the construction of explicit solutions remains the subject of current research. It turns out, from (Dacorogna et al., 2008), that piecewise $C^1$ rigid maps can provide such explicit solutions.

---

**Example 11.10** Let $\Omega = (-c, c) \times (-d, d)$ for $c, d > 0$ and $\varphi(x, y) = (\alpha x, \beta y)$ for $\alpha, \beta \in (0, 1)$ such that

$$\frac{d^2}{c^2} = \frac{1 - \alpha^2}{1 - \beta^2}.$$

The explicit solution for this Dirichlet problem will be a piecewise $C^1$ rigid fold whose crease pattern is made by combining the base module crease patterns shown in Figure 11.1. In these base modules, we define the various lengths to be as follows:

$$a' = \frac{a}{4}(1 + \alpha), \quad a'' = \frac{a}{2}(1 - \alpha), \quad b' = \frac{b}{4}(1 + \beta), \quad b'' = \frac{b}{2}(1 - \beta).$$

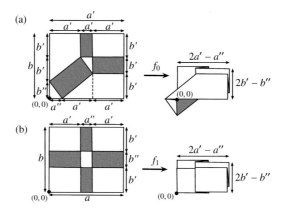

**Figure 11.1** The base module crease patterns for Example 11.10.

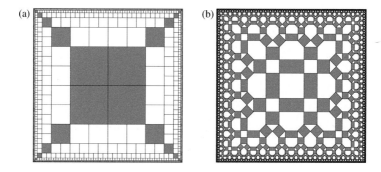

**Figure 11.2** The (a) rectangular tiling and (b) crease pattern for the rigid fold $f$ in Example 11.10 (with $a = b = 1$ and $\alpha = \beta = 0.3$).

---

**Diversion 11.1**    Prove that the crease patterns shown in Figure 11.1 satisfy the Kawasaki condition at each of their interior vertices. Then prove that the rigid maps $f_0$ and $f_1$ that result from the Recovery Theorem (Theorem 11.9) agree (up to an isometry) with the map $\varphi$ on the four corners of the rectangle $R = [0, a] \times [0, b]$.

---

Now we divide $\Omega$ into infinitely many rectangles similar to $\Omega$ as shown in Figure 11.2(a). We place the base module for $f_1$ into the rectangles on the diagonals of $\Omega$ and the base module for $f_0$ into the other rectangles (scaled and rotated appropriately) to form the crease pattern $\Sigma$ shown in Figure 11.2(b). By Diversion 11.1 we know that each interior vertex of $\Sigma$ satisfies the Kawasaki condition. Therefore, there exists a rigid map $f \colon \Omega \rightarrow \mathbb{R}^2$ with $\Sigma(f) = \Sigma$. We have that $J_p(f)$ is an isometry almost everywhere in $\Omega$ and thus is an orthogonal matrix. Also note that on our base modules we have

$$f_0(a, 0) = (2a' - a'') = (\alpha a, 0), f_0(0, b) = (0, 2b' - b'') = (0, \beta b),$$
$$f_1(a, 0) = (2a' - a'') = (\alpha a, 0), f_1(0, b) = (0, 2b' - b'') = (0, \beta b),$$

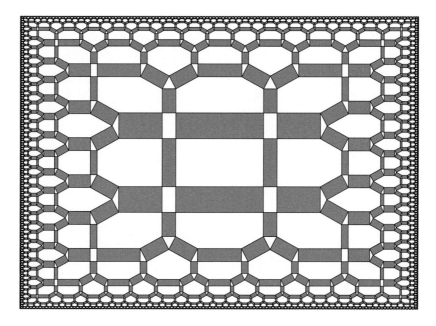

**Figure 11.3** The rigid fold $f$ in Example 11.10 (with $a = 1.4$, $b = 1$, and $\alpha = \beta = 0.3$).

and therefore $f(x, y) = \varphi(x, y)$ for all vertices of our rectangle tiling in Figure 11.2(a). Since the boundary $\partial\Omega$ is contained in the closure of all these rectangle vertices and $f$ is continuous, we thus have that $f = \varphi$ on $\partial\Omega$. We conclude that $f$ is a solution to this Dirichlet problem.

---

The rigid fold solution to the Dirichlet problem in Example 11.10 is interesting in a number of ways. First, notice that the definition of "rigid fold" that we are using here does not specify an MV assignment for the crease pattern $\Sigma$. For the purposes of a solution, all that is needed is a map from $\Omega$ to $\mathbb{R}^2$, and so mountains and valleys are irrelevant. However, that also means that for the purposes of a Dirichlet problem, we do not care if the resulting rigid fold is actually flat-foldable. Indeed, it might not be! The example shown in Figure 11.2(b) is foldable (this is not obvious; it is extremely difficult to fold), meaning that an MV assignment can be found that is globally flat-foldable.

However, for different choices of the parameters $a$ and $b$, the resulting crease pattern in this example might not be flat-foldable. Figure 11.3 shows one example where $a = 1.4$ and $b = 1$. Notice that in the crease pattern we see near-equilateral triangles that match the impossible crease pattern from Figure 6.1(a) in Chapter 6. Therefore, there is no valid MV assignment that can be assigned to these equilateral triangles, and thus this crease pattern is not flat-foldable. It is a phantom fold, however, and that is all the Dirichlet problem requires.

Second, the rigid folds generated by Example 11.10 give examples of an unusual collection of origami crease patterns known as **self-similar origami**. These are

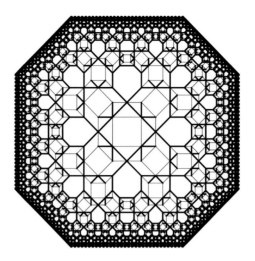

**Figure 11.4** A rigid fold $f$ that is an explicit solution to the Dirichlet problem where $\Omega$ is an octagon and $\varphi(x, y) = (x/2, y/2)$. This crease pattern was discovered by origami artist Frank van Kollem.

origami crease patterns that possess self-similar aspects and involve an infinite number of crease lines approaching the boundary or a single point on the paper. Of course, the case where we have an infinite number of crease lines cannot be physically folded, and so origami practitioners stop short and approximate the crease pattern by folding only a finite number of iterations of the self-similar pattern.

Figure 11.4 shows another example, discovered by Dutch origami artist Frank van Kollem. This crease pattern is flat-foldable, and a valid MV assignment is shown.

---

**Diversion 11.2**     Demonstrate that the rigid map singular set shown in Figure 11.4 provides a solution to the Dirichlet problem where $\Omega$ is a regular octagon and the boundary data is $\varphi(x, y) = (x/2, y/2)$.

---

These examples of crease patterns highlight a difference between the isometric folding model of Robertson (and Lawrence and Spingarn) and the more analytic model for piecewise $C^1$ rigid folds given by Dacorogna et al. Crease patterns with an infinite number of creases are not allowed in an isometric folding; they would allow a geodesic curve to map into an infinite zig-zag, which is explicitly forbidden in an isometric fold. But such crease patterns where the crease lines approach limits on the boundary of the domain $\Omega$ are allowed in piecewise $C^1$ rigid folds because $\Omega$ is open.

One can see the need to avoid limit points in the crease pattern, since any creases or vertices that approach a limit in the interior of the paper will cease to form a cellular decomposition of the domain at the limit point. That is, $\Sigma(f)$ will have an infinite number of strata, which causes problems for some of the proofs of Chapter 10.

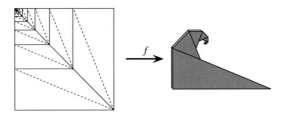

**Figure 11.5** The self-similar wave fold, from (Hull, 2012).

This raises the question of how we should incorporate such "infinite" crease patterns into a mathematical model of origami. Indeed, there are many such origami works in the origami literature. Another example that is not the solution to a Dirichlet problem is the self-similar wave shown in Figure 11.5 (this appears in (Hull, 2012)). There are many other examples from origami artists such as Paulo Barreto (1997), Shuzo Fujimoto (2011), Jun Maekawa (1997), and Chris Palmer (1997).

One way to include such origami models into the realm of flat-foldability and isometric foldings is to restrict ourselves to what we can physically fold. In reality, we cannot make an infinite number of creases, and any approximation of, say, the self-similar wave will actually be an isometric folding. Therefore, for the purposes of modeling paper folding, infinite crease patterns can be considered as a limit $f$ of a family $f_n$ of isometric foldings. The limit case is then only of esoteric interest, although it can have value in its own right, as shown by the Dirichlet problems discussed in this chapter.

# Part IV

## Non-flat Folding

Non-flat origami comes in many forms. It might describe a flat-foldable origami model that happens to be opened into a 3-dimensional state. Or it could describe the 3-dimensional motion of an origami model opening and closing. Or the subject might be a static origami model that simply cannot be pressed into a book without crumpling or adding new creases. For instance, we might fold a square sheet of paper into a cube or an octahedron.

The mathematics behind such models is both similar and completely different to that of flat-foldable origami. Maekawa's and Kawasaki's Theorems no longer hold, but there are analogs for non-flat vertex folds. The matrix model we saw for flat folds also works for non-flat folds, but in the 3-dimensional case it also captures information and relationships that simply weren't present in the 2-dimensional case, like the concept of a **folding angle** at a crease.

We start in Chapter 12 by considering **rigid origami**, or **rigid folds**, which are 3-dimensional origami models made of flat, polygonal faces, where we stay focused on static models, which do not change over time. But in Chapter 13 we consider models that can fold and unfold through a continuum of folding angles while the faces remain rigid, and we call these **rigid foldings**. We end with Chapter 14 on more theoretical aspects of rigid origami, specifically its computational complexity, configuration spaces, and self-foldability.

# 12  Rigid Origami

In this chapter we examine the properties of origami that has been folded into a 3-dimensional shape while still being a continuous, piecewise-isometric mapping from the paper $R$ into $\mathbb{R}^3$. We will still use Definition 5.2 from Chapter 5 of an origami on a crease pattern $G$, but since our end result is not a flat origami, we will need a new definition for our final, 3-dimensional shape.

A complicating factor is that issues of folding parts of the paper into a flat plane and layer order are still present in 3-dimensional origami. Figure 12.1 shows an example of a non-flat vertex that folds into a cube corner. Notice that the crease pattern contains more information than a mere MV assignment. In order to know what the proper, 3-dimensional configuration of the paper is supposed to be, we need to specify the **folding angles**, which are the displacement from the flat, unfolded state, at which each crease is to be folded (i.e., Definition 5.3). Recall that a positive folding angle will be a valley crease and a negative one will be a mountain crease. Thus folding angles may be in the range $[-\pi, \pi]$ with the endpoints of this interval representing creases that fold flat.

Also notice that in our cube corner vertex example, all of the faces of the crease pattern are kept planar in the folded image. We refer to this, where the faces are mapped isometrically into $\mathbb{R}^3$, as a **rigid origami**. Let us state this more precisely in a definition.

**Definition 12.1**   Given a crease pattern $G = (V, E)$ on a region $R$ and a function $\mu: E \to [-\pi, \pi]$ called the **folding angles** of the creases, a **rigid origami on** $G$ is an infinite sequence of origamis $\{\sigma_n\}_{n=1}^{\infty}$ on $G$ such that

- for each face $f$ of $G$, the images $\{\sigma_n(f)\}_{n=1}^{\infty}$ uniformly converge to a planar polygon congruent to $f$,
- for each crease $l \in E$, the folding angles of the images $\{\sigma_n(l)\}_{n=1}^{\infty}$ converge to $\mu(l)$.

If there exists a rigid origami on a given crease pattern $G$, then we say that $G$ is **rigid-foldable**.

As with the flat-foldable case, we will often refer to a rigid origami by its folding map $\sigma = \lim_{n\to\infty} \sigma_n$ that describes the actual 3-dimensional origami model.

In Figure 12.1 we see that the faces $F_3$, $F_4$, and $F_5$ all lie in the same plane in $\mathbb{R}^3$, although the cartoon picture in the figure shows the creases to be not quite folded flat

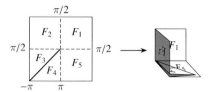

**Figure 12.1** A non-flat, rigid fold that makes a corner of a cube. The values shown along the boundary of the crease pattern are the folding angles at each vertex.

so that we may see the layering that the paper would make among these faces. Thus there may be parts of a non-flat origami that fold flat and could be modeled using the flat-fold definitions of Chapter 5, but we will not focus on this in our treatment in this chapter. Rather, we will concentrate on the novel aspects of non-flat, rigid origami, such as what conditions the crease pattern must possess in order for such a rigid origami to exist without tearing the paper.

In Section 12.1 we will see how a matrix model, similar to that of Section 5.7, can be used for rigid origami and give us a generalization of Kawasaki's Theorem. In Section 12.2 a Gaussian curvature approach will be applied to make a similar, if more theoretical, analysis. We will finish the chapter with an attempt to generalize Maekawa's Theorem to rigid, non-flat vertex folds.

## 12.1    Matrix Model and Necessary Conditions

We begin with a single-vertex crease pattern positioned with the vertex at the origin and our crease pattern lying in the $xy$-plane in $\mathbb{R}^3$. Denote the crease lines $l_1, \ldots, l_n$, in counterclockwise order with $l_1$ on the $x$-axis. For convenience, assign to each crease line a pair of angles $(\theta_i, \rho_i)$ where $\theta_i$ is the angle $l_i$ makes with the positive $x$-axis and $\rho_i$ is the folding angle of crease line $l_i$. (That is, $\mu(l_i) = \rho_i$.)

The following theorem was first discovered by Kawasaki (1987, 1997) and was formalized and proved by belcastro and Hull (2002a, b).

**Theorem 12.2**    *Given a single-vertex crease pattern G as described above that is rigid-foldable, let $\chi_i$ denote the orthogonal matrix that rotates $\mathbb{R}^3$ about the line containing crease $l_i$ in the positive direction by an angle of $\rho_i$. Then*

$$\chi_1 \cdots \chi_n = I. \tag{12.1}$$

This can be considered a non-flat version of Theorem 5.37, with the rather large difference that Theorem 12.2 is not a necessary and sufficient condition. Our proof will follow that of (belcastro and Hull, 2002a,b).

*Proof*    Imagine that we fold our rigid-foldable vertex so that the face of the paper between creases $l_1$ and $l_n$ lies fixed in the $xy$-plane (recall that $l_1$ lies on the $x$-axis).

Then we imagine a spider positioned just below the $x$-axis on the $xy$-plane crawling counterclockwise around the vertex on our folded piece of paper. Every time the spider crosses a folded crease line $l_i$ (moving from one face of the paper to another), it will rotate around the folded image of $l_i$ by the folding angle $\rho_i$. Because the folded paper is a piecewise linear object, we may express this action via a linear map (matrix), which we denote $L_i$.

We have that $L_1 = \chi_1$. We may describe the action of $L_2$ by first undoing operation $L_1$ to put crease $l_2$ back in the $xy$-plane, then performing $\chi_2$, and then redoing our first crease to put $l_2$ back into its folded image position. Thus $L_2 = \chi_1\chi_2\chi_1^{-1}$. Using this type of reasoning, we see in general that $L_i = $ (matrix to redo the previous $L$'s)$\cdot\chi_i \cdot$ (matrix to undo the previous $L$'s in reverse order), so that

$$L_i = (L_{i-1}\cdots L_1)\chi_i(L_1^{-1}\cdots L_{i-1}^{-1})$$
$$= \chi_1\cdots\chi_{i-1}\chi_i\chi_{i-1}^{-1}\cdots\chi_1^{-1},$$

where the last equality is achieved by telescoping and simplifying the recursion.

Now, because the spider must come back to where it started, we have $L_nL_{n-1}\cdots L_2L_1 = I$, which is simply a condition to ensure that we have continuity of the folding map along the $l_n$ crease line (i.e., no rips in the paper). Therefore, we have

$$I = L_n\cdots L_1$$
$$= (\chi_1\cdots\chi_{n-1}\chi_n\chi_{n-1}^{-1}\cdots\chi_1^{-1})(\chi_1\cdots\chi_{n-2}\chi_{n-1}\chi_{n-2}^{-1}\cdots\chi_1^{-1})\cdots(\chi_1\chi_2\chi_1^{-1})\chi_1$$
$$= \chi_1\cdots\chi_{n-1}\chi_n.$$

This concludes the proof.                                                    □

---

**Example 12.3** (Non-sufficiency)  To see that Theorem 12.2 cannot be a sufficient result for rigid origami, consider the single-vertex fold shown in Figure 12.2. As in the corner-of-a-cube example, we show the crease pattern with the folding angles denoted on the boundary. Notice that when these creases and folding angles are made, the faces $\{F_1, F_2, F_3\}$ and face $F_5$ end up on opposite sides of $F_4$, which means that face $F_1$ must intersect face $F_4$ as our sequence of origamis $\{\sigma_n\}$ approaches the limit fold $\sigma$ made by this crease pattern. Therefore, we cannot find a sequence of maps $\{\sigma_n\}$ that

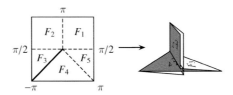

**Figure 12.2**  A piecewise isometric fold in $\mathbb{R}^3$ that self-intersects, yet also satisfies Equation (12.1).

are one-to-one as they approach $\sigma$, and thus this is not a rigid origami (i.e., it violates the self-intersection property).

Nonetheless, the crease pattern in Figure 12.2 satisfies the conclusion of Theorem 12.2, showing that this theorem cannot be made into a necessary and sufficient condition.

---

**Diversion 12.1**    Show that the crease pattern in Figure 12.2 satisfies Equation (12.1) in the statement of Theorem 12.2.

---

There are other ways to prove Theorem 12.2, as well as other ways to express it. Demaine and O'Rourke (2007) stated it in the following way:

**Theorem 12.4**    *Let $R_z(\theta)$ denote the matrix that rotates $\mathbb{R}^3$ about the z-axis counter-clockwise by angle $\theta$, and let $R_x(\theta)$ be the same except for rotating about the x-axis counterclockwise. Then given a rigid origami vertex with n creases and angles $\alpha_i$ between consecutive creases in the crease pattern and folding angle $\rho_i$ for the crease between faces with angles $\alpha_i$ and $\alpha_{i+1}$, we must have*

$$R_x(\rho_n)R_z(\alpha_n)\cdots R_x(\rho_2)R_z(\alpha_2)R_x(\rho_1)R_z(\alpha_1) = I. \qquad (12.2)$$

---

**Diversion 12.2**    Prove that Equations (12.1) and (12.2) are equivalent.

---

**Remark 12.5**    There are many ways to describe Equation (12.1) in terms of matrices. For example, using the notation in Theorem 12.4, given a single-vertex rigid origami with crease lines $l_1, \ldots, l_n$ with associated angles $(\theta_i, \rho_i)$ as defined at the beginning of this section, we can see that one way to describe $\chi_i$ is as

$$\chi_i = R_z(\theta_i)R_x(\rho_i)R_z^{-1}(\theta_i),$$

where

$$R_z(\theta_i) = \begin{pmatrix} \cos\theta_i & -\sin\theta_i & 0 \\ \sin\theta_i & \cos\theta_i & 0 \\ 0 & 0 & 1 \end{pmatrix} \quad \text{and} \quad R_x(\rho_i) = \begin{pmatrix} 1 & 0 & 0 \\ 0 & \cos\rho_i & -\sin\rho_i \\ 0 & \sin\rho_i & \cos\rho_i \end{pmatrix}.$$

Therefore, Equation (12.1) gives us a set of nine trigonometric equations in terms of the angles $\theta_i$ and $\rho_i$. These equations capture all of the relationships between these angles in a rigid origami. In fact, if we change, say, one of the folding angles a little bit, these equations can be used to determine how all the other folding angles will change (up to degrees of freedom) in an equivalent way to how kinematic equations from physics and engineering determine the angle relationships of rigid-body motions. In our case, the equations generated from Equation (12.1) can be thought of as equations from spherical trigonometry if we imagine the rigid origami vertex as being at the center of a sphere and the piecewise linear faces of the paper as cutting great circles on the sphere. This is an example of what William Thurston describes as spherical trigonometry being equivalent to the algebra of $3 \times 3$ orthogonal matrices (Thurston, 1997, p. 74).

We will develop this idea more in Chapter 13 when we discuss the folding angle relationships for continuous folding motions of rigid origami.

---

**Diversion 12.3** Use Theorem 12.2 (or Theorem 12.4) to prove the necessary direction of Kawasaki's Theorem (from Section 5.3).

---

**Remark 12.6** The proof of Theorem 12.2 rests on the fact that $L_n \cdots L_1 = I$, and we argued that this condition must hold in order for the paper to not rip. There is another reason why this condition must be true that will be crucial for generalizing Equation (12.1) to multiple-vertex crease patterns. It relies on the subtle difference between a physically folded rigid vertex origami model and the "mathematically folded" paper, which is the difference between our sequence of origamis $\{\sigma_n\}$ for high enough $n$ and the limit rigid origami $\sigma$. Consider, again, the example of the cube corner fold in Figure 12.1. If $R$ is the region of paper for this rigid fold, then $\sigma(R)$ has the geometry of a cone with cone angle $2\pi - \pi/2 = 3\pi/2$, whereas for large $n$, $\sigma_n(R)$ will have the geometry of a flat piece of paper, since each $\sigma_n$ is one-to-one, while their limit $\sigma$ is not.

Another way to specify this difference is to use the parallel transport of a vector around a closed curve on the folded paper around our vertex. On $\sigma(R)$ the parallel transport will change the orientation of the vector, but on $\sigma_n(R)$ the parallel transport will not change, since $\sigma_n(R)$ will be piecewise-isometric to the plane for large enough $n$.

Therefore, instead of considering a closed curve around the vertex on the folded image $\sigma(R)$, we can let our simple closed curve $\gamma$ be around the vertex on the unfolded paper and examine the image $\sigma(\gamma)$. Because $\gamma$ lies in the plane and $\sigma$ is composed only of rotations, $\sigma(\gamma)$ will be locally isometric to $\gamma$, and thus the parallel transport of a vector around $\sigma(\gamma)$ will not change its orientation. Therefore, because the $L_i$ are rotations, $L_n \cdots L_1$ encodes the effect on the orientation of a vector of parallel transport and thus is equivalent to $I$.

Our matrices $L_i$ also define the piecewise linear folding map $\sigma : \mathbb{R}^2 \subset \mathbb{R}^3 \to \mathbb{R}^3$ of our rigid fold as follows: Denote, in order, the faces of the single-vertex crease pattern in $\mathbb{R}^2$ (the $xy$-plane) as $F_1, \dots, F_n$ so that face $F_j$ lies between crease lines $l_j$ and $l_{j+1}$. Then we have

$$\sigma((x,y)) = \sigma((x,y,0)) = \begin{cases} L_1(x,y,0) & \text{for } (x,y) \in F_1, \\ L_2 L_1(x,y,0) & \text{for } (x,y) \in F_2, \\ \quad \vdots & \quad \vdots \\ L_{n-1} \cdots L_1(x,y,0) & \text{for } (x,y) \in F_{n-1}, \\ L_n L_{n-1} \cdots L_1(x,y,0) = I(x,y,0) & \text{for } (x,y) \in F_n. \end{cases}$$

Or, alternatively, assuming that crease $l_1$ between faces $F_1$ and $F_n$ is on the positive $x$-axis,

$$\sigma((x,y)) = \sigma((x,y,0)) = \begin{cases} \chi_1(x,y,0) & \text{for } (x,y) \in F_1, \\ \chi_1\chi_2(x,y,0) & \text{for } (x,y) \in F_2, \\ \quad\vdots & \quad\vdots \\ \chi_{n-1}\cdots\chi_1(x,y,0) & \text{for } (x,y) \in F_{n-1}, \\ \chi_n\chi_{n-1}\cdots\chi_1(x,y,0) = I(x,y,0) & \text{for } (x,y) \in F_n. \end{cases}$$

We now examine the case of rigid origami whose crease patterns have more than one vertex, following closely the treatment in (belcastro and Hull, 2002a,b) since this notation and level of detail has proven useful to engineers (as evidenced by the large number of citations these two papers have received in the engineering literature). Given a crease pattern $G = (V, E)$ on a piece of paper $R$ in the $xy$-plane, choose some vertex $v_0$ to be at the origin and denote the other vertices as $v_1, \ldots, v_b$. Some creases may terminate on the boundary of $R$; we call these **virtual vertices** and denote them by $(vv)_i$, enumerating them as $(vv)_{b+1}, \ldots, (vv)_c$, for indexing purposes only. Here we will assume that our paper $R$ is bounded, but in theory there is no problem with infinite paper as long as our folding map is defined on all points in $R$.

Now, label crease lines in $E$ with $l_{(i,j)}$ and $l_{(j,i)}$, where $i$ and $j$ are the indices of the vertices (which may be virtual) that define this edge in $G$. The order of $(i,j)$ will determine the vertex (the vertex whose index is $i$) from which we are considering the edge at a given time. Choose one face adjacent to $v_0$ to be fixed in the $xy$-plane, and denote it $F_0$. Then denote the other faces by $F_{(i_1,i_2,\ldots)}$, where $i_1, i_2, \ldots$ are the indices of the vertices, in order, that surround each face.

As before, we associate to each crease labeled $l_{(i,j)}$ an ordered pair $(\theta_{(i,j)}, \rho_{(i,j)})$, where $\theta_{(i,j)}$ is the angle by which $l_{(i,j)}$ deviates by a line from $v_i$ that is parallel to and in the direction of the positive $x$-axis in the $xy$-plane and where $\rho_{(i,j)}$ is the folding angle of the crease, measured by looking from vertex $v_i$ to $v_j$ where counterclockwise is considered the positive angle direction. That is, we have $\theta_{(j,i)} = \pi + \theta_{(i,j)}$ and $\rho_{(j,i)} = -\rho_{(i,j)}$.

Because we no longer have simply one vertex at the origin, we will use $4 \times 4$ matrices in homogeneous coordinates to describe our crease line rotations. We let $\chi_{(i,j)}$ denote the folding matrix that rotates $\mathbb{R}^3$ about the line containing $l_{(i,j)}$ by $\rho_{(i,j)}$. That is, $\chi_{(i,j)}$ rotates as if we consider $l_{(i,j)}$ to extend from the vertex with index $i$, whereas $\chi_{(j,i)}$ would rotate considering the crease extended from the vertex corresponding to index $j$.

Now consider a closed vertex-avoiding path $\gamma$ on the crease pattern that is not tangent to any creases and that begins and ends in $F_0$. Let $l_{(i_t,j_t)}$ be the crease lines, in order, that $\gamma$ crosses. Following the model from Theorem 12.2, let $L_{(i_t,j_t)}$ encode the rotation around the folded image of $l_{(i_t,j_t)}$ in $\sigma(R)$ by $\rho_{(i_t,j_t)}$. Note that $L_{(i_t,j_t)}$ is a $4 \times 4$ matrix in homogeneous coordinates and is dependent on $\gamma$. It is recursively defined, as follows: $L_{(i_1,j_1)} = \chi_{(i_1,j_1)}$, $L_{(i_2,j_2)} = L_{(i_1,j_1)}\chi_{(i_2,j_2)}L_{(i_1,j_1)}^{-1}$, and, in general,

$$L_{(i_t,j_t)} = L_{(i_{t-1},j_{t-1})} \cdots L_{(i_1,j_1)}\chi_{(i_t,j_t)}L_{(i_1,j_1)}^{-1} \cdots L_{(i_{t-1},j_{t-1})}^{-1}.$$

This simplifies to

$$L_{(i_t,j_t)} = \left(\chi_{(i_1,j_1)} \cdots \chi_{(i_{t-1},j_{t-1})}\right) \chi_{(i_t,j_t)} \left(\chi^{-1}_{(i_{t-1},j_{t-1})} \cdots \chi^{-1}_{(i_1,j_1)}\right).$$

The following theorem is the 3-dimensional, rigid origami analog to the flat-folding, multiple-vertex Theorem 6.4 (the generalized Kawasaki's Theorem).

**Theorem 12.7** (belcastro and Hull, 2002b)  *Consider a rigid origami multiple-vertex crease pattern, using the notation above. Let $\gamma$ be a closed, vertex-avoiding curve on the crease pattern, beginning and ending on face $F_{(i_p,j_p,\ldots)}$. Let $l_{(i_1,j_1)}$, $l_{(i_2,j_2)}$, $\ldots$ $l_{(i_n,j_n)}$ be the creases, in order, that $\gamma$ crosses, and let $\chi_{(i_t,j_t)}$ be the matrix that rotates counterclockwise around the crease line $l_{(i_t,j_t)}$ by $\rho_{(i_t,j_t)}$. Then*

$$\chi_{(i_1,j_1)}\chi_{(i_2,j_2)} \cdots \chi_{(i_n,j_n)} = I.$$

*Proof*  Let $\psi$ be a vertex-avoiding path from $F_0$ to $F_{(i_p,j_p,\ldots)}$. Then $\Gamma = \psi^{-1}\gamma\psi$ is a closed, vertex-avoiding curve beginning and ending on $F_0$. Let $L_{(d_1,e_1)}, \ldots, L_{(d_m,e_m)}$ be the folding matrices, as defined above, that correspond to the crease lines that $\Gamma$ crosses, in order. Let $\Gamma'$ be the image of $\Gamma$ under these matrices. Using similar reasoning as in Remark 12.6, we see that the parallel transport of a vector around $\Gamma$, and thus around $\Gamma'$, does not change its orientation. Therefore, as the accumulation of orientation changes along $\Gamma'$ is $L_{(d_m,e_m)} \cdots L_{(d_1,e_1)}$, that transformation must be equivalent to the identity:

$$L_{(d_m,e_m)}L_{(d_{m-1},e_{m-1})} \cdots L_{(d_1,e_1)} = I.$$

It then follows, by the same cancellation as in the proof of Theorem 12.2, that

$$\chi_{(d_1,e_1)}\chi_{(d_2,e_2)} \cdots \chi_{(d_m,e_m)} = I.$$

Let $\chi_{(a_1,b_1)}, \ldots, \chi_{(a_q,b_q)}$ be the matrices associated with the crease lines that $\psi$ crosses, in order. Then the matrices associated with $\psi^{-1}$ are $\chi^{-1}_{(a_q,b_q)}, \ldots, \chi^{-1}_{(a_1,b_1)}$, in order. We may now rewrite our identity as

$$\left(\chi_{(a_1,b_1)} \cdots \chi_{(a_q,b_q)}\right)\left(\chi_{(i_1,j_1)} \cdots \chi_{(i_n,j_n)}\right)\left(\chi^{-1}_{(a_q,b_q)} \cdots \chi^{-1}_{(a_1,b_1)}\right) = I.$$

This has the form $MNM^{-1} = I$, so that $N = I$, or $\chi_{(i_1,j_1)} \cdots \chi_{(i_n,j_n)} = I$. □

Theorem 12.7 can be proven in a number of different ways. The above proof has the advantage of corresponding to the definition of the rigid folding map $\sigma : \mathbb{R}^2 \subset \mathbb{R}^3 \rightarrow \mathbb{R}^3$ given previously, except that it has a fixed face $F_0$ and is generalized to multiple-vertex crease patterns (see also Definition 12.8 below).

---

**Diversion 12.4**  Create an inductive proof of Theorem 12.7, like the one used to prove the generalized Kawaski's Theorem (Theorem 6.4).

---

Again, there are different ways to express the matrices $\chi_{(i,j)}$. Let $T_i$ be the $4 \times 4$ homogeneous coordinates matrix that translates $\mathbb{R}^3$ by $v_i$ (or $(vv)_i$ if index $i$ represents a virtual vertex). Then we may write

$$\chi_{(i,j)} = T_i R_z(\theta_{(i,j)}) R_x(\rho_{(i,j)}) R_z^{-1}(\theta_{(i,j)}) T_i^{-1},$$

where here we let $R_z$ and $R_x$ be the rotation matrices as before but in homogeneous coordinates. That is, we have

$$R_z(\theta) = \begin{pmatrix} \cos\theta & -\sin\theta & 0 & 0 \\ \sin\theta & \cos\theta & 0 & 0 \\ 0 & 0 & 1 & 0 \\ 0 & 0 & 0 & 1 \end{pmatrix}, \quad R_x(\rho) = \begin{pmatrix} 1 & 0 & 0 & 0 \\ 0 & \cos\rho & -\sin\rho & 0 \\ 0 & \sin\rho & \cos\rho & 0 \\ 0 & 0 & 0 & 1 \end{pmatrix},$$

$$\text{and} \quad T_i = \begin{pmatrix} 1 & 0 & 0 & x_i \\ 0 & 1 & 0 & y_i \\ 0 & 0 & 1 & 0 \\ 0 & 0 & 0 & 1 \end{pmatrix},$$

where $v_i$ (or $(vv)_i$) has coordinates $(x_i, y_i, 0)$.

We may now define the folding map for general rigid origami. (This is the 3-dimensional analog of Definition 6.5 and Theorem 6.6.)

**Definition 12.8**   Let $F$ be the face that is fixed in the $xy$-plane of a rigid origami $G$ on a region $R$. Given any other face $F_{(i_p, j_p, \dots)}$, let $\gamma$ be any vertex-avoiding path in $R$ from a point in $F$ to a point in $F_{(i_p, j_p, \dots)}$. Let the crease lines that $\gamma$ crosses be, in order, $l_{(i_1, j_1)}, \dots, l_{(i_p, j_p)}$. Then the **general folding map** $\sigma_F : R \to \mathbb{R}^3$ is

$$\sigma_F(x, y, 0) = \sigma_F(x, y, 0, 1) = \chi_{(i_1, j_1)} \chi_{(i_2, j_2)} \cdots \chi_{(i_p, j_p)}(x, y, 0, 1)$$

for $(x, y, 0) \in F_{(i_p, j_p, \dots)}$.

**Theorem 12.9**   *The general folding map $\sigma_F$ of a crease pattern $G$ on a region $R$ is well-defined, that is, it is independent of the choice of the curve $\gamma$.*

*Proof*   Consider any two vertex-avoiding paths $\gamma_1$ and $\gamma_2$ from face $F$ to face $F_{(i_p, j_p, \dots)}$. Let $\gamma_1$ cross (in order) crease lines $l_{(i_1, j_1)}, l_{(i_2, j_2)}, \dots, l_{(i_m, j_m)}$, and let $\gamma_2$ cross (in order) crease lines $l_{(i_n, j_n)}, l_{(i_{n-1}, j_{n-1})}, \dots, l_{(i_{m+1}, j_{m+1})}$. The folding map on $F_{(i_p, j_p, \dots)}$ via $\gamma_1$ is $\chi_{(i_1, j_1)} \chi_{(i_2, j_2)} \cdots \chi_{(i_m, j_m)}$, and the folding map on $F_{(i_p, j_p, \dots)}$ via $\gamma_2$ is $\chi_{(i_n, j_n)} \chi_{(i_{n-1}, j_{n-1})} \cdots \chi_{(i_{m+1}, j_{m+1})}$. If we consider the path $\gamma_1 \gamma_2^{-1}$, we see that it is closed, and the proof of Theorem 12.7 shows that

$$\chi_{(i_1, j_1)} \chi_{(i_2, j_2)} \cdots \chi_{(i_m, j_m)} \chi_{(i_{m+1}, j_{m+1})}^{-1} \cdots \chi_{(i_n, j_n)}^{-1} = I.$$

It then follows that

$$\chi_{(i_1, j_1)} \chi_{(i_2, j_2)} \cdots \chi_{(i_m, j_m)} = \chi_{(i_n, j_n)} \chi_{(i_{n-1}, j_{n-1})} \cdots \chi_{(i_{m+1}, j_{m+1})},$$

and thus the folding map $\sigma_F$ will be the same using $\gamma_1$ or $\gamma_2$.   □

The general folding map $\sigma_F$, or variations on it, is the basic theoretical tool used for rigid origami computer simulation. More details of how this can be used in practice will be discussed in Chapter 13.

## 12.2    The Gauss Map

Another tool for studying rigid origami is the Gauss map from differential geometry. Its use for this purpose was first made in a seminal paper by David Huffman (1976), and it was subsequently used by Koryo Miura (1989) to describe his Miura-ori in the context of rigid origami.

**Definition 12.10**    Let $X$ be an oriented surface embedded in $\mathbb{R}^3$ and let $S^2 \subset \mathbb{R}^3$ denote the unit sphere. The **Gauss map** is the function $N\colon X \to S^2$ where for all $x \in X$, $N(x)$ is the unit vector normal to $X$ at $x$.

---

**Example 12.11**    In Figure 12.3(a) we see that if $X$ is a cylinder, then $N(X)$ is a great circle (an equator) $C$ on the sphere $S^2$.

In Figure 12.3(b) we have that $X$ is a cone. The Gauss map will not be defined at the apex of the cone, but everywhere else will give a unit normal vector whose tip, when translated to the center of $S^2$, will lie along a circle $C$ that will not be a great circle. In fact, the smaller the cone angle of our cone, the larger the circle $C$ will be, with a great circle being the limit. On the other hand, the larger the cone angle, the smaller the circle $C$ will be on $S^2$. If we have a cone angle of $2\pi$, then our cone is flat, like a piece of paper, and all the normal vectors will point in the same direction. In this case, the Gauss map image will be a single point on $S^2$.

What if we take a cone with cone angle greater than $2\pi$?

---

**Diversion 12.5**    Prove that the Gauss map of a cone $K$ whose cone angle is greater than $2\pi$ is a circle on $S^2$ whose orientation is opposite that of a circle drawn around the apex of the cone. That is, if $\Gamma\colon [0, 1] \to K$ is a circle oriented clockwise around the apex of the cone, then $N(\Gamma)$ will be a circle on $S^2$ oriented counterclockwise.

---

One simple take-away message from Example 12.11 is that the Gauss map of a flat piece of paper is a single point, perhaps the "North Pole" of $S^2$. Keep that in mind in the next definition.

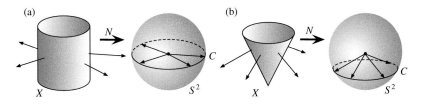

**Figure 12.3**  (a) The Gauss map of a cylinder is an equator of $S^2$. (b) The Gauss map of a cone is a circle on $S^2$.

**Definition 12.12**   Let $X$ be an oriented surface embedded in $\mathbb{R}^3$ and $p \in X$. Let $\Gamma$ be a closed curve in $X$ oriented clockwise around the point $p$, and we imagine shrinking $\Gamma$ and in the limit contracting to the point $p$. Then the **(intrinsic) Gaussian curvature of $X$ at $p$** is the real number $\kappa$ given by

$$\kappa = \lim_{\Gamma \to p} \frac{\text{Area}(N(\Gamma))}{\text{Area}(\Gamma)},$$

where $\text{Area}(S)$ denotes the intrinsic, signed area inside the curve $S$ on the surface (either $X$ or $S^2$).

By **signed area** we mean that an area enclosed by clockwise curves is positive and that enclosed by counterclockwise curves is negative. Also, the image of the Gauss map is sometimes called the **trace** of the curve $\Gamma$.

Gaussian curvature can also be defined by the determinant of the Jacobian of the Gauss map $N$, but for our purposes the geometric definition given above will suffice. In particular, we need to examine how the Gauss map behaves around **polyhedral vertices**, since this includes partially folded origami vertices whose faces remain planar (i.e., rigid).

**Remark 12.13**   It can be very difficult to visualize the Gauss map trace along a closed curve around a polyhedral vertex, and we will shortly see several examples. But first we call attention to one detail, illustrated in Figure 12.4. Let a face $F$ of the polyhedral vertex $v$ have plane angle $\beta_i$ at $v$ and be bordered by edges $e_i$ and $e_{i+1}$. Let $\vec{n}_i$ denote the unit normal to $F$ and the vertex of the spherical polygon made by the Gauss map trace corresponding to face $F$. **Then if $e_i$ and $e_{i+1}$ have the same convexity/concavity, the exterior angle of the spherical polygon trace at $\vec{n}_i$ will equal $\beta_i$. If $e_i$ and $e_{i+1}$ have different convexity, then the interior angle of the spherical polygon trace at $\vec{n}_i$ will equal $\beta_i$.** This is because as the closed curve $\Gamma$ travels around the polyhedral vertex $v$, its normal vectors will be constant on the faces surrounding $v$. As $\Gamma$ crosses $e_i$, the normal vectors will swing along an arc in a plane orthogonal to $e_i$, thus making an arc of the spherical polygon trace. When $\Gamma$ crosses the next edge $e_{i+1}$ of $v$, the normal vectors will swing along a plane orthogonal to $e_{i+1}$, and these two planes will meet at an angle of $\pi - \beta_i$. This will be the interior angle of the spherical polygon trace if both $e_i$ and $e_{i+1}$ have the same convexity/concavity, as shown in Figure 12.4(a). If $e_i$

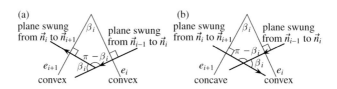

**Figure 12.4**   The (a) exterior or (b) interior angle of the Gauss map spherical polygon equals the plane angle of the face it spans.

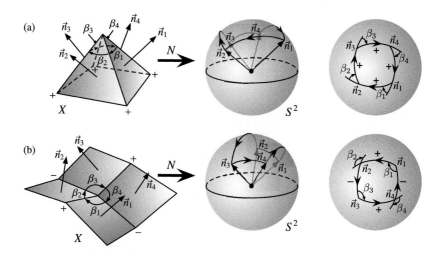

**Figure 12.5** (a) The Gauss map of a convex polyhedral vertex, where the plane angles $\beta_i$ correspond to the exterior angles of the spherical polygon. (b) The Gauss map of a non-convex polyhedral vertex.

and $e_{i+1}$ have different concavity/convexity, then $\pi - \beta_i$ will be an exterior angle of the spherical polygon trace, making $\beta_i$ the interior angle (Figure 12.4(b)).

---

**Example 12.14** (Gauss map examples)  Figure 12.5 shows two examples of the Gauss map around polyhedral vertices. In (a) we see a convex vertex with plane angles $\beta_1, \ldots, \beta_4$ around the vertex. If we travel around this vertex via a clockwise simple closed curve $\Gamma$, then $N(\Gamma)$ will be a spherical polygon oriented clockwise as well. Each corner of this spherical polygon corresponds to the unit normal vector $\vec{n}$ of one of the faces of our polyhedral vertex. The arc of our spherical polygon from $\vec{n}_1$ to $\vec{n}_2$ has length equal to the supplement of the dihedral angle between the faces corresponding to $\vec{n}_1$ and $\vec{n}_2$. Also shown in Figure 12.5 is how the plane angles of the polyhedral vertices $\beta_1, \ldots, \beta_4$ become the exterior angles of the spherical polygonal trace of the Gauss map.

If the polyhedral vertex has concave edges (marked with $-$) as well as convex edges (marked with $+$), such as the example shown in Figure 12.5(b), then on the Gauss map the convex edges will swing arcs in the same direction as $\Gamma$ while the concave edges will swing in the opposite direction. In our example, a concave polyhedral vertex is achieved by plane angles $\beta_i$ that are each greater than $90°$, and this results in the image curve $N(\Gamma)$ having opposite, counterclockwise orientation. Also, for a plane angle $\beta_i$ between $+$ and $-$ edges on the polyhedral vertex, we have by Remark 12.13 that the interior angle of the corresponding spherical polygon vertex will also equal $\beta_i$, as seen in Figure 12.5(b).

Now consider the examples shown in Figure 12.6, which illustrate what happens to the Gauss map trace around a convex vertex when one of its edges, labeled $x$ in

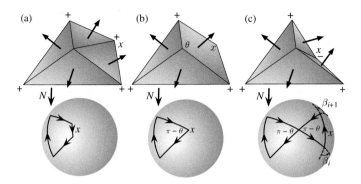

**Figure 12.6** (a) The Gauss map trace around a convex vertex of degree 4. (b) The same trace where the edge marked $x$ has been lowered so that its adjacent faces are planar. (c) The trace where the same edge $x$ has been lowered farther to be concave.

the figure, is lowered into a concave position. In our example, the convex vertex is of degree 4, and the trace is simply a quadrilateral, where one side of this quadrilateral represents the unit normals from the Gauss map swinging through $\pi - \delta$, where $\delta$ is the dihedral angle at $x$, and thus we label this side by $x$ in Figure 12.6(a). As the edge $x$ is lowered, it will reach a point where its adjacent faces become coplanar, and the side $x$ in the trace will shrink to a point, as seen in Figure 12.6(b). Most interestingly, as the edge $x$ becomes concave, the normals to the faces adjacent to $x$ will begin to point toward each other, whereas before they were pointing away from each other. This will make the trace form a **bow tie** region in the Gauss map, where one side of the bow tie is a triangle with counterclockwise orientation, corresponding to the normals of the sides around the concave edge $x$. The other side of the bow tie will remain the same clockwise triangle from (b), corresponding to the normals of the faces adjacent to the convex edge opposite of $x$. Such bow tie figures can occur in Gauss map traces of polyhedral vertices with a mix of convex and concave edges.

The previous examples lead to another definition and a theorem.

**Definition 12.15**   Let $v$ be a polyhedral vertex with plane angles $\beta_1, \ldots, \beta_n$. Then the **angle defect of** $v$, denoted $D(v)$, is

$$D(v) = 2\pi - \sum_{i=1}^{n} \beta_i.$$

**Theorem 12.16**   *Let $v$ be a polyhedral vertex and let $\Gamma$ be a clockwise simple closed curve around $v$. Then*

$$Area(N(\Gamma)) = D(v).$$

*Proof*  Let the plane angles of $v$ be $\beta_1, \ldots, \beta_n$. Our proof will be by induction on the number of **non-convex** edges of $v$. The base case is when $v$ has only convex edges, where we will have that the interior angles of the spherical $n$-gon that the Gauss map traces will all be $\pi - \beta_i$ (since as we saw in Remark 12.13, the exterior angles in this case will be $\beta_i$). Then by the formula for the area of a spherical $n$-gon, we have

$$\text{Area}(N(\Gamma)) = \sum_{i=1}^{n}(\pi - \beta_i) - (n-2)\pi = 2\pi - \sum_{i=1}^{n}\beta_i = D(v).$$

Now suppose that we are given a polyhedral vertex $v$ of degree $n$ with a mix of concave and convex edges. We will divide our induction step into three cases.

**Case 1:**  There exists a concave edge $x$ whose immediate neighbors around $v$ are both convex.

This is the situation shown in Figure 12.6(c), and if we remove the edge $x$ and add a plane face between $x$'s neighbors with plane angle $\theta$, we will create a new vertex $v'$ with $n-1$ edges. The difference in trace of $v$ and $v'$ will be the same as that shown between Figures 12.6(c) and (b); that is, a counterclockwise triangle whose interior angles $\pi - \theta$, $\beta_i$, and $\beta_{i+1}$ will be removed from the trace of $v$ to make the trace of $v'$. By the induction hypothesis, $\text{Area}(N(\Gamma)) = D(v')$ for a clockwise closed curve $\Gamma$ around $v'$. Extending $\Gamma$ to be such a curve around $v$ gives us

$$\text{Area}(N(\Gamma)) = D(v') - (\pi - \theta + \beta_i - \beta_{i+1} - \pi)$$

$$= 2\pi - \left( \sum_{k \neq i, i+1} \beta_k \right) - \theta + \theta - \beta_i - \beta_{i+1} = D(v).$$

**Case 2:**  There exists a concave edge $x$ whose immediate neighbors around $v$ have different concavity.

An example of this case is shown in Figure 12.7(a), where we label the angle between $x$ and its concave neighbor $\beta_{i+1}$ and the angle between $x$ and its convex neighbor $\beta_{i+2}$. The edge $x$ will correspond to the side of a counterclockwise spherical polygon $P$ in the Gauss map trace, and because of the hypothesis of this case, the path of the trace will make at one endpoint of the side $x$ a reflex angle that is inside $P$ and at the other endpoint a reflex angle that is outside of $P$ (see the first Gauss map in Figure 12.7(a)). Therefore, if we remove the edge $x$ from our polyhedral vertex $v$ and replace it with a plane face with plane angle $\alpha$, then the resulting new vertex $v'$ will have a concave and a convex edge surrounding the $\alpha$ face and the new spherical polygon $P'$ in the trace of $v'$ will have one fewer side than that of $P$ and an interior angle of $\alpha$. In particular, $P'$ will be the spherical polygon $P$ with one triangle piece sliced off of it. By the induction hypothesis, if $\Gamma$ is a clockwise closed curve around $v'$, we have $\text{Area}(N(\Gamma)) = D(v')$, and extending $\Gamma$ to be a curve around $v$ we obtain

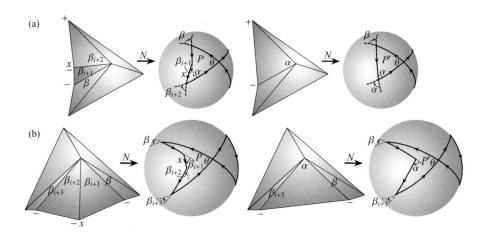

**Figure 12.7** Cases 2 and 3 in the proof of Theorem 12.16.

$$\text{Area}(N(\Gamma)) = D(v') - (\beta_{i+1} + \beta_{i+2} + \pi - \alpha - \pi)$$

$$= 2\pi - \left( \sum_{k \neq i+1, i+2} \beta_k \right) - \alpha - \beta_{i+1} - \beta_{i+2} + \alpha = D(v).$$

**Case 3:** There exists a concave edge $x$ whose immediate neighbors around $v$ are both concave.

See Figure 12.7(b), where we have labeled the plane angles of $v$ on either side of $x$ as $\beta_{i+1}$ and $\beta_{i+2}$. In the trace of $v$, the edge $x$ will correspond to a side of a counterclockwise spherical polygon $P$ where both endpoints of the side $x$ will have their reflex angles in the interior of $P$. When we create a new polyhedral vertex $v'$ by removing the edge $x$ and replacing it with a plane face with plane angle $\alpha$, the polygon $P$ will become a new polygon $P'$ with one less side, and the difference between these two spherical polygons is that $P'$ will be exactly $P$ with a small triangle, one side of which was $x$, cut out of it. The plane angle $\alpha$ will appear in $P'$ as shown in Figure 12.7(b). By the induction hypothesis, $\text{Area}(N(\Gamma)) = D(v')$ for a clockwise curve $\Gamma$ around $v'$, and extending this to $v$ gives us

$$\text{Area}(N(\Gamma)) = D(v') - (\beta_{i+1} + \beta_{i+2} + (\pi - \alpha) - \pi)$$

$$= 2\pi - \left( \sum_{k \neq i+1, i+2} \beta_k \right) - \alpha - \beta_{i+1} - \beta_{i+2} + \alpha = D(v).$$

These three cases cover all possible ways we could have a concave edge $x$ in our polyhedral vertex $v$, and the proof is complete. □

**Remark 12.17**   We chose to include the proof of Theorem 12.16 as well as a somewhat lengthy discussion of the Gauss map of polyhedral vertices because of the lack of good references on the matter. Hilbert and Cohn-Vossen's famous book *Geometry and the Imagination* contains an extensive introduction to the Gauss map (Hilbert and Cohn-Vossen, 1956, p. 193) but does not mention this theorem or provide many examples of the trace of polyhedral vertices. David Huffman's 1970s articles on paper folding and Gauss map representations of surfaces (Huffman, 1976, 1978) provide more examples and a statement of Theorem 12.16, but he only mentions that the proof involves cutting up the spherical polygon trace into triangles.

However, it can be very difficult to visualize the Gauss map trace for non-convex polyhedral vertices, and since rigid origami vertices fall into this category, it is useful to develop some intuition on such traces. In fact, the reader may find our proof of Theorem 12.16 unconvincing by speculating that the trace of a non-convex vertex could self-intersect or be convoluted in ways that our three cases do not cover.

An example of such a convoluted trace is shown in Figure 12.8(a), and we offer it as a way to illustrate that the cases in the proof of Theorem 12.16 are, in fact, exhaustive. We labeled the faces of our polyhedral vertex $F_1, \ldots, F_6$, and these correspond to the corners of the trace. A key observation is to note that when calculating the spherical area in such overlapping traces, we must allow "double-counting" of the area in order for Theorem 12.16 to hold. That is, to find the area of the trace in Figure 12.8(a), we must compute the positive area of the spherical quadrilateral $F_2 F_1 yx$, then add the negative area of the spherical triangles $xF_4F_3$ and $yF_6F_5$. In this way, all the angles of the trace, which relate to the plane angles of the polyhedral vertex, will be included in the trace area, and the angles inside the trace at the intersection points $x$ and $y$ will cancel. This is how the trace area will equal the angle defect. If we had tried to eliminate areas of the sphere that were double-counted in this computation, we would introduce more angles and possibly eliminate others, thereby obtaining an area that does not equal the angle defect.

Thus, in order to perform our induction process on our trace example in Figure 12.8(a), we could delete the concave edge between faces $F_5$ and $F_6$ to get the reduced polyhedral vertex and trace shown in Figure 12.8(b). The intersection labeled $y$ will now correspond to the new planar face between faces $F_1$ and $F_4$, and we see that the spherical triangle $yF_6F_5$ has been deleted from the trace, whereas all other parts of the trace remain intact. Therefore, our induction process will proceed smoothly, despite the convoluted self-intersection of our original trace.

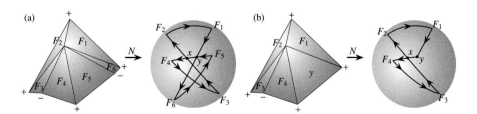

**Figure 12.8** A more complicated Gauss map trace.

---

**Diversion 12.6**    Prove that the sum of the angle defects of the vertices of a spherical (genus zero, not necessarily convex) polyhedron is always $4\pi$. (This is known as Descartes' Theorem and is a special, discrete case of the Gauss–Bonnet Theorem from differential geometry.)

---

## 12.3    The Gauss Map and Rigid Origami

An immediate consequence of Theorem 12.16 is the following:

**Corollary 12.18**    *Given a rigid origami $\sigma$ on a region $R \subset \mathbb{R}^2$ whose crease pattern has a single vertex in its interior, let $\Gamma$ be a simple clockwise closed curve around the polyhedral vertex in the image $\sigma(R)$. Then $Area(N(\Gamma)) = 0$.*

This follows because the angle defect of a rigid origami vertex folded from a flat sheet of paper is zero. This is an example of the more general fact (which we will not prove here) that the Gauss map and Gaussian curvature of a surface $X$ does not change if the surface is bent or manipulated without stretching or tearing. That is, the zero curvature of a flat piece of paper does not change as we fold it.

This observation, alone, can be used to prove some simple things about rigid origami vertices. For example, no rigid-foldable vertex can be of degree 3, since any degree-3 vertex will trace a spherical triangle as its Gauss map, which cannot have zero area.

---

**Diversion 12.7**    Prove that a degree-4 rigid origami vertex cannot have all its creases be mountains (or all valleys).

---

**Diversion 12.8**    Consider the degree-4 vertex whose crease pattern has 90° angles between all consecutive creases, where three of the creases are valleys and one is a mountain. Prove that the only way that this vertex can fold flat rigidly is to first fold the two opposite valley creases flat (this is called a **book fold** among origamists) and then fold the other two creases (the mountain and the valley) only when they're folded on top of each other.

---

**Diversion 12.9**    Prove that every degree-4 rigid-foldable vertex must obey Maekawa's Theorem: three of its creases must be mountains (convex) and one valley (concave) or vice versa. (Note that we are **not** assuming flat-foldability here!)

---

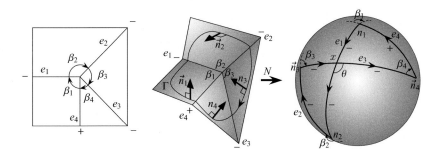

**Figure 12.9** A degree-4 rigid vertex crease pattern, partially folded, and its Gauss map trace.

**Example 12.19** (Degree-4 rigid vertex folds)   There are many properties of rigid-foldable degree-4 vertices that can be gleaned from the Gauss map trace. First of all, by Diversion 12.7, such a folded vertex will form a polyhedral vertex that is not convex, and thus its Gauss map trace will form a bow tie, as seen in Figure 12.9. Since the area of this bow tie must be zero, the two triangles that make up the bow tie must have equal areas, one positively and the other negatively oriented. This is easy to see on its own, since the sum of these signed areas is

$$(\pi - \beta_2 + \pi - \beta_3 + \pi - \theta - \pi) - (\beta_1 + \beta_2 + \pi - \theta - \pi) = 2\pi - \beta_1 - \beta_2 - \beta_3 - \beta_4,$$

which equals zero since we are folding a flat piece of paper.

By Diversion 12.9, every degree-4 rigid vertex fold will have one crease with convexity opposite of the other three creases. Using language introduced by Lang (2018), we define the **major creases** of a degree-4 rigid vertex fold to be the opposite pair of crease lines that have the same convexity. Similarly, the **minor creases** are the pair of opposite creases with different convexity.

Our next theorem lists relationships between degree-4 rigid-foldable vertices and their Gauss maps that were discovered by Huffman (1976). Refer to Figure 12.9 for our labelings of the various parts of our rigid vertex and the Gauss map trace. For convenience, we will use the labels $\beta_i$ to denote the plane angles surrounding the rigid-folded vertex as well as the face of the folded vertex that contains this angle.

**Theorem 12.20**   *Given a degree-4 rigid-foldable vertex as labeled in Figure 12.9 in a folded state,*

   (i) *the angle between the major creases in the folded state ($e_1$ and $e_3$ in our figure) is the angle $\theta$ in the Gauss map trace,*
  (ii) *the point $x$ on the Gauss map trace forms the normal vector to the plane containing the major creases,*
 (iii) *the common area $E$ of the two triangles that form the bow tie is equal to the decrease in angle between the major creases from the flat, unfolded state.*

*Proof*   First, note that in the Gauss map trace the arc from $\vec{n}_1$ (the normal to face $\beta_1$) to $\vec{n}_2$ (the normal to face $\beta_2$) is made up of vectors that are all perpendicular to the crease $e_1$, and thus $e_1$ is normal to the plane defined by the arc $\vec{n}_1\vec{n}_2$. Similarly, the plane defined by the arc $\vec{n}_3\vec{n}_4$ has $e_3$ as a normal vector. The angle between the $\vec{n}_1\vec{n}_2$ and $\vec{n}_3\vec{n}_4$ planes is $\theta$, and thus this is also the angle between $e_1$ and $e_3$, proving (i). Furthermore, the vector made by $x$ on the Gauss map trace is on the intersection of the $\vec{n}_1\vec{n}_2$ and $\vec{n}_3\vec{n}_4$ planes, and thus this $x$ vector is perpendicular to both $e_1$ and $e_3$, which proves (ii).

For (iii), note that in the flat, unfolded state we have that the angle between $e_1$ and $e_3$ is

$$\theta = \beta_1 + \beta_4 = 2\pi - \beta_2 - \beta_3.$$

Here we are assuming without loss of generality that $\beta_1 + \beta_4 \le \beta_2 + \beta_3$, since one of these angle-sum pairs must be less than or equal to the other. Now, we previously saw that the area $E$ of the two bow tie triangles is

$$(\pi - \beta_2) + (\pi - \beta_3) + (\pi - \theta) - \pi = 2\pi - \beta_2 - \beta_3 - \theta$$

and

$$\beta_1 + \beta_4 + (\pi - \theta) - \pi = \beta_1 + \beta_4 - \theta,$$

and thus as the vertex is rigidly folded, $\theta$ equals how much we subtract from the original angle between the major creases (in the unfolded state), and the area $E$ captures this decrease, as desired.   □

Part (iii) of Theorem 12.20 suggests that the angle $\theta$ in the trace would be a good parameter to use for modeling the rigid folding and unfolding of degree-4 vertices, and some more recent treatments of rigid origami vertices have done this. For example, Evans et al. (2015a) use the angle between the minor creases as a key parameter in their model.

Another result of Huffman having to do with the midpoints of the bow tie trace of (not necessarily flat-foldable) degree-4 rigid vertices is particularly elegant, but no simple proof (say, without using pages of spherical trigonometry) is known, and Huffman himself provides only an intuitive proof of it (Huffman, 1976). Therefore, we merely state it as a **very** challenging diversion.

---

**Diversion 12.10**   The midpoints of the four sides of the bow tie trace made from a degree-4 rigid origami vertex are collinear (i.e., lie on a common great circle on the sphere).

---

**Example 12.21** (Flat-foldable degree-4 rigid vertex folds)   Note that all the results from the previous example hold for arbitrary degree-4 rigid vertex folds. In the case where our degree-4 rigid-foldable origami vertex is also flat-foldable, the Gauss map has extra symmetry that provides some easy and surprising facts. Note that the example shown in Figure 12.9 is a flat-foldable vertex (although we did not exploit that fact in Example 12.19). Kawasaki's Theorem gives us that $\pi - \beta_2 = \beta_4$ and $\pi - \beta_3 = \beta_1$, and therefore the two triangles in the bow tie trace have exactly the same interior angles. (This is not necessarily true of rigid degree-4 vertices in general, although the two triangles must have the same area.) Further, in spherical geometry, two triangles with equal pairs of angles are always congruent.

Therefore, the two bow tie triangles are actually congruent in the flat-foldable degree-4 case. The converse is true as well; if the two bow tie triangles are congruent, then their angles are too and we have $\beta_1 = \pi - \beta_3$ and $\beta_4 = \pi - \beta_2$, so the vertex is flat-foldable. These triangles being congruent implies that their side lengths are equal, and thus so are the opposite folding angles of the folded vertex, since the lengths of the sides of the Gauss map trace equal the folding angles of the creases. Thus we immediately have the following:

**Theorem 12.22** *A rigid-foldable degree-4 origami vertex is flat-foldable if and only if the folding angles of the major creases are equal, as are those of the minor creases, throughout the rigid folding. That is, if $\rho_1$ and $\rho_3$ are the folding angles of the major creases and $\rho_2$ and $\rho_4$ the folding angles of the minor creases, then*

$$\rho_1 = \rho_3 \quad and \quad \rho_2 = 2\pi - \rho_4.$$

This theorem can also be proven using spherical trigonometry in a variety of ways. See, for example, (Hull, 2012).

A version of Theorem 12.22 also exists for non-flat-foldable degree-4 rigid origami vertices. The following theorem, and its proof, are due to Huffman (1976).

**Theorem 12.23** *Let $\rho_1, \ldots, \rho_4$ be the folding angles of a degree-4 rigid origami vertex in a rigidly folded state, where $\rho_1$ and $\rho_3$ are the folding angles of the major creases. Also let the plane angles $\beta_1, \ldots, \beta_4$ be labeled as in Figure 12.10. Then we have*

$$1 - \cos \rho_3 = \frac{\sin \beta_1 \sin \beta_2}{\sin \beta_3 \sin \beta_4}(1 - \cos \rho_1).$$

*Proof* A schematic of the Gauss map trace for such a degree-4 vertex is shown in Figure 12.10, where we note that the bow tie triangles labeled $E$ will have the same area but might not be congruent, since this vertex is not necessarily flat-foldable. We extend the arcs of the bow tie made by the folding angles $\rho_2$ and $\rho_4$ until they intersect at a point $\vec{n}_0$ at an angle of $\beta_0$. Then the spherical triangles $\vec{n}_0\vec{n}_1\vec{n}_2$ and $\vec{n}_0\vec{n}_3\vec{n}_4$ have equal areas. A variation of the spherical law of cosines (sometimes called the second spherical law of cosines) then yields

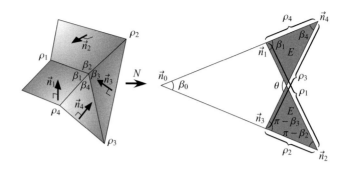

**Figure 12.10** The folding angles $\rho_i$ of a rigidly folded degree-4 vertex and extending the sides of the bow tie in the Gauss map trace.

$$\cos \beta_0 = -\cos(\pi - \beta_1)\cos(\pi - \beta_2) + \sin(\pi - \beta_1)\sin(\pi - \beta_2)\cos \rho_1$$

and

$$\cos \beta_0 = -\cos \beta_3 \cos \beta_4 + \sin \beta_3 \sin \beta_4 \cos \rho_3.$$

Equating these, adding $\sin \beta_1 \sin \beta_2 + \sin \beta_3 \sin \beta_4$ to both sides, and rearranging a little gives

$$-\cos \beta_1 \cos \beta_2 + \sin \beta_1 \sin \beta_2 + \sin \beta_3 \sin \beta_4 - \sin \beta_3 \sin \beta_4 \cos \rho_3$$
$$= -\cos \beta_3 \cos \beta_4 + \sin \beta_3 \sin \beta_4 + \sin \beta_1 \sin \beta_2 - \sin \beta_1 \sin \beta_2 \cos \rho_1$$
$$\Rightarrow -\cos(\beta_1 + \beta_2) + \sin \beta_3 \sin \beta_4(1 - \cos \rho_3)$$
$$= -\cos(\beta_3 + \beta_4) + \sin \beta_1 \sin \beta_2(1 - \cos \rho_1).$$

Since $\sum \beta_i = 2\pi$, $\cos(\beta_1 + \beta_2) = \cos(2\pi - (\beta_3 + \beta_4)) = \cos(\beta_3 + \beta_4)$. Therefore, the last equation gives the desired result. $\qquad\square$

---

**Diversion 12.11**   Prove that if $\rho_2$ and $\rho_4$ are the folding angles of the minor creases in a degree-4 origami vertex in a rigidly folded state, where $\rho_4$ has different sign than $\rho_1, \rho_2, \rho_3$, then we have

$$1 - \cos \rho_4 = \frac{\sin \beta_2 \sin \beta_3}{\sin \beta_1 \sin \beta_4}(1 - \cos \rho_2).$$

---

**Diversion 12.12**   Prove that Theorem 12.23 and Diversion 12.11 give us a proof of Theorem 12.22.

---

These results are applicable to not just a rigidly folded state of a vertex, but the whole parameterization of folded states, which we will refer to as **rigid foldings** in Chapter 13. The reader may look forward to seeing further discussion of rigid-foldable degree-4 vertices in Section 13.2. Also, what we present here on applying the Gauss

(a)

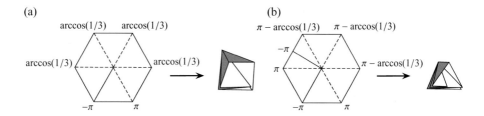

(b)

Figure 12.11 Solid angle vertex folds that form (a) the corner of an octahedron and (b) the corner of a tetrahedron. The labels on the creases are the folding angles.

map to rigid origami crease patterns is only the tip of the iceberg of what can be done. Many more results on rigidly folded origami based on the Gauss map can be found in (Lang, 2018).

## 12.4 Another Generalization of Maekawa

We now consider the question of whether or not Maekawa's Theorem can be generalized to non-flat, rigid origami vertices. General rigid vertex folds are so flexible in their design that one can make the quantity $M - V$ be a wide range of values, where $M$ is the number of convex creases and $V$ is the number of concave creases when looking at a specified side of the paper. Some observations can be made, however, if we restrict ourselves to a subset of rigid vertex folds.

**Definition 12.24**  A rigid origami vertex $\sigma$ on a piece of paper $R$ is called a **solid angle vertex fold** if $\sigma(R)$ forms a polyhedral vertex.

Examples of solid angle vertex folds are shown in Figures 12.1 and 12.11. Rigid folds that leave a flap of paper protruding from the model or that make a figure that contains zero volume are not considered solid angle folds, such as those in Figures 12.14 and 12.15. Note that solid angle folds need not be convex; Figure 12.12 shows an example of a non-convex solid angle fold.

Our solid angle folds that make a corner of a cube, a tetrahedron, and the non-convex example of Figure 12.12 all have $M - V = \pm 3$, while the corner of an octahedron has $M - V = \pm 4$.

---

**Diversion 12.13**  Make examples of solid angle vertex folds that have $M - V = \pm 1$ and $M - V = 0$.

---

These values of $M - V$ vary so wildly as to make one wonder if any pattern in them exists. If we restrict ourselves to solid angle folds that form a convex solid angle when folded, there actually is a pattern.

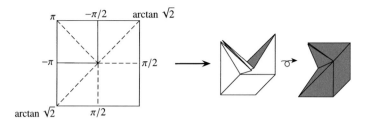

**Figure 12.12** A non-convex solid angle vertex fold.

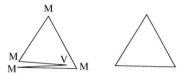

**Figure 12.13** Polygonal cross sections showing the M and V creases (left) and the convex hull (right).

**Theorem 12.25**  *Given a convex solid angle vertex fold $\sigma$ on a piece of paper R, if the valency of the polyhedral vertex that $\sigma(R)$ makes is d, then the crease pattern will have $M - V = \pm d$.*

*Proof*  Consider the polygonal cross section of the solid angle fold after it has been folded, as if we were truncating the solid angle. We want to also consider creases with folding angles of $-\pi$ and $\pi$ to be "flat" corners of this polygonal cross section. (See Figure 12.13.) Then the number of sides of this polygon is $M + V$.

We now consider the fact that the sum of the interior angles of the polygonal cross section equals $(M + V - 2)\pi$. We may assume that we are looking at this polygon "from above," so the flat mountain creases will contribute 0 radians to this sum and the flat valleys will each contribute $2\pi$. Since we're looking at this from above, the edges of the solid angle will all be mountain creases, and their interior angle sum will be $(d - 2)2\pi$. The other mountain creases will contribute 0 radians each. Finally, since the solid angle is convex, all the valley creases will be flat, contributing $2\pi$ each. Thus we may decompose the angle summation as follows:

$$\sum \text{interior angles of the cross section} = (M + V - 2)\pi$$
$$\Rightarrow (d - 2)2\pi + 2\pi V = (M + V - 2)\pi.$$

Dividing the last equality through by $\pi$ and simplifying gives the desired result.  □

An example of a non-convex solid angle fold can be seen in Figure 12.12. To make sense out of $M - V$ in such a case, we must keep track of which edges of the finished solid angle are convex and which are concave. More precisely, we may let $M_c$ and $V_c$ be the number of mountain and valley creases in the crease pattern and $M_s$, $V_s$ be the number of convex and concave edges of folded the solid angle, which is equivalent to

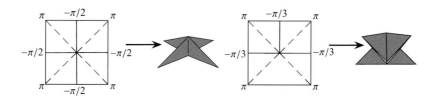

**Figure 12.14** Two non-flat vertex folds of zero volume with $M - V = 0$ (left) and 1 (right).

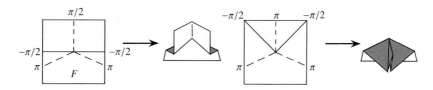

**Figure 12.15** Two other challenges to generalizing Maekawa.

the number of **non-flat** (i.e., folding angle $\neq \pm\pi$) convex and concave creases in the crease pattern. The proof of Theorem 12.25 can be modified to prove that, in general, $M_c - V_c = M_s - V_s$ for solid angle vertex folds.

Other pathological situations may occur. Figure 12.14 shows two non-flat folds that create a "solid angle" of zero volume and return different values of $M - V$. In Figure 12.15, left, not only are there "flat flaps" protruding from the solid angle, but their arrangement causes a face of the crease pattern, $F$, to span three faces, or what we might think of as faces, of the folded solid angle. Furthermore, by changing the crease with folding angle $\pi/2$ to a folding angle of $\pi$, we can convert this "solid angle" fold to one with zero volume without affecting $M - V$ (Figure 12.15, right). Such examples seem to imply that any meaningful, more general variation of Theorem 12.25 might not be possible.

What about extending this to multiple-vertex crease patterns? It is possible, and in fact quite common in origami art, to have multiple-vertex crease patterns where each vertex is a convex solid angle fold. For example, the origami technique of box pleating (see (Lang, 2011)) creates vertices that are, generally, convex solid angles of the type in Figure 12.1. Also see Figure 12.16 for two examples of how to use convex solid angle vertex folds to create an origami cube from one sheet of paper.

Theorem 12.25 can be generalized in a similar way to our generalization of Maekawa's Theorem in Section 6.3. For every interior vertex $v$ in a crease pattern made entirely of convex solid angle vertices, let $d(v)$ be the valency (i.e., degree, or number of sides) of the solid angle that $v$ creates when folded. Also let us define $U$ to be the set of **up vertices** (interior vertices with $M - V > 0$ locally) and $D$ to be the set of **down vertices** (interior vertices with $M - V < 0$ locally).

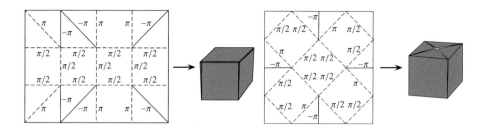

**Figure 12.16** Two multiple-vertex solid angle folds that make cubes. The left model can be made to lock together by inserting flaps into proper pockets.

**Theorem 12.26**  *Given a rigid-foldable crease pattern $G = (V, E)$ and a folding angle assignment $\mu: E \rightarrow [-\pi, \pi]$ that has M (resp. V) convex (resp. concave) creases, where every vertex of the crease pattern is a convex solid angle vertex fold, we have*

$$M - V = \sum_{v \in U} d(v) - \sum_{v \in D} d(v) - IM + IV,$$

*where IM and IV are the number of interior convex and concave creases in G, respectively.*

---

**Diversion 12.14**  Prove Theorem 12.26 using Theorem 12.25.

---

## 12.5    Open Problems

Much of the material in this chapter serves as required background for the next two chapters. As such, there are not a lot of open questions relating to origami in some of these areas. For example, the Gauss map is a great way to study rigid origami, and anyone wanting to investigate this further should study Robert J. Lang's work in this (Lang, 2018). But that ground has been explored pretty thoroughly.

The matrix model of rigid origami has become standard for engineering applications, and thus is now also well-understood. However, there is still room for exploration of this model when it comes to detecting when the paper will intersect itself. This was discussed briefly in (belcastro and Hull, 2002b), but the general problem still remains open, which we frame here in the single-vertex case.

**Open Problem 12.1**  Given a single-vertex crease pattern $G = (V, E)$ and folding angle function $\mu: E \rightarrow [-\pi, \pi]$ that results in a rigid origami $\sigma$, is there an easy way, possibly using the matrix model for $\sigma$, to detect if $\sigma$ has any self-intersections?

There are ways to do this kind of intersection detection for specific cases, such as in (Liu et al., 2018). We will also see in Chapter 14 how configuration spaces of rigid foldings can help determine when rigid origami goes awry. But for the open problem stated here, we seek a general, all-purpose method.

## 12.6  Historical Remarks

While the idea of specifically using matrices to model rigid origami seems to originate from the work of Toshikazu Kawasaki (1987, 1997), the concept is quite a bit older in the engineering kinematics literature. For example, in 1955 Denavit and Hartenberg described how mechanical linkage systems that form a closed loop can be modeled by a matrix product being the identity (Denavit and Hartenberg, 1955). Because of this, some more recent engineering papers on rigid origami mechanisms cite Denavit and Hartenberg as the originators of the matrix product result of Theorem 12.2 and Equation (12.1). However, such interpretations of the history of this work misses an important, if rather subtle, point: In Equation (12.1),

$$\chi_1 \cdots \chi_n = I,$$

the matrices $\chi_i$ are rotations about the crease lines of the **unfolded** crease pattern. That is, each $\chi_i$ is a rotation matrix whose axis of rotation lies in the $xy$-plane. This is very different from the matrix products from (Denavit and Hartenberg, 1955), where each matrix represents a rotation based at a different point in space (in fact, each matrix is a change of coordinates transformation). Denavit and Hartenberg's matrix model is more in line with the product $L_n \cdots L_1 = I$ from the proof of Theorem 12.2, where each matrix $L_i$ is a rotation of $\mathbb{R}^3$ whose rotation axis is the image of the crease line $l_i$ after the paper is folded. In the case of rigid origami, this $L_i$ matrix product simplifies to the $\chi_i$ matrix product, which is much more simple. Such a reduction is not possible for general, closed mechanical linkages as studied in (Denavit and Hartenberg, 1955).

The contribution of belcastro and Hull to Kawasaki's work was to add enough formalism to allow Equation (12.1) to be proven (belcastro and Hull, 2002a,b), as opposed to stating the equation as a definitional property of rigid origami (which was Kawasaki's approach).

Applying the Gauss map to rigid paper folding was first done by David A. Huffman in a series of two papers (Huffman, 1976, 1978). It is worth noting that this is the same person who created Huffman codes for data compression in 1952. Despite the fact that his only published papers on paper folding were from the 1970s, Huffman continued to work in this area. After Huffman's death in 1999, his research notes on paper folding were shared by his family with researchers in mathematical and computational origami. This resulted in a number of posthumous publications on origami research for Huffman, such as (Demaine et al., 2016).

# 13 Rigid Foldings

In Chapter 12 we introduced rigid origami, which we may think of as a continuous mapping $\sigma$ from a planar sheet of paper region $R$ into $\mathbb{R}^3$ such that the faces of the crease pattern $G = (V, E)$ on $R$ map to planar polygons in $\sigma(R)$. As such, any adjacent crease pattern faces must have a fixed folding angle between them in $\sigma(R)$. In this chapter we explore the relationships between folding angles in a rigid origami so as to understand how they change with respect to each other. This will allow us to model how a rigid origami $\sigma_1$ can change into another rigid origami $\sigma_2$ on the same crease pattern via a continuous change of its folding angles. By taking the folding angles (or a subset of them) as parameters, we can define a **rigid folding** to be a parameterized family of rigid origamis that model how a rigid-foldable crease pattern can flex, say, from the unfolded state to a folded state.

There is a strong connection between rigid foldings and rigid-body kinematics and rigidity in general. Thus we start our investigation with infinitesimal rigid foldability, also known as first-order rigid foldability, which is analogous to infinitesimal rigidity of graph (bar and joint) structures. We then, in Section 13.2, turn to second-order (or finite) rigid foldability, which implies a range of motion of the folding angles and thus leads, in some cases, to angle relationships between the folding angles and some interesting observations about the **relative speed** at which folding angles change. We will then, in Section 13.3, look at a simple way to determine if a single-vertex crease pattern, with an assigned MV assignment, will have a rigid folding from the flat state.

## 13.1 Infinitesimal Rigid Foldability

The idea behind infinitesimal rigidity is to quantify when a mechanical system can flex by an infinitely minuscule amount, as opposed to a full range of motion. In a rigid-bar-and-joint structure, as one might see in architecture, this would mean that a potential flexing motion exists where the tangent vectors of the joint motions agree with the constraints imposed by the rigid bars. This can happen sometimes even if the bar-and-joint structure is actually rigid.

The approach to studying infinitesimally rigid origami structures pioneered by Naohiko Watanabe and Kenichi Kawaguchi (Watanabe and Kawaguchi, 2009) and Tomohiro Tachi (2010a, b, 2012) is similar in approach to that of ball-and-joint

rigidity. Readers may consult (Graver et al., 1993) for an introduction to the latter. Infinitesimal rigidity in an origami context is seldom explicitly defined in the
engineering literature. Therefore, we will attempt to define things carefully.

**Definition 13.1** Let $f\colon \mathbb{R} \to \mathbb{R}$ be a real-valued function. We say that $f$ is **constant
in the first order of** $t$ **at** $t = t_0$ if the linear part of the Taylor expansion of $f$ at $t = t_0$
is a constant function of $t$. That is, if the Taylor expansion is

$$f(t) = a_0 + a_1(t - t_0) + \cdots,$$

then $a_0 + a_1(t - t_0) = c$ for all $t \in \mathbb{R}$, where $c \in \mathbb{R}$ is a constant. Similarly, we say that
$f$ is **constant in the $k$th order of** $t$ **at** $t = t_0$ if the sum of the terms up to and including
the $(t - t_0)^k$ term of the Taylor expansion of $f$ at $t = t_0$ is a constant function of $t$.
  We may also extend these definitions for functions $f\colon \mathbb{R} \to \mathbb{R}^n$ or $f\colon \mathbb{R} \to M_{n,n}$,
where $M_{n,n}$ is the set of all $n \times n$ real-valued matrices.

  We will initially consider infinitesimally rigid origami as a special case of infinitesimal deformations of discrete surfaces, following the treatment of Schief, Bobenko,
and Hoffmann in (Schief et al., 2008). Discrete surfaces arise out of the developing
field of discrete differential geometry.
  Recall that if $G$ is a planar graph with vertices $V(G)$ and edges $E(G)$, then a planar
embedding of $G$ will define a set of faces, $R(G)$. The **geometric dual graph** $G^*$ of
$G$ is the planar graph whose vertices are $R(G)$ and whose edges $E(G^*)$ are defined as
follows: for every edge adjacent to faces $f_1$ and $f_2$ in $G$, we include an edge $\{f_1, f_2\} \in
R(G)$. Note that this allows $G^*$ to, possibly, have multiple edges.

**Definition 13.2** A **discrete surface** $F$ is the image of a mapping

$$F\colon V(G) \to \mathbb{R}^3,$$

where $V(G)$ denotes the vertices of a planar graph $G$. A **dual** discrete surface $F^*$ is a
mapping that is defined on the vertices of the geometric dual graph $G^*$ of $G$, that is,
$F^*\colon V(G^*) \to \mathbb{R}^3$.

  A discrete surface could be a polyhedron, a triangulated surface, or a partially
folded rigid origami crease pattern, to name a few examples.

**Definition 13.3** Given a discrete surface $F\colon V(G) \to \mathbb{R}^3$, we define a **deformed
surface** $F^\varepsilon$ to be

$$F^\varepsilon = F + \varepsilon \overline{F},$$

where the constant $\varepsilon$ is called the **deformation parameter** and is normally considered
to be small ($|\varepsilon| \ll 1$) and $\overline{F}\colon V(G) \to \mathbb{R}^3$ describes the displacement of the vertices
of $F$.
  An **infinitesimal isometric deformation** $F^\varepsilon$ of a discrete surface $F$ is a deformed
surface whose faces are not changed to the first order, which means that their side
lengths and interior angles are constant in the first order of $\varepsilon$.

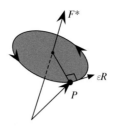

**Figure 13.1** An infinitesimal rotation of a point $P$ about an axis $F^*$ can be described by a small vector $\varepsilon R$ that is orthogonal to the plane made by $P$ and $F^*$.

**Remark 13.4** It will be useful to remark on what, exactly, the infinitesimal isometries implied in the definition of infinitesimal isometric deformations are in $\mathbb{R}^3$.

Every orientation-preserving isometry of $\mathbb{R}^3$ that preserves the shape of a polygon must be a composition of a translation and a rotation, either of which may be trivial. An infinitesimal translation can be specified simply by a scalar $\varepsilon$ times a translation vector $T$, where $|\varepsilon|$ is very small. An infinitesimal rotation, however, can be specified more simply than by a rotation matrix. Such a small rotation will need an axis of rotation, which we will denote by a vector $F^*$. (The reason for this choice of notation will be apparent later.) Suppose, then, that we want to rotate a point $P$ around this axis by a very small amount. Then $P$ will be nudged in a direction $\varepsilon R$ that is orthogonal to both the vector $F^*$ and the vector $P$, as illustrated in Figure 13.1. Following the right-hand rule, we may describe the vector $R$ as a cross product $F^* \times P$.

Therefore, if we let $P$ be a point in a face $f$ of a discrete surface $F$ and perform an infinitesimal isometric deformation $F^\varepsilon$, letting $P^\varepsilon$ be the point in $F^\varepsilon$ corresponding to $P$, then the displacement between $P^\varepsilon$ and $P$ can be described as $P^\varepsilon - P = \varepsilon \overline{P}$ where

$$\overline{P} = T + F^* \times P, \tag{13.1}$$

where $T$ is the translation and $F^*$ the axis of rotation for the displacement of the face $f$, both of which are independent of the choice of $P \in f$. Thus for our discrete surface $F$, an infinitesimal isometric deformation $F^\varepsilon$ defines a translation $T$ for each face, which is a function $T \colon V(G^*) \to \mathbb{R}^3$, and a rotation vector $F^*$ for each face, which we may also think of as a function $F^* \colon V(G^*) \to \mathbb{R}^3$.

We chose the notation $F^*$ for the vector axes of rotation of the faces in $F$ under an infinitesimal isometric deformation because they literally form a dual discrete surface of $F$. This dual is quite hard to visualize. The domains of $F$ and $F^*$ are the vertices of $G$ and $G^*$, respectively, which is easy to visualize. But under their respective mappings of $F$ and $F^*$, these vertices will map to points in $\mathbb{R}^3$ that could easily not resemble a dual structure.

The point, however, is that the dual surface $F^*$ **does** have a nice geometric property that makes it an easy-to-check visual tool in the case where the surface $F$ is a rigid origami and $F^\varepsilon$ an infinitesimal isometric deformation of the rigid origami.

**Definition 13.5** Dual discrete surfaces $F$ and $F^*$ are called **reciprocal-parallel** if the dual edges in $F^*$ are parallel to their corresponding edges in $F$.

**Theorem 13.6** *If F is a discrete surface and $F^{\varepsilon}$ an infinitesimal isometric deformation with rotation vectors $F^*$ corresponding to the faces of F, then the discrete surfaces F and $F^*$ are reciprocal-parallel.*

*Proof* Let $f_r$ and $f_l$ be two adjacent faces of $F$ with their shared edge denoted $\{P_1, P_2\}$ in the image of $F$. These two vertices $P_1$ and $P_2$ belong to both of the faces $f_r$ and $f_l$, and so their displacements, using Equation (13.1), can be calculated in two different ways. Let $T_r$ and $F_r^*$ be the translation and rotation vectors, respectively, for the displacement of face $f_r$ and similarly for $f_l$. Then we have

$$\overline{P}_1 = T_r + F_r^* \times P_1, \qquad \overline{P}_2 = T_r + F_r^* \times P_2,$$
$$\overline{P}_1 = T_l + F_l^* \times P_1, \qquad \overline{P}_2 = T_l + F_l^* \times P_2. \qquad (13.2)$$

Now if we consider the edge $\{P_1, P_2\}$ in the surface $F$ and $\{F_r^*, F_l^*\}$ in the surface $F^*$, which are dual to each other, we have that

$$(F_r^* - F_l^*) \times (P_1 - P_2) = F_r^* \times P_1 - F_r^* \times P_2 - F_l^* \times P_1 + F_l^* \times P_2$$
$$= (\overline{P}_1 - T_r) - (\overline{P}_2 - T_r) - (\overline{P}_1 - T_l) + (\overline{P}_2 - T_l) = 0.$$

Therefore, $\{P_1, P_2\}$ and $\{F_r^*, F_l^*\}$ are parallel. $\qquad \square$

A major question is whether or not a discrete surface $F$ can be infinitesimally isometrically deformed, that is, whether there exists an infinitesimal isometric deformation $F^{\varepsilon}$ for the surface. This would mean the surface could be flexed "in the first order" keeping the faces of $F$ planar and rigid and bending only along the edges of $F$. Surprisingly, the existence of a reciprocal-parallel dual of $F$ is enough to guarantee that such an $F^{\varepsilon}$ exists. The following theorem is due to Schief, Bobenko, and Hoffmann (Schief et al., 2008).

**Theorem 13.7** (Schief et al., 2008) *A discrete surface F admits an infinitesimal isometric deformation if and only if there exists a reciprocal-parallel discrete surface $F^*$.*

*Proof* Theorem 13.6 gives us one direction of the proof. For the other, suppose that we have a discrete surface $F$ and a reciprocal-parallel discrete surface $F^*$. Our aim is to show that we may generate an infinitesimal isometric deformation of $F$ from $F^*$.

The vectors in $F^*$ will be our rotation vectors for every face of $F$, but we still need to define the translation vectors. Given an edge $\{P_1, P_2\}$ of $F$ and the faces $f_r$ and $f_l$ to the left and right, say, of this edge, we want to find translations $T_r$ and $T_l$ for these two faces that agree with the necessary requirements from Equations (13.2) of the rotation vectors $F^*$ from our reciprocal-parallel surface. That is, we want to have

$$T_r - T_l = -(F_r^* - F_l^*) \times P_1 \quad \text{and} \quad T_r - T_l = -(F_r^* - F_l^*) \times P_2.$$

These two equations in two unknowns $T_r$ and $T_l$ will uniquely define a function $T: V(G^*) \to \mathbb{R}^3$ provided that we fix an initial translation $T_0$ for a fixed face $f_0$. However, we need to make sure that this function $T$ is well-defined, since any translation

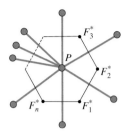

**Figure 13.2** Proving well-defined-ness of translation map $T \colon V(G^*) \to \mathbb{R}^3$ around a vertex $P$ of $F$.

$T_r$ could, say, be defined by any of its boundary edges that are internal to the surface $F$. That is, consider a vertex $P$ of the surface $F$ and the faces $F_1^*, \ldots, F_n^*$ surrounding $P$, which are vertices in the dual surface $F^*$, as depicted in Figure 13.2. Since $F^*$ is reciprocal-parallel to $F$, we know that the Equations (13.2) hold, and we would define the translations on these dual faces using

$$T_{k+1} - T_k = -(F_{k+1}^* - F_k^*) \times P, \quad \text{for } k = 1, \ldots, n,$$

where we identify $T_{n+1} = T_1$ and $F_{n+1}^* = F_1^*$. We want these equations to be consistent in the closed loop around the vertex $P$, and thus we compute

$$\sum_{k=1}^{n}(T_{k+1} - T_k) = -\sum_{k=1}^{n}(F_{k+1}^* - F_k^*) \times P$$

$$= -\sum_{k=1}^{n} F_{k+1}^* \times P + \sum_{k=1}^{n} F_k^* \times P = -F_{n+1}^* \times P + F_1^* \times P = 0.$$

Thus the vectors $T_{k+1} - T_k$ for $k = 1, \ldots, n$ form a closed loop and are therefore consistent and well-defined.

Then we may define the displacement of a vertex $P$ of $F$ as $\overline{F} = T + F^* \times P$, where $T$ is defined by any of the dual edges surrounding the face dual to $P$. We obtain an infinitesimal isometric deformation $F^\varepsilon = F + \varepsilon \overline{F}$ of $F$. ☐

**Example 13.8** Suppose that we take a vertex of an octahedron together with its four adjacent faces, $f_1, \ldots, f_4$, viewed as a discrete surface $F$, and we take an infinitesimal isometric deformation $F^\varepsilon$ of it. In Figure 13.3 we show this, where we have placed the vertex at the origin and fixed the face $f_1$ (whose other two corners are $(1, 0, 0)$ and $(1/2, -\sqrt{3}/2, 0)$) so that it does not move in our infinitesimal deformation. Then we may compute the infinitesimal rotation vectors $F_i^*$. We see that $F_1^* = (0, 0, 0)$, since face $f_1$ does not move, and that $F_2^* = (1, 0, 0)$ and $F_4^* = (1/2, -\sqrt{3}/2, 0)$ because the faces $f_2$ and $f_4$ merely rotate about their edges adjoining the fixed face $f_1$. The coordinates of the vector $F_3^*$ will depend on the value of $\varepsilon$; in Figure 13.3 we calculated the position of $F_3^*$ using Mathematica with $\varepsilon = 0.02$.

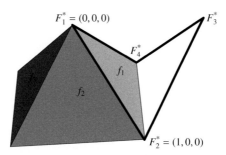

**Figure 13.3** An infinitesimal isometric deformation of a corner of an octahedron with its reciprocal-parallel surface $F^*$ shown.

Notice that the edges $F_2^* - F_1^*$, $F_3^* - F_2^*$, $F_4^* - F_3^*$, and $F_1^* - F_4^*$ form a 4-cycle, which is the graph of the dual of the octahedron vertex figure. Also, it does appear that the edges of the surface $F^*$ are parallel to the edges of the octahedron vertex.

Of specific interest, of course, is the application of Theorem 13.7 to origami. We immediately obtain that if we view a rigidly folded origami crease pattern $C$ as a discrete surface, then an infinitesimal isometric deformation exists for this rigid origami if and only if there is a reciprocal-parallel discrete surface $C^*$. But more can be said in the rigid folding case, especially when we are considering the paper as it folds from the flat, unfolded state.

First, some definitions are in order, following (Demaine et al., 2016).

**Definition 13.9**   Given a crease pattern $G = (V, E)$ on a region $R$ of the plane, the **parameter space** of $G$ is $[-\pi, \pi]^{|E|}$, where each point $\rho \in [-\pi, \pi]^{|E|}$ has coordinates equal to the possible folding angles of the crease line edges of $G$. The **configuration space** $\mathcal{C}(G)$ of $G$ is the subset of points $\rho = (\rho_1, \ldots, \rho_{|E|})$ of the parameter space that give rise to a rigid origami $\sigma$ whose folding angle function is $\mu(l_i) = \rho_i$, where $E = \{l_1, \ldots, l_{|E|}\}$. That is, the points in $\mathcal{C}(G)$ correspond to valid sets of folding angles that allow $\sigma$ to fold rigidly and without self-intersections.

We then say that an origami crease pattern $G$ on a region $R$ of the plane has a **rigid folding** if there exist a finite-length path $\gamma : [a, b] \to \mathcal{C}(G)$ and a parameterized family of rigid origamis $\sigma(t)$ on $G$ such that the folding angle function for $\sigma(t)$ is given by $\gamma(t)$.

We know that a rigid origami $\sigma$ exists for a crease pattern $G$ on a region $R$ if and only if the image $\sigma(R)$ has no self-intersections (by which we mean for large enough $k$ the sequence of origamis $\sigma_k$ that define $\sigma$ are injective) and around each interior vertex of the crease pattern we have that Theorem 12.2 holds, that is $\chi_1 \cdots \chi_n = I$. Let us rewrite $R(l_i, \rho_i) = \chi_i$, where $l_i$ is the crease line forming the rotation axis of $\chi_i$ and $\rho_i$ is the folding angle at $l_i$, where $l_1, \ldots, l_n$ are the crease lines meeting at our vertex. We then define a function $F$ on the parameter space $[-\pi, \pi]^n$ by

$$F(\rho) = R(l_1, \rho_1)R(l_2, \rho_2)\cdots R(l_n, \rho_n),$$

where $\rho = (\rho_1, \ldots, \rho_n)$. Then a necessary condition for $\rho$ to be in our configuration space is that $F(\rho) = I$, the identity matrix.

**Definition 13.10**    An origami crease pattern $G$ is **$k$th-order rigid-foldable** if there is a parameterized $C^1$ curve $\rho(t)$ in the parameter space with nontrivial initial velocity $\rho'(0)$ such that around each vertex $F(\rho(t))$ is constant in the $k$th order of $t$ at $t = 0$.

First-order rigid foldability is also called **infinitesimal rigid foldability**.

This definition of $k$th-order rigid foldability is convenient because it is concerned only with folding near the flat (all folding angles $\rho_i = 0$) state. As such, all that is needed is for $F(\rho) = I$ to hold at each vertex, because (at least in the case where the region of paper $R$ is bounded) near the unfolded state there will be no chance of self-intersections. Also note that first-order (infinitesimal) rigid foldability of origami corresponds to the definition of infinitesimal isometric deformations of discrete surfaces.

From Chapter 12 we already saw an explicit formula for the matrices $R(l_i, \rho_i)$, but they can also be computed by Rodrigues' rotation matrix formula (Murray, 1994, p. 28). Let $l_i$ be unit vectors in the direction of our crease lines with coordinates $(l_i^x, l_i^y, l_i^z)$ (where $l_i^z = 0$ when we are in the flat, unfolded state), then define the cross product operator to be the skew-symmetric matrix

$$[l_{i\times}] = \begin{pmatrix} 0 & -l_i^z & l_i^y \\ l_i^z & 0 & -l_i^x \\ -l_i^y & l_i^x & 0 \end{pmatrix}.$$

That is, $[l_{i\times}]v = l_i \times v$. Then Rodrigues' formula states

$$R(l_i, \rho_i) = I + \sin \rho_i [l_{i\times}] + (1 - \cos \rho_i)[l_{i\times}]^2. \tag{13.3}$$

**Theorem 13.11**    *A crease pattern $G$ is first-order rigid-foldable if and only if there exists a reciprocal-parallel surface of $G$. Furthermore, around each vertex $v$ in $G$ the reciprocal-parallel surface in the unfolded state is a loop made by connecting, tip to tail, scaled vectors in the direction of the crease lines at $v$ that are oriented toward $v$ if they are mountains and away from $v$ if they are valleys.*

The first part of this theorem is proven by the more general Theorem 13.7 for discrete surfaces, but the second part requires more work. Therefore, we will present a self-contained proof of Theorem 13.11 from (Tachi, 2012) and (Demaine et al., 2016).

*Proof*    First, note that Equation (13.3) gives us

$$\frac{\partial R(l_i, \rho_i)}{\partial \rho_i} = \cos \rho_i [l_{i\times}] + \sin \rho_i [l_{i\times}]^2$$

$$= [l_{i\times}] + \sin \rho_i [l_{i\times}]^2 + [l_{i\times}]^3 - \cos \rho_i [l_{i\times}]^3 = R(l_i, \rho_i)[l_{i\times}],$$

where we used that fact that $[l_{i\times}]^3 = -[l_{i\times}]$ since $l_i$ is a unit vector. This means that

$$\left.\frac{\partial F}{\partial \rho_i}\right|_{\rho=\vec{0}} = R(l_1, \rho_1) \cdots R(l_i\rho_i)[l_{i\times}]R(l_{i+1}\rho_{i+1}) \cdots R(l_n, \rho_n)\big|_{\rho=\vec{0}} = [l_{i\times}].$$

Therefore, if we take the Taylor expansion of $F$ around $t = 0$ along a $C^1$ curve $\rho(t)$ (where $\rho(0) = \vec{0}$ is the unfolded state), we get

$$F(\rho(t)) = I + \left.\frac{dF}{dt}\right|_{t=0} t + o(t^2).$$

By the chain rule,

$$\frac{dF}{dt} = \frac{\partial F}{\partial \rho_1}\frac{d\rho_1}{dt} + \cdots + \frac{\partial F}{\partial \rho_n}\frac{d\rho_n}{dt}.$$

Thus,

$$F(\rho(t)) = I + \sum_{i=1}^{n} \left.\frac{\partial F}{\partial \rho_i}\rho_i'(t)\right|_{t=0} t + o(t^2).$$

We then see that a necessary and sufficient condition for first-order rigid foldability is to find nontrivial values for the $\rho_i'(0)$ terms so that

$$\sum_{i=1}^{n}[l_{i\times}]\rho_i'(0) = Z$$

(where $Z$ is the zero matrix). Looking at the individual entries of this matrix equation, we see that the sum of the x-, y-, and z-coordinates of the vectors $l_i$ (multiplied by the respective $\rho_i'(0)$ numbers) equals 0. Thus an equivalent way to write this is

$$\sum_{i=1}^{n} \rho_i'(0)l_i = \vec{0}. \tag{13.4}$$

Note that since we are folding infinitesimally from the flat, unfolded state, we will have $\rho_i'(0) > 0$ if $l_i$ is a valley and $\rho_i'(0) < 0$ if $l_i$ is a mountain. Equation (13.4) thus tells us that the vectors $l_i$, scaled appropriately and given direction according to their MV parity, will form a closed loop.

Thus if we are given an infinitesimal rigid fold, it will satisfy Equation (13.4) around each vertex, and thus a dual surface can be made with the vectors $\rho_i'(0)l_i$, giving a reciprocal-parallel closed figure around each vertex, which can be tiled to make the full dual surface because their shared edges have compatible directions and lengths. For sufficiency, if we are given a reciprocal-parallel diagram for each vertex of our crease pattern, then each edge $e_i$ of this diagram will correspond to an edge $l_i$ of the crease pattern's vertex. If we assign $\rho_i'(0)$ to be the signed length of the edge of $e_i$ (where the sign switches, say, if the vector directions switch as we travel counterclockwise around the reciprocal-parallel diagram for the vertex), then we will obtain folding speeds $\rho_i'(0)$ satisfying Equation (13.4), thus giving us a first-order rigid folding for each vertex. □

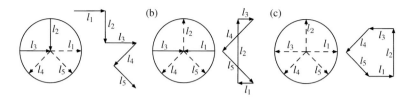

**Figure 13.4** Examples of single-vertex crease patterns and their possible reciprocal-parallel diagrams to test for infinitesimal rigid foldability. (Based on examples from (Watanabe and Kawaguchi, 2009).)

**Example 13.12**   Figure 13.4 shows three examples, from (Watanabe and Kawaguchi, 2009), of how Theorem 13.11 can be used to see if a given single-vertex crease pattern is infinitesimally rigid-foldable. They have the same crease pattern but different MV assignments. Note that, as we saw in the proof of the theorem, the lengths of the edges in the reciprocal-parallel diagram equal the speed $\rho_i'(t)$ of the folding angle $\rho_i(t)$ at $t = 0$. In (a), we see that no matter how we change the lengths of the edges, we cannot make the reciprocal-parallel diagram form a closed loop, and thus this MV assignment is not infinitesimally (first-order) rigid-foldable.

The MV assignment in (b) is first-order rigid-foldable, since the directions of the reciprocal-parallel edges will allow it to form a closed loop. We will later see that this example also exhibits second-order rigid foldability, meaning that it will rigidly fold (i.e., flex) more than just an infinitesimal amount.

The MV assignment in (c), which is all valley creases, also forms a closed loop in the reciprocal-parallel diagram. Thus it is also infinitesimally rigid-foldable. However, we saw in Diversion 12.7 that a degree-4 rigid origami vertex cannot have all of its creases be valleys, and the same is true for a degree-5 vertex. Therefore, the MV assignment in Figure 13.4(c) is only infinitesimally rigid-foldable; it will not flex rigidly by a finite (i.e., non-infinitesimal) amount. We will soon see that finite rigid foldability corresponds to second-order rigid foldability, which the example in (c) does not satisfy.

**Example 13.13**   A 2-vertex rigid-foldable crease pattern is shown in Figure 13.5(a), together with its reciprocal-parallel discrete surface. The two reciprocal-parallel diagrams for each vertex are joined together along the sides corresponding to the crease line $l_4 = l_5$ to make the full dual surface.

Some references on this topic use an equivalent, but slightly different, structure to determine first-order (and as we will see, second-order) rigid foldability called the reciprocal diagram (Tachi, 2012; Demaine et al., 2016). The **reciprocal diagram** of an infinitesimally rigid origami vertex is the reciprocal-parallel diagram but with each of the edges rotated 90° clockwise. Such rotated edges will still satisfy Equation (13.4) and thus can be used interchangeably with the reciprocal-parallel diagram. Also, while

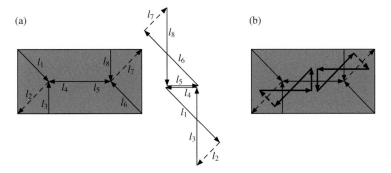

**Figure 13.5** (a) A 2-vertex rigid fold and its reciprocal-parallel diagram. (b) The same 2-vertex crease pattern with reciprocal diagram overlaid.

the edges in the reciprocal-parallel diagram are parallel to the crease lines, the edges of the reciprocal diagram will be orthogonal to the crease lines, which can give it a more dual-like appearance when compared to the crease pattern. Figure 13.5(b) shows the reciprocal diagram for our two-vertex example, which does have the advantage of looking more like a dual of the original crease pattern.

**Remark 13.14** While the reciprocal and reciprocal-parallel diagrams are thus far only meaningful for infinitesimal rigid origami, there is a further connection with finite rigid foldability, where the fold is guaranteed to flex rigidly by some nonzero measure set in the parameter space. One intuitive way to see this is through the Gauss map trace from Sections 12.2 and 12.3. Suppose that we are given a rigid vertex that is folded only a little bit from the flat state. Then the Gauss map trace of a closed curve around this folded vertex will be a small, zero-area spherical polygon. An example of this is shown in Figure 13.6, where a degree-4 vertex generates a spherical bow tie as the trace. We then project the spherical trace onto a plane tangent to the sphere, where the point of tangency is near the center of the trace. If the trace is small enough, this

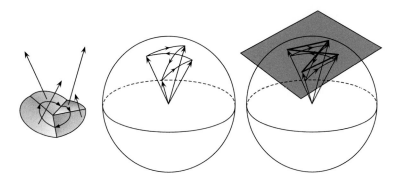

**Figure 13.6** Projecting the Gauss map trace of a rigid vertex onto a tangent plane to obtain a reciprocal diagram of the vertex discrete surface.

projection will form a very good approximation of the reciprocal (or equivalently, the reciprocal-parallel) diagram of the vertex.

Since the spherical polygon trace must have zero area by Corollary 12.18, we see that the reciprocal diagram should have zero signed area as well, at least for very small Gauss map traces. But even a small Gauss map trace will represent a finite, not infinitesimal, rigid folding, and therefore this zero-area property should be tied to more than first-order rigid foldability. It turns out that second-order rigid foldability does the trick.

**Definition 13.15**    A reciprocal (or reciprocal-parallel) diagram of a crease pattern $G$ is said to be **zero-area** if the signed area of the polygons it encloses is zero, where the signed area is measured by following the diagram around each vertex of $G$ and counting its polygonal areas as positive if they are traversed clockwise and negative if they are traversed counterclockwise.

**Theorem 13.16** (Demaine et al., 2016)    *A crease pattern is second-order rigid-foldable if and only if there exists a nontrivial zero-area reciprocal diagram for the crease pattern.*

*Proof*   We begin with the second-order Taylor expansion about $t = 0$ of our rigid folding function $F$ around a vertex, using notation as previously seen in the proof of Theorem 13.11. Note that

$$\frac{d^2F}{dt^2} = \frac{d}{dt}\left(\frac{\partial F}{\partial \rho_1}\rho_1'(t) + \cdots + \frac{\partial F}{\partial \rho_n}\rho_n'(t)\right)$$

$$= \sum_{i=1}^{n}\sum_{j=1}^{n}\frac{\partial^2 F}{\partial \rho_i \partial \rho_j}\rho_i'(t)\rho_j'(t) + \sum_{i=1}^{n}\frac{\partial F}{\partial \rho_i}\rho_i''(t),$$

where in the last step we employed the product rule and the multivariable chain rule $n$ times. Thus our second-order Taylor expansion is

$$F = I + \sum_{i=1}^{n}\frac{\partial F}{\partial \rho_i}\rho_i'(t)\bigg|_{t=0} t$$

$$+ \frac{1}{2}\left(\sum_{i=1}^{n}\sum_{j=1}^{n}\frac{\partial^2 F}{\partial \rho_i \partial \rho_j}\rho_i'(t)\rho_j'(t) + \sum_{i=1}^{n}\frac{\partial F}{\partial \rho_i}\rho_i''(t)\right)\bigg|_{t=0} t^2 + o(t^3). \tag{13.5}$$

The second-order partial derivatives are

$$\frac{\partial^2 F}{\partial \rho_i \partial \rho_j} = R(l_1, \rho_1)\cdots R(l_i, \rho_i)[l_{i\times}]\cdots R(l_j, \rho_j)[l_{j,\times}]\cdots R(l_n, \rho_n)$$

when $i < j$. Let us denote by $H_{i,j}$ the value of this at the flat, unfolded state:

$$H_{i,j} = \frac{\partial^2 F}{\partial \rho_i \partial \rho_j}\bigg|_{t=0} = \begin{cases} [l_{i\times}][l_{j\times}] & \text{for } i \leq j, \\ [l_{j\times}][l_{i\times}] & \text{for } i > j. \end{cases}$$

Letting $l_i = (l_i^x, l_i^y, 0)$ and $l_j = (l_j^x, l_j^y, 0)$, the $i \leq j$ case is

$$
[l_{i\times}][l_{j\times}] = \begin{pmatrix} -l_i^y l_j^y & l_i^y l_j^x & 0 \\ l_i^x l_j^y & -l_i^x l_j^x & 0 \\ 0 & 0 & -l_i^x l_j^x - l_i^y l_j^y \end{pmatrix}.
$$

The $i > j$ case gives $H_{i,j}$ to be just the transpose of the above matrix.

We have that a crease pattern is second-order rigid-foldable if and only if the derivatives $\rho'(0)$ and $\rho''(0)$ can be found that make the third term of Equation (13.5) vanish (where we want $\rho'(0)$ to not be the zero vector). That is, we want

$$
\sum_{i,j} H_{i,j}\rho_i'(0)\rho_j'(0) + \sum_i [l_{i\times}]\rho_i''(0) =
$$

$$
\sum_{i,j} \begin{pmatrix} -l_i^y l_j^y & l_a^y l_b^x & 0 \\ l_a^x l_b^y & -l_i^x l_j^x & 0 \\ 0 & 0 & -l_i^x l_j^x - l_i^y l_j^y \end{pmatrix} \rho_i'(0)\rho_j'(0) + \sum_i \begin{pmatrix} 0 & 0 & l_i^y \\ 0 & 0 & -l_i^x \\ -l_i^y & l_i^x & 0 \end{pmatrix} \rho_i''(0) = Z,
$$

where $Z$ is the zero matrix, $a = \min\{i,j\}$, and $b = \max\{i,j\}$. Now, notice that the two matrices $H_{i,j}$ and $[l_{i\times}]$ are "complementary" in the sense that wherever $H_{i,j}$ is nonzero, $[l_{i\times}]$ is zero, and vice versa. Thus in order to get the zero matrix, both of the summations must be zero. The second summation can be made to be zero by setting the acceleration $\rho''(0) = \vec{0}$, which means that $\rho'(t)$ needs to be a constant vector in a neighborhood of zero. This is always possible because $t = 0$ corresponds to the flat, unfolded state, and a path $\rho(t)$ through the origin will always be symmetric in a rigid folding, that is, $\rho(t) = -\rho(-t)$ for $t$ in some neighborhood of zero. The interpretation of this is that for $t < 0$ we have that some creases are mountains and some are valleys, and as we pass through $t = 0$, the flat state, all the MV parities will switch when $t > 0$, giving us the symmetry. Therefore, it is natural to assume that our folding angle curve $\rho(t)$ is MV-symmetric (i.e., symmetric about the origin) in a neighborhood of $t = 0$.

Then we need to see when the first summation, $H = \sum H_{i,j}\rho_i'(0)\rho_j'(0)$ will become the zero matrix. Let us denote $\omega_i = \rho_i'(0)l_i = (\omega_i^x, \omega_i^y, \omega_i^z)$. Then the diagonal entries of $H$ are all zero because, for example, $\sum_{i,j} l_i^x l_j^x \rho_i'(0)\rho_j'(0) = (\sum_i \omega_i^x)(\sum_j \omega_j^x) = 0$ by Equation (13.4) (that is, because second-order rigid-foldability implies first-order rigid-foldability, and so the reciprocal diagram about a vertex will form a closed loop).

Now, let $H_{i,j}(a,b)$ denote the $(a,b)$ entry of the matrix $H_{i,j}$. We have that

$$
2\sum_{i,j} H_{i,j}(2,1)\rho_i'(0)\rho_j'(0) =
$$

$$
\sum_{i,j} (H_{i,j}(2,1) + H_{i,j}(1,2))\rho_i'(0)\rho_j'(0) + \sum_{i,j} (H_{i,j}(2,1) - H_{i,j}(1,2))\rho_i'(0)\rho_j'(0).
$$

Furthermore,

$$
\sum_{i,j} (H_{i,j}(2,1) + H_{i,j}(1,2))\rho_i'(0)\rho_j'(0) = \sum_{i,j} (\omega_i^x \omega_j^y + \omega_i^y \omega_j^x) = 0
$$

by the closed-loop Equation (13.4). For the difference summation, the diagonal ($i = j$) entries cancel and the off-diagonal ($i \neq j$) entries become symmetric, giving us

$$\sum_{i,j}(H_{i,j}(2,1) - H_{i,j}(1,2))\rho_i'(0)\rho_j'(0) = 2\sum_{i<j}(\omega_i^x\omega_j^y - \omega_i^y\omega_j^x) = 2\sum_{i<j}\|\omega_i \times \omega_j\|,$$

where we take $\|\omega_i \times \omega_j\|$ to be the signed length (or just the $z$-coordinate) of the vector $\omega_i \times \omega_j$. Thus we have that $\rho(t)$ gives a second-order rigid-foldable motion of our crease pattern at $t = 0$ if and only if it satisfies Equation (13.4) and

$$\sum_{i<j}\|\omega_i \times \omega_j\| = 0. \tag{13.6}$$

Each of the vectors $\omega_i \times \omega_j$ are parallel to the $z$-axis, and thus to each other. Therefore, equality will hold for the triangle inequality, and inside the summation of Equation (13.6) we have (for example)

$$\|\omega_1 \times \omega_n\| + \|\omega_2 \times \omega_n\| + \cdots + \|\omega_{n-1} \times \omega_n\| =$$
$$\|(\omega_1 \times \omega_n) + (\omega_2 \times \omega_n) + \cdots + (\omega_{n-1} \times \omega_n)\| = \|(\omega_1 + \omega_2 + \cdots + \omega_{n-1}) \times \omega_n\|.$$

Then the left-hand side of Equation (13.6) equals

$$\|(\omega_1 + \cdots + \omega_{n-1}) \times \omega_n\| + \|(\omega_1 + \cdots + \omega_{n-2}) \times \omega_{n-1}\| + \cdots + \|\omega_1 \times \omega_2\|$$
$$= \sum_{i=0}^{n}\|(x_i - x_1) \times (x_{i+1} - x_i)\|, \text{(with } x_{n+1} = x_1)$$

where the $x_i$ are the coordinates of the vertices of the reciprocal diagram, so that $\omega_i = x_{i+1} - x_i$. But $\|(x_i - x_1) \times (x_{i+1} - x_i)\|$ is twice the signed area of the triangle with vertices $x_1, x_i,$ and $x_{i+1}$. Therefore, Equation (13.6) equals twice the signed area of the reciprocal diagram. In other words, a crease pattern will be second-order rigid-foldable if and only if there exists a folding path $\rho(t)$, flat at $t = 0$, whose reciprocal diagram is zero-area at each vertex (and thus for the whole diagram as well), as desired. □

---

**Example 13.17**    An example of Equation (13.6) in action is shown in Figure 13.7, where one of the vertices from Figure 13.5 and its reciprocal-parallel diagram are

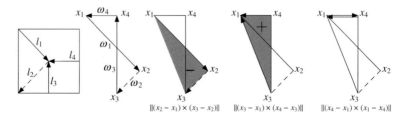

**Figure 13.7**  An example of how Equation 13.6 gives that the signed area of a reciprocal-parallel diagram is zero.

depicted. With the vertices $x_i$ of the reciprocal-parallel diagram labeled as shown, we see that the three terms of Equation (13.6) give a negative-area triangle $x_1x_2x_3$, a positive-area triangle $x_1x_2x_3$, and a zero-area, degenerate triangle $x_1x_4x_1$.

We now have everything we need, via Theorem 13.16, to prove that in the single-vertex case, the existence of a rigid folding from the flat state is equivalent to the existence of a zero-area reciprocal diagram.

**Theorem 13.18** (Demaine et al., 2016)   *A single-vertex origami crease pattern has a rigid folding from the flat state if and only if it has a nontrivial zero-area reciprocal (or reciprocal-parallel) diagram.*

*Proof*   First, if a single-vertex crease pattern $G$ has a rigid folding from the flat state, then for every rigid fold $\sigma(t)$ around $t = 0$, we have that the matrix function $F(\rho(t))$ equals the identity matrix around our vertex. Thus $F(\rho(t))$ is constant in the $k$th order for any $k$, and in particular $G$ is second-order rigid-foldable, which by Theorem 13.16 means that it has a nontrivial zero-area reciprocal diagram.

Conversely, suppose that our single-vertex crease pattern $G$ has a nontrivial zero-area reciprocal diagram. We want to show that we can create a rigid origami $\sigma$ of $G$ using some, to-be-determined, folding angles. What we can do is make a cut along one of the creases $l_k$ of $G$ and then define folding angles $\rho_i$ for all the other creases $l_i$ ($i \neq k$) by making them proportional to the folding velocities $\rho_i'$ given by the lengths of the sides of the reciprocal diagram. We then claim that since our vertex is second-order rigid-foldable, the resulting gap left between the cut ends of crease $l_k$ will be $o(t^3)$ in length, where $t$ is one of the folding parameters, and that this will be small enough to glue the ends of $l_k$ back together by modifying the other folding angles a small amount.

To prove this last claim more rigorously, first note that we may assume that $G$ has at least four creases. (If there are only two creases, they must be along a straight line, and the proportional folding velocities given by the reciprocal diagram provide the rigid folding angles exactly. If there are only three creases in $G$, then the reciprocal diagram forms a triangle, which cannot have zero area.) Let $l_i$, $l_j$, and $l_k$ be consecutive creases where the angle $\theta_{ij}$ between $l_i$ and $l_j$ plus the angle $\theta_{jk}$ between $l_j$ and $l_k$ is less than $\pi$. (Such a triple of creases can always be found in a degree-4 or greater vertex.) We slit the crease pattern along $l_k$ to create two edges in the paper, $l_k$ and $l_{k'}$, where the $l_k$ side is still adjacent to $l_j$. We may then fold the rest of the folding angles according to the reciprocal diagram. To visualize this, we think of our folded vertex (thus far) as being at the center of a small sphere, so that the planar faces of the crease pattern cut the sphere with arcs of great circles. In this way, we may view our partially folded vertex as a spherical linkage, as illustrated in Figure 13.8, where the crease lines $l_i$, $l_j$, $l_k$, and $l_{k'}$ are now just points, which we will refer to as $i$, $j$, $k$, and $k'$ for convenience.

We define the parameter $t$ to be the folding angle at vertex $j$, so $t = \rho_j$.

**Claim:** For a range of sufficiently small $t$, $\theta_{ik'} < \theta_{ij} + \theta_{jk}$.

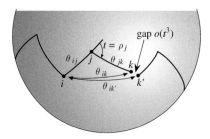

**Figure 13.8** The partially folded vertex, viewed as a spherical linkage.

If we can prove this claim, then we know that we can modify the folding angles at $i, j$, and $k$ in order to make $k = k'$ and thus glue the edges $l_k$ and $l_{k'}$ back together, giving us a rigid origami near the flat, unfolded state.

**Proof of claim:** Since our vertex has a zero-area reciprocal diagram, it is also second-order rigid-foldable by Theorem 13.16. Therefore, as a function of the parameter $t$, the distance $\theta_{ik'}$ on the sphere between $i$ and $k'$ is second-order approximate to $\theta_{ik}$, that is, $\theta_{ik} = \theta_{ik'} + o(t^3)$. In order to see how these spherical arc-lengths change with respect to one another as we change $t$, we need to appeal to spherical geometry. By the spherical law of cosines we have

$$\cos(\theta_{ik'} + o(t^3)) = \cos\theta_{ik} = \cos\theta_{ij}\cos\theta_{jk} + \sin\theta_{ij}\sin\theta_{jk}\cos(\pi - t). \quad (13.7)$$

On the other hand, using the sum of angle cosine formula and the Taylor series for sine and cosine,

$$\begin{aligned}
\cos(\theta_{ik'} + o(t^3)) &= \cos\theta_{ik'}\cos(o(t^3)) - \sin\theta_{ik'}\sin(o(t^3)) \\
&= \cos\theta_{ik'}(1 - o(t^6)) - \sin\theta_{ik'}o(t^3) \\
&= \cos\theta_{ik'} + o(t^3),
\end{aligned}$$

where this last $o(t^3)$ term also contains non-constant sine and cosine terms, but since they are bounded they will not affect the asymptotic behavior. Combining this with Equation (13.7) we obtain

$$\begin{aligned}
\cos\theta_{ik'} - \cos(\theta_{ij} + \theta_{jk}) &= \cos\theta_{ik'} - \cos\theta_{ij}\cos\theta_{jk} + \sin\theta_{ij}\sin\theta_{jk} \\
&= \cos(\theta_{ik'} + o(t^3)) - \cos\theta_{ij}\cos\theta_{jk} + \sin\theta_{ij}\sin\theta_{jk} + o(t^3) \\
&= \sin\theta_{ij}\sin\theta_{jk}\cos(\pi - t) + \sin\theta_{ij}\sin\theta_{jk} + o(t^3) \\
&= \sin\theta_{ij}\sin\theta_{jk}\left(1 - 1 + \frac{t^2}{2} + o(t^3)\right) + o(t^3) \\
&= \frac{t^2}{2}\sin\theta_{ij}\sin\theta_{jk} + o(t^3),
\end{aligned}$$

which is greater than 0 for sufficiently small $t$. Since cosine is a decreasing function over $[0, \pi]$, this implies that $\theta_{ik'} < \theta_{ij} + \theta_{jk}$ holds for a range of sufficiently small $t$, as desired. □

Therefore, the reciprocal and reciprocal-parallel diagrams provide a good, geometric way to determine if a single-vertex crease pattern has a rigid folding. These diagrams also capture the geometry of the rigid folding, making them good tools for rigid folding analysis.

It turns out, however, that there is an even simpler way to determine if a single-vertex crease pattern with an MV assignment has a rigid folding from the unfolded state. We will see this in Section 13.3.

## 13.2   Angle Relationships for Single-Vertex Rigid Foldability

In order to model a rigid folding completely, say for the purposes of making a computer animation of the rigid folding and unfolding process, one needs to know the relationships between the folding angles $\rho_i$ at the creases. In this section we will derive some nice, explicit equations for these folding angle relationships for degree-4 flat-foldable vertices. We will also describe how to use basic kinematics to derive relationships, which often involve very complicated equations, for general rigid vertices.

First, we need to establish the concept of the degrees of freedom, which comes from physics and mechanical engineering. We provide the most general definition possible and then apply it to rigid origami.

**Definition 13.19**   The **degree of freedom (DOF)** of a mechanical system is the number of independent parameters that uniquely determine its configuration.

In our rigid origami case, we have that the parameters are the folding angles $\rho_i$, which determine the parameter space of the crease pattern. The question then becomes, "What is the fewest number of folding angles needed to uniquely determine the remaining folding angles?" This fewest number of folding angles will be the degree of freedom of the rigid origami model.

**Theorem 13.20**   *A single-vertex crease pattern G of degree n with a rigid folding will have n − 3 degrees of freedom.*

*Proof*   As in the proof of Theorem 13.18, we think of our rigid-foldable vertex as being in the center of a small sphere, so that the paper cuts the sphere in spherical arcs and our crease lines $l_i$ project to points $p_i$ on the sphere. With no constraints whatsoever, each point $p_i$ has two parameters that determine its location on the sphere, say the angle from the $x$-axis around the $xy$-plane and the angle from the positive $z$-axis, as in standard spherical coordinates. That gives us $2n$ parameters. But each angle between consecutive creases in the paper introduces a constraint that removes one degree of freedom (i.e., the point $p_i$ must lie on the circle drawn on the sphere centered at $p_{i-1}$ with spherical radius equal to the angle between $p_i$ and $p_{i-1}$ on the crease pattern, assuming our sphere has radius one). Then we need to subtract the three rigid motions possible on the surface of a sphere (say, rotating about the $x$-axis and the $z$-axis to position a given point $p$, and then rotating about the point $p$ to specify a rotation). This gives us a total of $2n - n - 3 = n - 3$ degrees of freedom.    □

**Example 13.21** (Degree-4, flat-foldable vertices)    As we have previously seen in Section 12.3 and, in particular, Theorem 12.22, there are especially nice relationships between some of the folding angles in a rigid folding of a degree-4, flat-foldable vertex. That is, if we label the folding angles $\rho_1, \ldots, \rho_4$ where we assume that $\rho_2$ and $\rho_4$ are both valley creases (the major creases, using Lang's terminology presented in Example 12.19) while $\rho_1$ and $\rho_3$ have opposite MV parity (the minor creases), then by the symmetry of the Gauss map trace as the rigid folding opens and closes, we will have $\rho_2 = \rho_4$ and $\rho_1 = -\rho_3$.

Recall from Diversion 12.8 that if the angles between creases in a degree-4 vertex are all $\pi/2$, then the creases cannot all be rigidly folded at the same time; rather, the two major creases need to be folded completely flat first, and then the minor creases may be folded. (This follows from the geometry of the Gauss map trace in this instance.) Therefore, if our degree-4 vertex is to have a rigid folding from the flat, unfolded state, then there must be two consecutive angles that are both less than $\pi/2$ radians, say $0 < \alpha \le \beta < \pi/2$. Since this is a flat-foldable vertex, the other angles will be $\pi - \alpha$ and $\pi - \beta$. Let us assume that our degree-4 vertex will have three valleys and one mountain. By the Big-Little-Big Lemma from Chapter 5, the two creases surrounding angle $\alpha$ must have different MV parity (unless $\alpha = \beta$, but in that symmetric case we may assume that $\alpha$ is the angle with a mountain and a valley bordering it). This gives us two **modes** to consider for our rigid folding, as shown and labeled in Figure 13.9. Modes 1 and 2 will have different relationships between their folding angles.

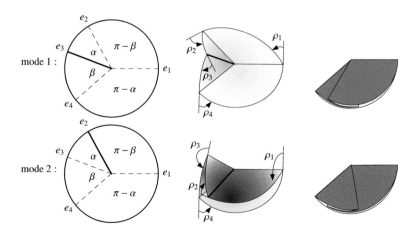

**Figure 13.9** The two modes of a rigid folding of a degree-4 flat-foldable vertex.

In particular, for mode 1 we would like to know the relationship between the folding angles $\rho_1$ and $\rho_2$. We know by Theorem 13.20 that a degree-4 rigid folding will have one degree of freedom, and so if we set the parameter $\rho_1 = t$ to be our degree of freedom, then knowing the formula for $\rho_2(t)$ together with $\rho_3 = -t$ and $\rho_4(t) = \rho_2(t)$ will tell is all of the folding angles needed.

There are many ways to derive a formula for $\rho_2(t)$. The most illuminating formula, however, reparameterizes the folding angles with $\tan(\rho_i/2)$. This approach has its origins outside of rigid origami in the field of flexible polyhedra from the late 1800s. Raoul Bricard's flexible octahedron analysis (Bricard, 1897) uses the tangent of half the dihedral angles of a corner of an octahedron with, possibly, different plane angles around the vertex. As such, this is exactly like a rigid folding of a degree-4 origami vertex, but on a cone instead of a flat piece of paper. The tangent of the half-angle reparameterization has since been rediscovered by other researchers. Huffman (1976, 1978) for example, described this approach in the context of paper folding. Our presentation of this result will follow that given in the rigid origami literature, in particular the work of Tomohiro Tachi (2010a,b 2012).

**Theorem 13.22** *Given a degree-4, flat-foldable vertex with plane angles $\alpha$, $\beta$, $\pi - \alpha$, and $\pi - \beta$, with $0 < \alpha < \pi/2$, and creases $e_1, \ldots, e_4$ with corresponding folding angles $\rho_1, \ldots, \rho_4$ as arranged in Figure 13.9, let $t_i = \tan(\rho_i/2)$ for $i = 1, \ldots, 4$. Then the $t_i$ values are proportional to each other according to the following relationship, depending on the mode of the vertex's MV assignment:*

$$(t_1, t_2, t_3, t_4) = \begin{cases} (c_1 t, t, -c_1 t, t) & \text{for mode 1,} \\ (t, -c_2 t, t, c_2 t) & \text{for mode 2,} \end{cases}$$

*where modes 1 and 2 are as shown in Figure 13.9, our reparameterized folding angle is $t = t_2$ for mode 1 and $t = t_1$ for mode 2, and the constants $c_1$ and $c_2$ are defined by*

$$c_1 = \frac{\cos \frac{\alpha+\beta}{2}}{\cos \frac{\alpha-\beta}{2}} = \frac{1 - \tan \frac{\alpha}{2} \tan \frac{\beta}{2}}{1 + \tan \frac{\alpha}{2} \tan \frac{\beta}{2}} \quad \text{and} \quad c_2 = -\frac{\sin \frac{\alpha-\beta}{2}}{\sin \frac{\alpha+\beta}{2}} = -\frac{\tan \frac{\alpha}{2} - \tan \frac{\beta}{2}}{\tan \frac{\alpha}{2} + \tan \frac{\beta}{2}}.$$

*Proof* We will prove the formula for $c_1$ in mode 1; the mode 2 case follows similarly.

We will use the rigid folding matrix Equation (12.1) from Theorem 12.2.

Let our unfolded vertex be positioned at the origin with the paper in the $xy$-plane in $\mathbb{R}^3$. From Figure 13.9 we have that $e_i$ are our crease lines with folding angles $\rho_i$, for $i = 1, \ldots, 4$, with $e_1$ on the positive $x$-axis and $e_2$, $e_3$, and $e_4$ proceeding counterclockwise from there. Let us further assume that the sector of paper between $e_4$ and $e_1$ remains fixed in the $xy$-plane as we fold all the creases.

Then we can calculate the matrices $\chi_i = R(e_i, \rho_i)$ for Equation (12.1). Specifically, if we let $R_x(\theta)$ and $R_z(\theta)$ denote the rotation matrices about the $x$- and $z$-axes by $\theta$, then the matrix product

$$A = \chi_4 \chi_3 \chi_2 \chi_1 = R(e_4, \rho_4) R(e_3, \rho_3) R(e_2, \rho_2) R(e_1, \rho_1)$$

becomes, using the fact that $\rho_3 = -\rho_1$ and $\rho_4 = \rho_2$ in mode 1,

$$R_z(\pi + \alpha)R_x(\rho_2)R_z(-(\pi + \alpha))R_z(\pi + \alpha - \beta)R_x(-\rho_1)R_z(-(\pi + \alpha - \beta))$$
$$\cdot R_z(\pi - \beta)R_x(\rho_2)R_z(-(\pi - \beta))R_x(\rho_1).$$

This matrix $A$, by Equation (12.1), must equal the identity matrix, and therefore the third row, third column entry, $A_{3,3}$, must equal 1. Simplifying this equation gives

$$\left( \sin \frac{\rho_2}{2} \cos \frac{\rho_1}{2} \cos \frac{\alpha + \beta}{2} - \cos \frac{\rho_2}{2} \sin \frac{\rho_1}{2} \cos \frac{\alpha - \beta}{2} \right)$$
$$\cdot (1 + \cos \rho_1 \cos \rho_2 + \cos \beta \sin \rho_1 \sin \rho_2) = 0. \quad (13.8)$$

Now, since

$$|\cos \rho_1 \cos \rho_2 + \cos \beta \sin \rho_1 \sin \rho_2| < |\cos \rho_1 \cos \rho_2| + |\sin \rho_1 \sin \rho_2|$$
$$\leq |\cos(\rho_1 \pm \rho_2)|,$$

where we choose the sign of the $\pm$ depending on the sign of $\cos \rho_1$ and $\cos \rho_2$, we have that the right-hand parenthetical part of Equation (13.8) is never zero. Thus for any non-flat rigid folding we must have that the left-hand factor of Equation (13.8) is zero, which simplifies to

$$\tan \frac{\rho_1}{2} = \frac{\cos \frac{\alpha + \beta}{2}}{\cos \frac{\alpha - \beta}{2}} \tan \frac{\rho_2}{2},$$

as desired.     □

Note that since $0 < \alpha \leq \beta < \pi/2$, we have that $c_1 > 1$ and $0 < |c_2| < 1$. This proves the following corollary, which explains the motivation behind Lang's terms **major** and **minor creases** (see Example 12.19).

**Corollary 13.23**  *In a flat-foldable, degree-4 rigid folding, the absolute value of the folding angles along the major creases will be greater than that of the minor creases.*

---

**Diversion 13.1**   There are many other ways to prove Theorem 13.22.

(i) Prove the mode 1 case by using the fact that as we fold the valley creases at $e_1$ and $e_4$, the endpoints of $e_2$ and $e_3$ (assuming our folded paper is a unit disc) will remain an arc-distance of $\alpha$ apart.
(ii) Prove the mode 1 case by using the fact from the proof of Theorem 13.16 that the second-order matrix

$$\sum_{i=1}^{n} \sum_{j=1}^{n} \frac{\partial^2 F}{\partial \rho_i \partial \rho_j} \rho_i'(0)\rho_j'(0)$$

must equal the zero matrix (applying this to the flat-foldable, degree-4 case, of course).

---

While the folding angle relationships discussed here are only for degree-4, flat-foldable vertices, they can be used to draw conclusions about larger crease patterns. The following two diversions are useful and fun examples. For more examples see (Evans et al., 2015b,c; Tolman et al., 2017; Hull and Tachi, 2017; Lang, 2018).

---

**Diversion 13.2**    Use Corollary 13.23 to prove that the square twist crease pattern with the MV assignment shown in Figure 6.12 does not have a rigid folding.

---

**Diversion 13.3**    Use Corollary 13.23 to prove that the MV assignment of the triangle twist crease pattern (see Figure 6.20) has a rigid folding.

---

The change-of-variable $t_i = \tan(\rho_i/2)$ is an interesting transformation in its own right. It is sometimes called the **Weierstrass substitution**, although it seems that it was first employed by Euler (1768, paragraph 261). We may think of it as a mapping $\Psi : \mathbb{R}/2\pi\mathbb{R} \to \mathbb{RP}^1$, where the domain is the real line mod $2\pi$ and the codomain is the real projective line $\mathbb{R} \cup \{\infty\}$. An illustration for the justification of this is shown in Figure 13.10, where we consider $\mathbb{RP}^1$ to be wrapped around a circle.

This transformation gives us a way to develop a concept of the **speed**, also known as the **folding multiplier**, of the folding angles as we open and close a degree-4, flat-foldable rigid folding. Specifically, Theorem 13.22 gives us that $t_1 = c_1 t_2$ for the mode 1 case, and this means that

$$\frac{dt_1}{dt_2} = c_1 \quad \text{and} \quad \frac{dt_2}{dt_2} = 1.$$

That is, in the mode 1 case with our reparameterization $t_i$, we may think of the crease $e_1$ as having speed 1 and the crease $e_2$ as having speed $c_1$. Choosing nice values for the angles $\alpha$ and $\beta$ between the creases, we can make these speeds proportional.

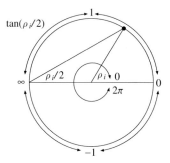

**Figure 13.10** The transformation $\Psi : \mathbb{R}/2\pi\mathbb{R} \to \mathbb{RP}^1$ given by $\Psi(\rho_i) = \tan(\rho_i/2)$.

For example, $\alpha = \arctan(3/4)$ and $\beta = \pi/2$ gives us $c_1 = 1/2$, which means that the speed of the major creases will be twice that of the minor creases. We will make use of this in Section 14.1 when we study the complexity of rigid foldings.

We conclude this lengthy example by explicitly noting our result: the relationships between the folding angles in a flat-foldable, degree-4 rigid folding are especially nice. To wit, in the mode 1 case in Figure 13.9, if we let $t$ be our folding parameter with $\rho_2(t) = t$, then

$$\rho_1(t) = 2 \arctan \left( \frac{\cos \frac{\alpha+\beta}{2}}{\cos \frac{\alpha-\beta}{2}} \tan \frac{t}{2} \right), \quad \rho_3(t) = \rho_1(t), \quad \text{and} \quad \rho_4(t) = -\rho_2(t).$$

The mode 2 case can be handled similarly, in accordance with the result of Theorem 13.22.

---

**Diversion 13.4**    Prove that if, using the notation of Figure 13.9, we have $\alpha = \beta$ then there is only one mode that allows all the creases at the degree-4 vertex to be folded (that is, having $\rho_i \neq 0$ for all $i$).

---

Outside of the degree-4 case, it is difficult to find explicit, general relationships between the folding angles $\rho_i$ in a rigid folding. However, standard procedures from mechanical engineering on understanding the kinematics of rigid-body mechanisms can be used to find such folding angle relationships for a given single-vertex crease pattern.

Devin Balkcom has written explicit directions for translating this kinematic process to rigid foldings of origami vertices (Balkcom, 2002, pp. 126–128; Balkcom and Mason, 2008). The method presented here is based on his approach.

Suppose that our vertex has creases labeled $e_1, \ldots, e_n$, where our unfolded paper lies in the $xy$-plane with the vertex at the origin and crease line $e_1$ along the positive $x$-axis. We label the sector angles of the paper by $\alpha_i$ and $\beta_i$ as shown in Figure 13.11. Let $\rho_i$ be the folding angle at crease $e_i$ once we start folding the vertex rigidly.

Our strategy will be to keep the region of the paper between creases $e_1$ and $e_n$ fixed and cut along the crease $e_3$. This way we can let the folding angles $\rho_1$ and $\rho_2$ be the **dependent** folding angles and the remaining $n - 3$ folding angles be the **independent** degrees of freedom of the vertex (c.f. Theorem 13.20). The endpoint of the crease $e_3$, which we label $p$ in Figure 13.11, can then be followed to its new location, $p_F$, after folding the independent folding angles; we refer to this as the **forward kinematics** direction of $p$'s folding. Then we can calculate $p$'s position, $p_I$, under the folding of the dependent folding angles, or the **inverse kinematics** direction. Setting $p_F = p_I$ will allow us to determine the values of $\rho_1$ and $\rho_2$ based on the independent variables $\rho_4, \ldots, \rho_n$.

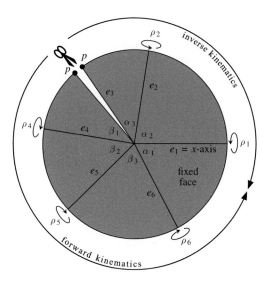

**Figure 13.11** Kinematic analysis of a rigid folding of a single-vertex crease pattern.

Using the notation of Section 13.1, we may calculate the forward kinematics as follows:

$$p_F = R(e_n, -\rho_n) \cdots R(e_5, -\rho_5)R(e_4, -\rho_4) \begin{pmatrix} \cos(\alpha_2 + \alpha_3) \\ \sin(\alpha_2 + \alpha_3) \\ 0 \end{pmatrix}. \qquad (13.9)$$

Note that due to the different orientations of the forward and inverse directions relative to the fixed face, the directions of the folding angles in the forward direction will be opposite those of the inverse direction. We are choosing to make the forward $\rho_i$ values negative in Equation (13.9), and in the inverse direction we will make them positive.

For the inverse kinematics direction, we have

$$p_I = R(e_1, \rho_1)R(e_2, \rho_2) \begin{pmatrix} \cos(\alpha_2 + \alpha_3) \\ \sin(\alpha_2 + \alpha_3) \\ 0 \end{pmatrix}$$

$$= R_x(\rho_1)R_z(\alpha_2)R_x(\rho_2)R_z(-\alpha_2) \begin{pmatrix} \cos(\alpha_2 + \alpha_3) \\ \sin(\alpha_2 + \alpha_3) \\ 0 \end{pmatrix}$$

$$= R_x(\rho_1)R_z(\alpha_2)R_x(\rho_2) \begin{pmatrix} \cos \alpha_3 \\ \sin \alpha_3 \\ 0 \end{pmatrix}$$

$$= R_x(\rho_1) \begin{pmatrix} \cos \alpha_2 \cos \alpha_3 - \sin \alpha_2 \sin \alpha_3 \cos \rho_2 \\ \sin \alpha_2 \cos \alpha_3 + \cos \alpha_2 \sin \alpha_3 \cos \rho_2 \\ \sin \alpha_3 \sin \rho_2 \end{pmatrix}. \qquad (13.10)$$

The $x$-coordinate shown in (13.10) will not change after multiplication by $R_x(\rho_1)$, and therefore it can be used to find an expression for $\rho_2$ in terms of the independent variables and the (fixed) sector angles. That is, if $x_F$ equals the $x$-coordinate of $p_F = p_I$ from Equation (13.9), we have

$$\cos \rho_2 = \frac{\cos \alpha_2 \cos \alpha_3 - x_F}{\sin \alpha_2 \sin \alpha_3}. \tag{13.11}$$

The particulars of Equation (13.11) can result in multiple possible values for $\rho_2$. Let $q$ be the value of the right-hand side of Equation (13.11). If $|q| < 1$ then there are two possible values for $\rho_2$, namely because $-\pi \leq \rho_2 \leq \pi$ and thus two different values for $\rho_2$, one positive and the other negative, will return the same value for $\cos \rho_2$. This corresponds to the crease $e_2$ being either a mountain (if $\rho_2 < 0$) or a valley (if $\rho_2 > 0$), where these two different cases will result in different values for $\rho_1$.

However, if $|q| = 1$ then there will be only one solution for $\rho_2$. This case corresponds to either $\rho_1 = 0$ (so $e_1$ is unfolded) or $\rho_1 = \pm\pi$ (so $e_1$ is completely folded). If $|q| > 1$ then there is no solution for $\rho_2$, which probably means that the choices for $\rho_4, \ldots, \rho_n$ are forcing the paper to rip.

Given a solution for $\rho_2$, a corresponding value for $\rho_1$ can be calculated by solving an equation from $p_F = p_I$, but doing this can lead to complicated trigonometric expressions from which it can be tricky to massage the solution for the full range of $\rho_2$. A better approach is to use the fact that the angle between the vector made from the $y$- and $z$-coordinates of (13.9) and those of the right-most part of (13.10) will be $\rho_1$. Thus if we let $p_F = (x_F, y_F, z_F)$, then we have

$$\rho_1 = \arctan\left(\frac{z_F}{y_F}\right) - \arctan\left(\frac{\sin \alpha_3 \sin \rho_2}{\sin \alpha_2 \cos \alpha_3 + \cos \alpha_2 \sin \alpha_3 \cos \rho_2}\right). \tag{13.12}$$

Equations (13.11) and (13.12) give parameterizations of the dependent folding angles for a rigidly folded vertex. By chaining such equations across multiple vertices, they can be used to create algebraically precise simulations of complex rigid-foldable crease patterns flexing. (See (Abel et al., 2015a,b) for examples.)

However, even in seemingly simple cases the equations that this kinematic method generates can be very convoluted. Sometimes clever manipulations can simplify things, but not usually. The next example illustrates this.

---

**Example 13.24** (General degree-4 vertices)  As a final example, suppose that we aim to model the rigid folding of a general degree-4 vertex, which is not necessarily flat-foldable. Consider such a vertex as labeled in Figure 13.12. If we let the folding angles be functions of $-\pi \leq t \leq \pi$ and set $\rho_4(t) = t$, then Equation (13.11) gives us

$$\rho_2(t) = \pm \arccos\left(\frac{\cos \beta_1 \cos \beta_2 - x_F}{\sin \beta_1 \sin \beta_2}\right),$$

where the $x_F$ value given by the forward kinematics is a rather lengthy expression involving the sine and cosine of various sums of the plane angles. If, however, we take the forward and inverse kinematics $x$-coordinates from Equations (13.9) and (13.10) and re-parameterize with $\cos t = (1 - t^2)/(1 + t^2)$ and $\cos(\rho_2) = (1 - \rho_2^2)/(1 + \rho_2^2)$, we obtain the somewhat less complex expression

$$\rho_2(t) = -2\arctan$$

$$\left(\frac{\sqrt{2}\tan\frac{t}{2}\sqrt{-\sin\beta_3\sin(\beta_1 + \beta_2 + \beta_3)}}{\sqrt{(1 + \tan^2\frac{t}{2})\cos(\beta_1 - \beta_2) - \cos(\beta_1 + \beta_2) - \tan^2\frac{t}{2}\cos(\beta_1 + \beta_2 + 2\beta_3)}}\right).$$

Then Equation (13.12) gives us

$$\rho_1(t) = \arctan\left(\frac{z_F}{y_F}\right) - \arctan\left(\frac{\sin\beta_2\sin(\rho_2(t))}{\sin\beta_2\cos\beta_1\cos(\rho_2(t)) + \sin\beta_1\cos\beta_2}\right),$$

where the forward kinematics gives us $y_F$ and $z_F$:

$$y_F = \sin(\beta_1 + \beta_2 + \beta_3)\cos\beta_3 - \cos(\beta_1 + \beta_2 + \beta_3)\sin\beta_3\cos t$$

and

$$z_F = \sin\beta_3\sin t.$$

This illustrates a few things. First of all, as Huffman notes in (Huffman, 1976), we can see that the relationship between the folding angles of adjacent creases in the non-flat-foldable degree-4 case does not have a simple formulation. Second, the kinematics approach does not necessarily give us the most simple expressions. Compare the kinematic-derived formula for $\rho_2(t)$ (even as simplified by our tangent of the half-angle re-parameterization) with Huffman's formula from Theorem 12.23. In our case we have that the major creases are those with the folding angles $\rho_4(t) = t$ and $\rho_2(t)$. Thus,

$$1 - \cos(\rho_2(t)) = \frac{\sin(2\pi - \beta_1 - \beta_2 - \beta_3)\sin\beta_3}{\sin\beta_1\sin\beta_2}(1 - \cos t),$$

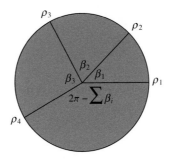

**Figure 13.12** A general degree-4 vertex.

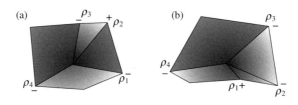

**Figure 13.13** A general degree-4 vertex rigid folding. (a) Mode 1. (b) Mode 2.

which is much more simple than what we previously obtained. (Although, in order to use this in a simulation, we need to use the arccos function, which is always positive. Thus if $t \leq 0$ we need to take the negative of the arccos, and if $t > 0$ we let it remain positive.)

For completeness, the parametrizations of $\rho_1(t)$ and $\rho_2(t)$ given thus far are not the only possibilities. As we saw in the flat-foldable degree-4 case, there are two modes that such a rigid folding can attain. The solution we've seen for the folding angles gives mode 1, as seen in Figure 13.13(a). To move the opposite-concavity crease to the appropriate adjacent crease, and attain mode 2 as seen in Figure 13.13(b), we need to change the sign of $\rho_2(t)$; this will also change the value of $\rho_1(t)$.

**Diversion 13.5**   Prove that in a rigid folding of a general degree-4 vertex, if $\rho_1$ is the folding angle whose sign is opposite that of the other folding angles, and if $\beta_1$ and $\beta_4$ are the plane angles adjacent to the $\rho_1$ crease, then we must have $\beta_1 + \beta_4 < \pi$.

As a final remark for this section, there are many other ways to describe folding angle relationships. In particular, different forms of the folding angle equations to those presented here in the non-flat-foldable, degree-4 case can be found in (Lang, 2018, Section 8.3.7).

## 13.3     An Intrinsic Condition for Rigid Vertices

In this chapter we have explored ways to model the rigid folding of materials along straight crease lines, and in doing so we developed tools to find relationships between the folding angles in a rigid folding. In Chapter 12 we saw a matrix-based necessary condition for a crease pattern to be rigid-foldable (Theorem 12.7), and such matrix equations are a major tool for finding folding angle relationships.

The question remains, however, whether there are sufficient conditions for a crease pattern to be rigid-foldable. As we saw in Chapter 12 (i.e., in Figure 12.2), satisfying the matrix model of rigid origami does not preclude the paper from self-intersecting. Further, one might wonder if there is a more simple condition, without resorting to computing rotation matrices and the proper folding angles, to see if a given crease pattern could rigidly fold from the unfolded state.

In this section we present a necessary and sufficient condition for a single-vertex crease pattern to have a rigid folding from the flat, unfolded state that relies only on the intrinsic geometry of the crease pattern. This result originally appeared in (Abel et al., 2016), and we follow the development and proofs from that paper.

We begin with some definitions. Let $\angle(c, d)$ denote the oriented angle going counterclockwise between two creases $c$ and $d$ that meet at a vertex in a crease pattern $C$. (That is, $c$ and $d$ might not be consecutive creases going around the vertex; there could be other creases between them.)

**Definition 13.25** Given a single-vertex crease pattern $C$ together with an MV assignment $\mu : C \to \{\pm 1\}$, we say that the pair $(C, \mu)$ **has a tripod** (or a **mountain tripod**) if there exist three creases $c_1, c_2, c_3 \in C$ in counterclockwise order, not necessarily consecutive, with the property that $\mu(c_i) = -1$ for $i = 1, 2, 3$ and $0 < \angle(c_i, c_{i+1}) < \pi$ for $i = 1, 2, 3$ (mod 3). A **valley tripod** is a tripod but with $\mu(c_i) = 1$ for $i = 1, 2, 3$. See Figure 13.14.

We also say that $(C, \mu)$ contains a **mountain (resp. valley) cross** if there exists four creases in counterclockwise order $c_1, c_2, c_3, c_4 \in C$, not necessarily consecutive, where $c_1$ and $c_3$ form a straight line, as do $c_2$ and $c_4$, and $\mu(c_i) = -1$ (resp. $\mu(c_i) = 1$) for $i = 1, 2, 3, 4$.

We say that $(C, \mu)$ contains a **bird's foot** if it contains either a tripod or a cross together with another crease that has the opposite MV parity of the tripod/cross. Figure 13.14 shows bird's feet made from (a) a mountain tripod (creases $c_1$–$c_4$) and (b) a mountain cross (creases $c_1$–$c_5$).

**Theorem 13.26** *A single-vertex crease pattern $C$ with an MV assignment $\mu$ has a rigid folding from the flat state if and only if $(C, \mu)$ contains a bird's foot.*

To prove Theorem 13.26 we will first need a sequence of lemmas, and one major result, that utilize the following idea: Consider our vertex to be at the center of a sphere sufficiently small so that every crease adjacent to the vertex will intersect the sphere (although by rescaling we may assume that this sphere has radius 1). The regions of paper between the creases will intersect the sphere along arcs of great circles, like we

 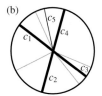

**Figure 13.14** Single-vertex crease patterns with (a) a mountain tripod and (b) a mountain cross. These are also bird's feet.

saw in the proof of Theorem 13.18. Then an equivalent way to think about a rigid origami vertex is as a **closed spherical linkage** that forms a non-intersecting closed loop on the sphere. Such a linkage will have vertices $c_i$ corresponding to the creases of our rigid origami. Let $\alpha_i$ be the angle on the sphere at vertex $c_i$ that arcs $c_i c_{i-1}$ and $c_i c_{i+1}$ make, so $\alpha_i$ is the supplement of the folding angle $\rho_i$ at crease $c_i$. We may also carry the MV assignment $\mu$ from the creases to the vertices of our spherical linkage.

Let $\text{arc}(c_i, c_j)$ denote the length of the shortest spherical arc between vertices $c_i$ and $c_j$ in a spherical linkage. Now, for flat-foldable vertices we have been using the fact that $\text{arc}(c_i, c_{i+1}) < \pi$ since this is an easy consequence of Kawasaki's Theorem. However, it is perhaps not immediately obvious that this is also true for non-flat rigid origami vertices. Indeed, this fact is sometimes assumed (e.g., see (Kapovich and Millson, 1997; Streinu and Whiteley, 2005)). For completeness we will include a proof of this fact.

**Lemma 13.27**  *Let $c_i$ and $c_{i+1}$ be consecutive creases of a rigid origami vertex viewed as vertices on a corresponding closed spherical linkage. Then $\angle(c_i, c_{i+1}) < \pi$, and thus $\text{arc}(c_i, c_{i+1})$ is the length of the linkage on the sphere from $c_i$ to $c_{i+1}$.*

*Proof*  Suppose that there are consecutive creases $c_i$ and $c_{i+1}$ with $a = \angle(c_i, c_{i+1}) \geq \pi$, and so $\text{arc}(c_i, c_{i+1}) = 2\pi - \angle(c_i, c_{i+1})$ on the sphere. Then consider the circles $C_i$ and $C_{i+1}$ drawn on the sphere centered at $c_i$ and $c_{i+1}$ with radii $\angle(c_{i-1}, c_i)$ and $\angle(c_{i+1}, c_{i+2})$, respectively. If we imagine holding the arc between $c_i$ and $c_{i+1}$, representing the sector of paper between $c_i$ and $c_{i+1}$, fixed as we fold our spherical linkage, then $C_i$ (resp. $C_{i+1}$) represents the set of points where $c_{i-1}$ (resp. $c_{i+2}$) can be folded. (Note that $c_{i-1} \neq c_{i+2}$ because a rigid origami vertex must have at least four creases, as we saw in Section 12.3.) There may be other points $c_j$ on the spherical linkage, but the distance $d = \text{arc}(c_{i-1}, c_{i+2})$ when the linkage/paper is unfolded represents the maximum distance $c_{i-1}$ and $c_{i+2}$ can be from each other as the linkage/paper is rigidly folded, for otherwise the linkage/paper would rip.

If $a > \pi$ then the only points on $C_i$ and $C_{i+1}$ for which the distance between the folded images of $c_{i-1}$ and $c_{i+2}$ is less than or equal to $d$ are along the great circle containing the arc between $c_i$ and $c_{i+1}$, which is the unfolded state. This means that the points $c_{i-1}$ and $c_{i+2}$ cannot be folded from the unfolded state without ripping the spherical linkage.

If $a = \pi$ then $c_i$ and $c_{i+1}$ are antipodal points, which means that the circles $C_i$ and $C_{i+1}$ lie on parallel planes through the sphere. In this case the only way for $c_{i-1}$ and $c_{i+2}$ to not fold away from each other would be for them to travel along the circles $C_i$ and $C_{i+1}$ in tandem, preserving the distance $d$ between them. But this would mean that the linkage would be straight at the points $c_{i-1}$ and $c_{i+2}$, that is, these points on the linkage would not be folded. This violates the condition that a rigid folding be actually folded along all of its creases (a consequence of the folding being non-differentiable on the crease pattern). $\qquad\square$

We now mention briefly a major result on rigid foldings of single-vertex crease patterns due to Ileana Streinu and Walter Whiteley (Streinu and Whiteley, 2005).

**Theorem 13.28** (Streinu and Whiteley, 2005)    *For every single-vertex rigid origami fold $\sigma : G \to \mathbb{R}^3$, there exists a path $\gamma : [0, 1] \to C(G)$ in the configuration space of $G$ that gives a rigid folding $\sigma_\gamma(t)$ where $\sigma_\gamma(1) = \sigma$ and $\sigma_\gamma(0)$ is the unfolded state. In other words, every single-vertex rigid origami can be unfolded to the flat, unfolded state without the paper self-intersecting.*

Streinu and Whiteley obtain this result by proving that every spherical polygon on the unit sphere with perimeter $2\pi$ can be, as a spherical linkage, unfolded to a great circle without crossing itself. Because this is an important result on the configuration spaces of single-vertex rigid foldings, we will discuss its proof later, in Chapter 14 (specifically in Section 14.2). For our current purposes we will need to use a corollary of the Streinu–Whiteley Theorem.

**Corollary 13.29**    *If a single-vertex crease pattern $G$ has a rigid folding $\sigma : G \to \mathbb{R}^3$, then there exists a rigid folding for $G$ from the flat, unfolded state.*

The proof of this corollary is trivial by Theorem 13.28, but the fact that one rigid folding of a single-vertex crease pattern implies the existence of a parameterized family of rigid foldings along a continuous curve in the crease pattern's configuration space to the unfolded state deserves special mention, as it is a useful tool in proving the existence of rigid foldings.

**Lemma 13.30**    *Given a closed spherical linkage $C = \{c_1, \ldots, c_k\}$ where the $c_i$ are the vertices, in order, embedded on a sphere in some configuration (i.e., with angles $\alpha_i \notin \{0, \pi\}$ at each vertex $c_i$ for $i = 1, \ldots, k$ and with $0 < \mathrm{arc}(c_i, c_{i+1}) < \pi$ for each arc), we may change the linkage $C$ by shrinking the length of any $\mathrm{arc}(c_i, c_{i+1})$ by some small amount while only changing the angles $\alpha_{i-1}, \alpha_i$, and $\alpha_{i+1}$ of our configuration by a small amount and without changing the MV parity of the vertices.*

*Proof*    Draw a circle $C_a$ on the sphere whose center is the midpoint of $\mathrm{arc}(c_i, c_{i+1})$ such that the circle contains the points $c_i$ and $c_{i+1}$. Note that $C_a$ will lie in a hemisphere because $\mathrm{arc}(c_i, c_{i+1}) < \pi$. Draw a circle $C_i$ centered at $c_{i-1}$ with radius $\mathrm{arc}(c_{i-1}, c_i)$. (See Figure 13.15; the fact that $\mathrm{arc}(c_{i-1}, c_i) < \pi$ prevents $C_i$ from being degenerate.) Then $C_i$ represents the region on the sphere where vertex $c_i$ can be if we rotate it about $c_{i-1}$. We claim that $C_a$ and $C_i$ are not tangent, for if they were then we'd have $\alpha_i = \pi$ and the vertex $c_i$ would not be folded. Thus $C_a$ and $C_i$ overlap, and so we can shrink the length of $\mathrm{arc}(c_i, c_{i+1})$ a little bit by rotating $c_i$ around $c_{i-1}$ a little bit to have vertex $c_i$ still be connected to the rest of the linkage. Doing this only changes the angles $\alpha_{i-1}$, $\alpha_i$, and $\alpha_{i+1}$ of the spherical linkage configuration.    $\square$

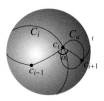

**Figure 13.15** Illustration of the spherical linkage for Lemma 13.30.

The next lemma is interesting in its own right as a way to add creases to a rigid origami without changing its rigid foldability and preserving the MV assignment of the original creases.

**Lemma 13.31**   *The existence of a rigid folding for single-vertex crease patterns is closed under the adding of new creases. That is, if a single-vertex crease pattern $C$ has a rigid folding under an MV assignment $\mu$ and $c'$ is a new crease added to the vertex in $C$, then $C \cup \{c'\}$ also has a rigid folding under the same $\mu$ (where we may choose $\mu(c')$ arbitrarily).*

*Proof*   Let $C = \{c_1, \ldots, c_k\}$ be a single-vertex crease pattern that has a rigid folding. Suppose that the new crease $c'$ we are adding is between creases $c_i$ and $c_{i+1}$. Consider the vertex to be at the center of a sphere and fold $C$ into a rigid configuration. We know $\angle(c_i, c_{i+1}) < \pi$, by Lemma 13.27. Use Lemma 13.30 to shrink $\text{arc}(c_i, c_{i+1})$ a little bit. Then replace $\text{arc}(c_i, c_{i+1})$ with a small "tent" (if we want a mountain) or "trough" (for a valley) at the position where we want $c'$ **and** so that the original $\text{arc}(c_i, c_{i+1}) = \text{arc}(c_i, c') + \text{arc}(c', c_{i+1})$. See Figure 13.16. Since we first decreased the length of $\text{arc}(c_i, c_{i+1})$, we will have that the folding angle at $c'$ will not be zero, and so the crease $c'$ will be folded and all the other creases are at most only changed slightly from their previous folding angles. The result is the new crease pattern $C \cup \{c'\}$ that is in a rigidly folded configuration. Furthermore, since this demonstrates a rigidly folded state of the single-vertex crease pattern $C \cup \{c'\}$, we know it has a rigid folding by the Streinu–Whiteley Theorem (that is, Corollary 13.29).   □

**Lemma 13.32**   *Degree-4 and -5 vertices that are bird's feet have a rigid folding from the unfolded state.*

*Proof*   Suppose that the bird's foot contains a tripod with three valley creases $c_1$, $c_2$, and $c_3$ and one mountain crease $c_4$, which we may assume is between $c_1$ and $c_3$, like the one shown in Figure 13.17(a). We aim to build a closed spherical linkage for

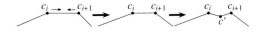

**Figure 13.16** Adding a crease to a single-vertex crease pattern, as witnessed in the proof of Lemma 13.31.

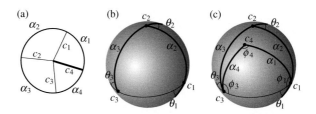

**Figure 13.17** Three creases of a tripod bird's foot make a spherical triangle when folded, and then the fourth crease may be added.

this vertex and thus prove that it has a rigidly folded state. Place $c_1$ and $c_3$ on the equator of a sphere so that $arc(c_1, c_3) < \alpha_1 + \alpha_4 < \pi$. (We know that $\alpha_1 + \alpha_4 < \pi$ by the definition of a bird's foot.) Then the arcs $c_1c_2$ and $c_2c_3$ can be drawn on the northern hemisphere of the sphere so that $c_1c_2c_3$ forms a spherical triangle (see Figure 13.17(b)) whose exterior angles $\theta_1, \theta_2, \theta_3 > 0$. (Here, $\theta_2$ is the fold angle of crease $c_2$.)

Consider the arcs $c_3c_4$ and $c_4c_1$ to be drawn on the sphere. There will be two candidate positions for $c_4$ to do this because $arc(c_1, c_3) < \alpha_1 + \alpha_4$. We chose the one in the northern hemisphere. Label the interior angles of triangle $c_1c_3c_4$ with $\phi_1, \phi_4, \phi_3 > 0$. (Here, $\phi_4 - \pi$ is the fold angle of crease $c_4$.)

The fold angles of $c_1$ and $c_3$ are then $\theta_1 + \phi_1 > 0$ and $\theta_3 + \phi_3 > 0$, respectively; see Figure 13.17(c). The motions of $c_1c_2c_3$ and $c_1c_3c_4$ are both continuous from a flat state. This means that we can assume that $\theta_1, \theta_3$ and $\phi_1, \phi_3$ are all small enough so that $\theta_1 + \phi_1$ and $\theta_3 + \phi_3$ do not exceed $\pi$, thus ensuring that $c_1$ and $c_3$ will be valley creases.

Therefore, this produces a folding motion with $c_1, c_2, c_3$ being valleys and $c_4$ being a mountain, again by Corollary 13.29.

For the cross case, just replace triangle $c_1c_2c_3$ with a $c_1c_2c'_2c_3$ spherical quadrilateral, and the rest follows the same argument. This gives us a rigidly folded instantiation of our bird's foot, which we know implies the existence of a rigid folding by Corollary 13.29. □

*Proof of Theorem 13.26* We start with the converse direction. We know by Lemma 13.32 that a degree-4 single-vertex crease pattern that is a bird's foot, with tripod, will rigidly fold, as will a degree-5 vertex that is a bird's foot with cross. By Lemma 13.31 if we add creases to this, it will still rigidly fold. Therefore, any single-vertex crease pattern with an MV assignment that contains a bird's foot will have a rigid folding from the unfolded state.

Of course, the other direction is harder. We assume that $(C, \mu)$ (a single-vertex fold $C$ with an MV assignment $\mu$) has a rigid folding (by at least some small amount) and suppose that no bird's foot (tripod or cross) exists in $(C, \mu)$. We may also assume that the MV assignment $\mu$ is not all mountains or valleys since, as we saw in Diversion 12.7, a degree-4 rigid origami vertex cannot have all mountains or valleys, and the same Gauss map argument used there works for arbitrary-degree rigid vertices.

**Figure 13.18** Overlapping semicircles $S_1$ and $S_2$, and the sector $S$ that will remain fixed in the folding of the vertex.

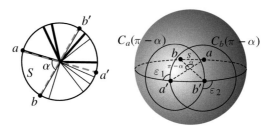

**Figure 13.19** In the proof of Theorem 13.26, when the creases are all folded a little, $a'$ and $b'$ can only move into the regions $\varepsilon_1$ and $\varepsilon_2$, respectively.

So what must $(C, \mu)$ look like? Well, the set of all mountains in $C$ must be confined to a half-plane (or a semicircle, if we think of $C$ as being drawn on a disk). This is because no tripods and no crosses implies that all the mountains are contained in a half-circle. The same is true for all of the valley creases in $C$.

The half-circle of all Ms could overlap with the half-circle of all Vs, however. In fact, we can always pick two semicircles $S_1$ and $S_2$ such that $S_1$ contains all the Ms, $S_2$ contains all the Vs, and $S_1 \cap S_2 \neq \emptyset$. (Even if the Ms and Vs split the disk evenly, we can modify one of the semicircles to make them have overlap.) This will mean that there is a sector of the paper disk, call it $S$, that is completely in neither $S_1$ nor $S_2$. (See Figure 13.18.) The sector $S$ defines a plane, and we imagine this plane being fixed, say in the equator disk of a sphere $B$, while we fold the rest of the disk inside $B$.

The sector $S$ is bounded by an M crease and a V crease. Let $a$ (resp. $b$) be the endpoint on the circle of the M (resp. V) crease bounding $S$, and let $\alpha$ be the angle between $a$'s mountain crease and $b$'s valley crease. Let $a'$ (resp. $b'$) be the antipodal point of $a$ (resp. $b$). Then $a'$ is on the boundary of the semicircle $S_1$, and so between $b$ and $a'$ there are only valley creases. Similarly, between $a$ and $b'$ there are only mountain creases. Our goal is to show that $a'$ and $b'$ move farther apart from each other in the folding, which is a contradiction. (See Figure 13.19 for visuals of this.)

We track regions on the surface of the sphere $B$ to which $a'$ and $b'$ can move under the rigid folding. Denote $C_p(r)$ to be a circle, with interior, drawn on the sphere with center $p$ on the sphere and arc-radius $r$. That is, $C_p(r) = \{x \in B : \text{arc}(x, p) \leq r\}$. Then the region where point $a'$ can be folded is precisely $C_b(\pi - \alpha)$; the boundary of $C_b(\pi - \alpha)$ is achieved by rotating $a'$ about line $b$-$b'$, which is the same as folding the crease whose endpoint is $b$, and folding any of the other valley creases between $a'$ and

$b$ will only move $a'$ inside $C_b(\pi - \alpha)$. Similarly, the image of the point $b'$ under the rigid folding (while keeping sector $S$ fixed) is $C_a(\pi - \alpha)$.

Now, if all the valley creases between $a'$ and $b$ are folded, the point $a'$ will move above the equator of $B$ in the region marked $\varepsilon_1$ in Figure 13.19. Similarly, if all the mountain creases between $b'$ and $a$ are folded, the point $b'$ will move below the equator in the region labeled $\varepsilon_2$ in Figure 13.19. But the shortest distance between the regions $\varepsilon_1$ and $\varepsilon_2$ is $\text{arc}(a', b')$, which is only achieved along the equator when the crease pattern $C$ is in the unfolded position. Thus when the creases are folded, the points $a'$ and $b'$ must increase their distance from each other, which is impossible because it would require the paper to rip.      □

An interesting corollary follows from Theorem 13.26.

**Definition 13.33**   Consider a rigid-foldable single-vertex fold $C$ placed on the equatorial plane of a sphere, with the vertex at the sphere's center, when in the unfolded state. We say that $C$ **pops up** (resp. **pops down**) when rigidly folded if all the creases fold to the lower (resp. upper) half-sphere or remain on the equator. (Note: Compare with the definitions of up vertex and down vertex from Sections 6.3 and 12.4.)

**Corollary 13.34**   *If a single-vertex crease pattern has an M tripod or cross, then it can be folded to pop up. If a single-vertex crease pattern has a V tripod or cross, then it can be folded to pop down.*

The proof of this simply follows from the construction given in Lemma 13.31. A consequence of Corollary 13.34 is that if a single vertex has both an M and a V tripod (or cross), then it can be made to snap from a pops-up folded state to a pops-down one. This characterizes formally a phenomenon seen in practice when rigidly folding single-vertex crease patterns.

A more general version of Theorem 13.26 can be made by considering a single-vertex crease pattern without an MV assignment $\mu$ specified. That is, we are considering the question of determining if an arrangement of creases meeting at a vertex can be rigidly folded using all of the creases.

To do this, we need to augment our definitions slightly. By an **unspecified cross** in a single-vertex crease pattern $C$ we mean a cross without an MV assignment. An unspecified cross is four creases $c_1, c_2, c_3, c_4 \in C$ such that $c_1$ and $c_3$ form a straight line, as do $c_2$ and $c_4$.

**Theorem 13.35**   *A single-vertex crease pattern $C$ is rigid-foldable if and only if $C$ has at least four creases, all sector angles of $C$ are strictly less than $\pi$, and $C$ is not an unspecified cross.*

*Proof*   If $C$ is rigid-foldable, then Theorem 13.26 gives us that $C$ contains a bird's foot. Thus $C$ cannot be an unspecified cross, and the bird's foot sector angles are all less than $\pi$, implying that the sector angles for all of $C$ are less than $\pi$ as well.

In the other direction, suppose that we have a single-vertex crease pattern $C$ with no MV assignment and at least four creases, not an unspecified cross, with all sector angles less than $\pi$. Let $W$ be the wedge between consecutive creases with the largest sector angle. It is possible for there to be another wedge diametrically opposite $W$, with the same angle, so that the four creases bounding this wedge form an unspecified cross, but in order for the whole pattern to not be an unspecified cross, there must be an additional crease somewhere; labeling the cross mountain and the additional crease valley creates a bird's foot and fulfills the requirements of a rigid-foldable vertex. On the other hand, if the creases do not form an equal wedge diametrically opposite $W$, then there is a crease within that opposite wedge (since $W$ has the largest sector angle) that, together with $W$, forms a tripod. Again, we can assign this tripod to be a mountain fold and assign any other crease to be a valley fold and thus satisfy Theorem 13.26.                                                                                      □

One might hope that the bird's foot condition of Theorem 13.26 or the unspecified cross condition of Theorem 13.35 might extend to crease patterns with multiple vertices. This is not the case. As we will see in Section 14.1, determining if a general crease pattern has a rigid folding from the unfolded state is NP-hard, so a condition that amounts to simply looking for bird's feet chained together is wishful thinking. In fact, consider the square twist crease pattern with the standard MV assignment (where the central square has all Ms or all Vs, as in Figure 6.12). In Diversion 13.2 we saw that this crease pattern does not have a rigid folding, let alone from the unfolded state. (In case you did not actually do this diversion, the reason is because the folding multipliers along the major creases around the central square form a strictly increasing sequence, making the crease pattern impossible to rigidly fold unless all the folding angles are zero.) Yet every vertex of this crease pattern has a bird's foot. A simple generalization of Theorem 13.26 to a global property is not possible.

## 13.4     Open Problems

The material in this chapter (and the next) is the subject of much research in science and engineering. Using rigid origami to design structures like solar panels or large telescopes lenses to unfold and deploy in outer space requires understanding the folding angle relationships and their folding speeds in order to control the mechanics of the system as it folds and unfolds. There are many research papers in mechanical engineering that employ (and oftentimes re-derive) the degree-4 folding angle equations from Theorem 13.22. In general, however, calculating the exact folding angle relationships between creases is a complicated algebra problem. We will touch on this more in the next chapter when we discuss configuration spaces of rigid origami.

The bird's foot condition of Section 13.3 is so simple that one can't help but wonder if there is more that can be done with this result. We state one possible open problem that could be pursued.

**Open Problem 13.1**   Are there any families of multiple-vertex crease patterns whose rigid foldability under a given MV assignment can be determined by simply applying Theorem 13.26 to every vertex (that is, checking that each vertex has a bird's foot)?

## 13.5    Historical Remarks

Rigidly folding origami can be considered a subset of flexible polyhedral surfaces, and there is a long history of studies in flexible polyhedra. Cauchy's rigidity theorem, which states that no convex polyhedra can be rigidly flexed along its edges while leaving its faces rigid (i.e., has a rigid folding) goes back to 1812. Raoul Bricard discovered rigidly flexible octahedra that do not contradict Cauchy's result because they self-intersect (Bricard, 1897). Work such as Cauchy's and Bricard's created the foundation for the mathematical subject of **rigidity theory** (see, for example, the book *Combinatorial Rigidity* (Graver et al., 1993)). While rigidity theory focuses more on ball-and-joint mechanisms (i.e., trying to determine when a graph with specified lengths for its edges, or a **linkage**, is rigid or flexible in $\mathbb{R}^2$ or $\mathbb{R}^3$), the connection to rigid folding origami is clear (see the book *Geometric Folding Algorithms* (Demaine and O'Rourke, 2007) for more details on linkage folding).

As we have seen, rigid foldings of single-vertex crease patterns are equivalent to foldings of closed spherical linkages that have a total length of $2\pi$ (assuming the linkage is on a sphere of radius 1). As such, many single-vertex rigid folding problems reduce to what engineers refer to as the kinematics of spherical linkages. See (Chiang, 2000) for a whole book on the subject of spherical mechanism kinematics. Still, single-vertex rigid foldings of 2-dimensional paper (i.e., origami) is a special case with its own advantages, which we've exploited in this chapter.

From the flexible polyhedra literature there has been substantial work that can apply to rigid origami foldings. In particular, Ivan Izmestiev produced a complete classification of the folding angle relationships for quadrangular **Kokotsakis polyhedra** (Izmestiev, 2016). Kokotsakis polyhedra are polyhedral surfaces made by a convex polygon $P$, called the **base**, where the faces adjacent to $P$ are quadrilaterals and between consecutive quadrilaterals around $P$ we have triangle faces, resulting in a polyhedral surface where every vertex has degree 4. Such a polyhedral surface will generally be flexible, and Izmestiev generated the folding angle equations for these Kokotsakis polyhedra where the base $P$ is a quadrilateral. In the special case where each vertex of the Kokotsakis polyhedron is **developable**, meaning that the sum of its plane angles is $2\pi$, we have that the unfolded polyhedral surface is an origami crease pattern. When the vertices also satisfy Kawasaki's Theorem (what Izmestiev calls the **antiisogram** case), then Izmestiev's equations are exactly those we found in Theorem 13.22. For developable vertices that do not satisfy Kawasaki's Theorem, Izmestiev gives folding angle relations that differ from the ones we derived kinematically in Example 13.24, but which do resemble similar formulas derived by Huffman (1976).

# 14 Rigid Origami Theory

Rigid origami is still an active area of research. As mentioned in Section 13.1, rigidly folding origami can be considered a special case of deformations of discrete surfaces, also known as flexible polyhedral surfaces.

In this chapter we will outline some of what is known about the theoretical side of rigid origami and foldings. We start, in Section 14.1, by describing the computational complexity of rigid foldings. In Section 14.2 we will explore configuration spaces of rigid foldings as determined by the folding angles (as opposed to the plane angles around a single vertex, as we saw in Section 5.6). We already defined configuration spaces in Chapter 13 (see Definition 13.9), but there is more to be said about their general and specific properties. We end the chapter with a theoretical development of an aspect of rigid foldings that grew out of applications in mechanical engineering—describing when a rigid-foldable crease pattern is **self-foldable**, meaning that it can somehow be programmed to fold to a certain state using simple functions, called **actuators**, for the crease folding angles. This is meant to model how mechanical folding devices can drive the folding angles to force the material to rigidly fold, without the aid of human hands, to a target folded state.

## 14.1 Complexity of Rigid Foldability with Optional Creases

In Section 13.3 we saw a simple method for determining whether or not a single-vertex crease pattern, with or without an MV assignment, will have a rigid folding from the unfolded state. For multiple-vertex crease patterns this question becomes much harder. In fact, determining if an arbitrary crease pattern has a rigid folding from the flat, unfolded state has been proven to be NP-hard (Akitaya et al., 2019). The full details of this proof are rather complex and will be discussed at the end of this section.

Instead, we will prove a slightly different result: suppose that we are given a crease pattern $C$, without an MV assignment, and we want to know if a **subset** of the creases in $C$ can be rigidly folded from the flat, unfolded state. We will outline the proof that this problem, which we will call the **rigid foldability with optional creases** problem, is NP-hard. For more thorough details, see (Akitaya et al., 2019).

Conceptually our proof will be similar to the one we used to show that flat-foldability is NP-hard in Section 6.6. We will show that the rigid foldability with

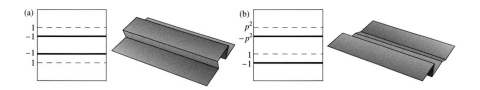

**Figure 14.1** The two types of wires signaling TRUE. (a) A gutter wire. (b) A staircase wire. The numbers indicate the relative folding angle multiplier (speed).

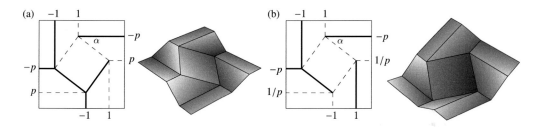

**Figure 14.2** Square twists where the twisting angle $\alpha = \arctan(3/4)$. This has two different rigid folding modes that propagate folding speeds as shown.

optional creases problem can be reduced to the **1-in-3 SAT problem**: given a collection of $n$ boolean variables $x_i$ and clauses $C_m$ of the form $(x_i \vee x_j \vee x_k)$, we ask if there exists a TRUE/FALSE assignment to the variables $x_i$ so that each clause has **exactly one** TRUE variable.

To do this we will use origami wires and gadgets. Wires will act as signals for each variable $x_i$, where TRUE will indicate that the wire is folded and FALSE will mean that the wire is not folded. (This is one way in which some of the creases being optional comes into play.) As in the flat-foldability proof, gadgets will allow us to split variable signals, change their direction, handle when wires cross each other, and perform the action of the clauses.

For wires we use sets of four parallel crease lines, as seen in Figure 14.1. For a TRUE signal we will allow either a **gutter**, where the creases form a VMMV pattern, or a **staircase**, with an MVMV pattern (Figure 14.1(a) and (b), respectively). In a gutter we want the folding speeds of the four creases to all be the same. But in a staircase we want one of the pairs to fold at a slower rate $p^2$. (The exact value of $p$ will be explained later.)

In fact, in order to argue that our eventual crease pattern is rigid-foldable, we will only show that it has a small rigid folding motion from the unfolded state. That way we won't have to worry about any potential self-intersections of the paper. In order to guarantee that a rigid folding motion exists, we need to make sure that the folding angles, or rather their speeds, are all synchronized in our folding motion.

A basic building block of our gadgets will be the square twists shown in Figure 14.2, where the twist angle $\alpha$ (as shown) is taken to be $\arctan(3/4)$. Then

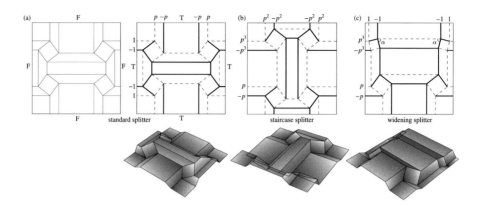

**Figure 14.3** The (a) standard, (b) staircase, and (c) widening splitter gadgets, with folding speeds indicated.

Theorem 13.22 tells us that if the major creases at each vertex have folding speed 1, then the minor creases have speed $p = 1/2$.

We will arrange such square twists in grids to form a variety of gadgets, the most basic of which is a **splitter gadget**, which transfers a horizontal signal on a wire to a vertical signal, or vice versa, by tiling four $\alpha = \arctan(3/4)$ square twists into a $2 \times 2$ grid. This is shown in Figure 14.3(a). Since every crease in a square twist must be folded, or none folded at all, a horizontal (or vertical) FALSE unfolded wire can exist only if the vertical (or horizontal) wire of the splitter is also FALSE. Similarly, a TRUE horizontal wire can only be folded if the vertical wire is also TRUE (and vice versa). Note, however, how the folding speed of 1 in a horizontal TRUE signal transfers to a TRUE vertical signal with speed $p$. Of course, if we rotate a splitter by $90°$, then we can reverse this relationship.

We refer to the splitter gadget in Figure 14.3(a) as the **standard splitter** to distinguish it from the **staircase splitter** (b) and the **widening splitter** (c). The staircase splitter is used to send staircase wires into the clause gadgets, and the widening splitter is a building block of the clause gadgets, as we will see. Note that in the widening splitter we use two $\arctan(3/4)$ square twists at the bottom and two copies of a different square twist with angle $\alpha = 2\arctan(7/9)$ at the top. This $\alpha$ value gives us minor crease folding speeds of $p^3 = 1/8$ if the speed of the major creases is 1 (again, this is given by Theorem 13.22).

These folding multiplier speeds are chosen to allow us to tile the gadgets together into a larger pattern. For example, a **1-in-3 clause gadget** is shown in Figure 14.4. This is made by stacking three widening splitter gadgets into a vertical column, overlaying the vertical creases between them, and adding a fourth widening splitter on the top flipped upside-down. The idea is that with the vertically oriented creases overlaid, only one of the three bottom-most widening splitters can be folded and thus act as the literals for the clause; if the bottom-most one is folded, then the wide pleats at its top will extend vertically and block the second and third literal clauses. Similarly, if

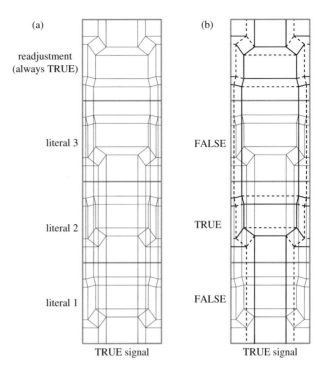

**Figure 14.4** (a) The 1-in-3 SAT clause gadget, made of overlaying widening splitter gadgets. (b) The clause being activated (folded) with literal 1 = FALSE, literal 2 = TRUE, and literal 3 = FALSE.

the second one is folded, then its vertical pleats will block the first and third literal clauses from folding. Furthermore, as long as the gutter wire creases at the bottom of the gadget are folded, then one of the literals must be folded. This is because the center-most vertical creases do not extend to the top-most gadget, forcing one of the other widening splitters to actually fold. The top-most widening splitter is only there to reset the vertical wire to be a standard gutter wire, so that another clause gadget can be stacked on top of it.

How we put this all together is shown in Figure 14.5. The scheme is to consider the boolean variable wires to be entering the paper along the lower left edge, in this case the literals $x_1, \ldots x_4$. If a literal is TRUE then its horizontal gutter wires are folded, which then forces the standard splitter next to the left edge of the paper to fold and thus send a vertical TRUE signal to the clause gadgets that are stacked along the left edge of the paper above the literal wire inputs.

To the right of the literal wire inputs and the clause gadgets is a grid of standard splitters (next to the wire inputs) and staircase splitters (next to the clauses). This allows TRUE signals from the inputs to be diverted up by a standard splitter and then to the left by a staircase splitter. The example shown in Figure 14.5 is very simple, with only one clause gadget, and therefore there are many possible rigid foldings we can create with TRUE/FALSE valuations of the $x_1, \ldots, x_4$ variables to make the clause

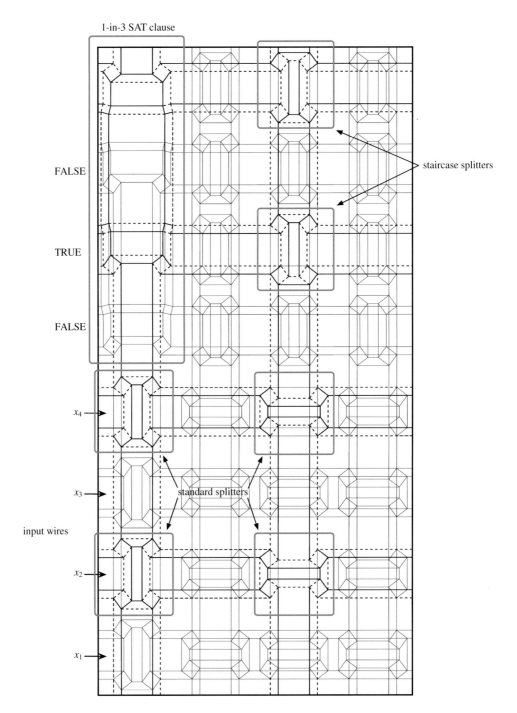

**Figure 14.5** Putting it all together: Four input wires $x_1, \ldots, x_4$ enter at the lower-left and are fed to the 1-in-3 SAT clause gadget above them using splitters. Copius use of optional creases is made to allow for all possibilities of the wires being TRUE or FALSE and being routed to the various parts of the clause.

be TRUE (that is, folded in some way); one such way is shown in Figure 14.5. A more complex collection of clauses can be represented with a similar, albeit much larger, crease pattern: More clause gadgets can be stacked vertically along the left edge of the (larger) sheet of paper in order $C_1, \ldots, C_k$, and more $3 \times 3$ grids of splitter gadgets to the right and above of those shown in Figure 14.5. The first $3 \times 3$ grid of splitters to the right of the wire inputs is used to divert wire signals to clause $C_1$, and the next $3 \times 3$ grid is used to divert signals to clause $C_2$, and so on. Such a crease pattern will have a rigid folding if and only if there is a solution to the corresponding 1-in-3 SAT problem for the clauses $C_1, \ldots, C_n$.

More details can be found in (Akitaya et al., 2019), but we have outlined the proof of the following.

**Theorem 14.1**    *Rigid foldability with optional creases is NP-hard.*

Actually, in (Akitaya et al., 2019) it is proved that the rigid foldability with optional creases problem is **strongly** NP-hard, meaning that it is NP-hard even when all the input parameters are bounded by a polynomial in the length of the input. Also in that reference is a proof that a restricted rigid-foldability problem using all creases in the given crease pattern is NP-complete, where the restrictions are that: all vertices must have rational coordinates, be flat-foldable, and have degree 4 and that the rigid folding is allowed to be **finite precision**, meaning that the folding multipliers must be within a certain tolerance of their required values.

These results indicate an aspect of complexity in origami that is not present in our previous results on the complexity of flat-foldability. That is, when we are rigidly folding initially from the flat (all fold angles equal zero) state, there is no risk of the paper intersecting itself. Thus, the computational problems of layer ordering that led to the NP-hardness of flat-foldability results in Section 6.6 do not come into play. Rather, there are other combinatorial issues present in rigidly folding from the unfolded state where the paper needs to decide which creases can be mountains or valleys in order to have folding angle ranges and speeds that allow the faces of the crease pattern to remain rigid.

These rigid-folding complexity issues are encountered in the work of physicists and engineers who deal with challenges of folding a sheet of material along a pre-scribed crease pattern. See, for example, (Evans et al., 2015a; Chen and Santangelo, 2018; Kang et al., 2019), which discuss how a physically folded material will have multiple ways to fold from the unfolded state and how one might incorporate design elements into the folding medium to bias folding in a certain way. (More will be said on this in Section 14.3.) Theorem 14.1 provides insight into why such care must be taken when designing mechanisms that fold from a flat sheet of some material.

These applications of origami in mechanics design require more information as to the various states into which a crease pattern can fold, especially when folding from the unfolded state. For this we need to first gain a better understanding of the configuration spaces of rigid foldings.

## 14.2     Configuration Spaces of Rigid Foldings

As previously seen in Section 13.1 (Definition 13.9), the **configuration space** $C(G)$ of a crease pattern $G = (V, E)$ on a region $R$ of the plane is the set of points $\rho = (\rho_1 \ldots, \rho_n)$, where $n = |E|$ and $\rho$ represents a valid set of fold angles for a rigid folding map $\sigma$. That is, if $E = \{l_1, \ldots, l_n\}$ is the set of creases, then the folding angle function $\mu$ for $\sigma$ is given by $\mu(l_i) = \rho_i$. In this section we will present a few examples of configuration spaces of rigid foldings, discuss some results on the general form of rigid folding configuration spaces, and see how they can be used in applications.

---

**Example 14.2** (Degree-4 flat-foldable vertex configuration spaces)  As seen in Section 13.2, the folding angles $\rho_1$ and $\rho_2$ of a degree-4, flat-foldable vertex with sector angles $\pi - \beta$ and $\alpha$ between the creases with folding angles $\rho_1, \rho_2$ and between $\rho_2, \rho_3$, respectively (see Figure 13.9), follow the relations

$$\tan \frac{\rho_1}{2} = \frac{\cos \frac{\alpha+\beta}{2}}{\cos \frac{\alpha-\beta}{2}} \tan \frac{\rho_2}{2} \text{ for mode 1, } \tan \frac{\rho_2}{2} = -\frac{\sin \frac{\alpha-\beta}{2}}{\sin \frac{\alpha+\beta}{2}} \tan \frac{\rho_1}{2} \text{ for mode 2,}$$

as per Theorem 13.22. Now, degree-4 single-vertex crease patterns have one degree of freedom, and thus the configuration space $C(G)$ will be the union of two curves (one for each mode) contained in the 4-dimensional cube $[-\pi, \pi]^4 \subset \mathbb{R}^4$. Furthermore, we know that $\rho_1 = -\rho_3$ and $\rho_2 = \rho_4$ in mode 1, while $\rho_1 = \rho_3$ and $\rho_2 = -\rho_4$ in mode 2. Therefore, to visualize the configuration space, we may consider only the axes corresponding the the folding angles $\rho_1$ and $\rho_2$. Figure 14.6(a) shows this 2-dimensional slice of the 4-dimensional $C(G)$, where we took $\alpha = \pi/4$ and $\beta = \pi/3$. Of course, both of the mode 1 and mode 2 curves intersect at the origin, which

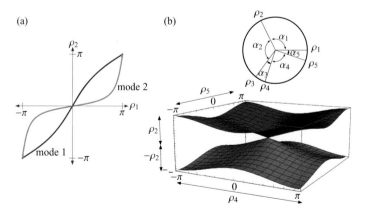

**Figure 14.6**  (a) A 2-dimensional slice of $(G)$ where $G$ is a flat-foldable degree-4 single-vertex crease pattern. (b) A 3-dimensional slice of $C(G)$ where $G$ is the degree-5 vertex shown.

corresponds to the unfolded state, and they also intersect at $(\pi, \pi)$ and $(-\pi, -\pi)$, which correspond to the two flat-folded states.

**Example 14.3** (Degree-5 vertex configuration space) Figure 14.6(b) shows a degree-5 single-vertex crease pattern. Degree-5 vertices have two degrees of freedom (Theorem 13.20), and therefore its configuration space $\mathcal{C}(G)$ will be defined by two parameters, giving rise to a 2-manifold in $\mathbb{R}^5$. In order to provide a visual of the configuration space, we pick three of the folding angles and depict their 3-dimensional slice of $\mathcal{C}(G)$. In Figure 14.6(b) we picked $\rho_4$ and $\rho_5$ to be the free parameters and used Balkcom's rigid folding kinematic process (as described in Section 13.2) to find a formula for $\rho_2$. This produced

$$\cos \rho_2 = \frac{1}{8}(2 + 3 \cos \rho_4 + 3 \cos \rho_5 - \sin \rho_4 \sin \rho_5).$$

For $\rho_4, \rho_5 \in [-\pi, \pi]$ this gives two values, a plus/minus pair, for $\rho_2$. The resulting $(\rho_2, \rho_4, \rho_5)$ slice of $\mathcal{C}(G)$ is shown in Figure 14.6(b).

Notice that in both of these examples we have that the origin is a singular point of the configuration space, if we view $\mathcal{C}(G)$ as the union of manifolds. This is always the case: $\mathcal{C}(G)$ **will always have a singular point at the origin.** (Here by **singular point** we mean a point $x \in \mathcal{C}(G)$ where no neighborhood of $x$ in $\mathcal{C}(G)$ is homeomorphic to a patch of $\mathbb{R}^k$ for some $k$.) The reason for this is simple. First, if $\rho \in \mathcal{C}(G)$ then $-\rho$ will also be in $\mathcal{C}(G)$ because we can reverse the mountain and valley creases of any valid folding angle function and obtain another valid rigid origami configuration. Second, we generally assume that if a crease line $l_i$ is part of a rigid-foldable crease pattern, then some rigid foldings must actually use $l_i$. Thus every coordinate $\rho_i$ must have at least a small range $\rho_i \in (-\varepsilon, \varepsilon)$ (for some $\varepsilon > 0$) in $\mathcal{C}(G)$. Therefore, near the origin the only way that $\mathcal{C}(G)$ could be homeomorphic to a patch of $\mathbb{R}^k$ would be if $k = n$, where $n$ is the number of creases in $G$. But $\mathcal{C}(G)$ will never be full-dimensional near the origin (or at any point) because $\mathcal{C}(G)$ is an affine variety determined by the constraint equations of the folding angles.

Another important elementary fact about $\mathcal{C}(G)$ is that in the context of rigid origami, configuration spaces are always closed.

**Diversion 14.1** Prove that if $G$ is a rigid-foldable crease pattern, then $\mathcal{C}(G)$ is a closed set. (Hint: Recall that each folding angle is restricted to a closed interval $\rho_i \in [-\pi, \pi]$.)

In fact, there is a structural statement that we can assert for the configuration spaces of rigid-foldable vertices near the origin. Recall that the **germ** of a manifold $\mathcal{M}$ at a point $x$ is the equivalence class of analytic functions that are (pairwise) identical on a neighborhood of $x$ in $\mathcal{M}$.

**Theorem 14.4**  *Let G be a single-vertex crease pattern with a rigid folding at the unfolded state. Then the germ of $\mathcal{C}(G)$ at the origin is isomorphic to the germ of a quadratic form at the origin with nullity 3.*

*Proof*  We have basically already proven this theorem – it follows directly from Theorems 13.16 and 13.18. That is, germs are a generalization of Taylor series, and the second-order term in the Taylor series of $F = \prod_{i=1}^{n} R(l_i, \rho_i)$ from the proof of Theorem 13.16 gave us

$$\left( \sum_{i=1}^{n} \sum_{j=1}^{n} \frac{\partial^2 F}{\partial \rho_i \partial \rho_j} \rho_i'(t) \rho_j'(t) + \sum_{i=1}^{n} \frac{\partial F}{\partial \rho_i} \rho_i''(t) \right) \Bigg|_{t=0} t^2 = 0,$$

which is the desired quadratic form. The nullity of 3 corresponds to the fact that the number of degrees of freedom of a rigidly folding vertex of degree $n$ is $n - 3$. $\qquad\square$

Theorem 14.4 is actually a special case of a more general and detailed result by Michael Kapovich and John J. Millson in a very interesting paper titled, "Hodge theory and the art of paper folding" (Kapovich and Millson, 1997). This paper explores the configuration spaces of rigid foldings of single-vertex crease patterns, which they equate to the configuration spaces of spherical linkages (as we did in Chapters 12 and 13).

To describe the result of Kapovich and Millson, let $\Pi$ denote a closed spherical linkage on the unit sphere with $n$ sides with lengths $r = (r_1, \ldots r_n)$. Let $C_r$ denote the configuration space of $n$-gon spherical linkages with side lengths $r$. We will say that $\Pi$ is **degenerate** if it lies in a great circle $S^1$ on the sphere. Thus, $\Pi$ might be a great circle itself if the linkage is unfolded completely (and has length sum equal to $2\pi$, which we are not necessarily assuming here), or $\Pi$ might be folded with folding angles all equal to $\pm\pi$. Give this great circle $S^1$ an orientation and define $\varepsilon_i \in \{\pm 1\}$ to be 1 if the orientation of the $i$-th edge of $\Pi$ agrees with that of $S^1$ (where we call the edge a **forward** edge) and $-1$ otherwise (where we call the edge a **backward** edge). Let $f$ and $b$ equal the number of forward and backward edges of $\Pi$, respectively (thus $f + b = n$). Let $w = turn(\Pi)$ be the turning number of $\Pi$, which we defined in Section 6.4, but in this context also satisfies $\sum_{i=1}^{n} \varepsilon_i r_i = 2\pi w$.

If $\Pi$ represents the projection of a rigid origami vertex fold $\sigma$ onto the unit sphere (meaning $\sum r_i = 2\pi$) and $\Pi$ is degenerate, then either $\Pi$ is a great circle, in which case $\sigma$ is unfolded, $f = n$, $b = 0$, and $w = 1$, or $\Pi$ is a proper subset of a great circle, in which case $\sigma$ is a flat origami fold, $n$ is even, $f = b = n/2$, and $w = 0$. However, the following theorem includes cases where $\sum r_i < 2\pi$ and our piece of paper is a cone as well as where $\sum r_i > 2\pi$ and we are folding hyperbolic paper. In those cases, $\Pi$ being degenerate means that the corresponding flat single-vertex fold will be a cone fold as described in Section 5.4.

**Theorem 14.5** (Kapovich and Millson, 1997)

(i) *If $\Pi$ is nondegenerate then $C_r$ is smooth at $\Pi$.*

(ii) *If $\Pi$ is degenerate then the germ of $C_r$ at $\Pi$ is analytically isomorphic to the germ of the null cone of a quadratic form on $\mathbb{R}^{n+1}$ of nullity 3 and signature $(f - 2w - 1, b + 2w - 1)$.*

The proof of Theorem 14.5 given in (Kapovich and Millson, 1997) is a complicated excursion into the deformation theory of representations of SO(3); we refer interested readers to their paper. A different proof of the signature part of (ii) can be found in (Chen and Santangelo, 2018).

These results describe the configuration space for single-vertex rigid foldings: They have a singular point at the origin, where $\mathcal{C}(G)$ can be described by a quadratic form. And away from the origin $\mathcal{C}(G)$ is a smooth manifold.

More can be said, however.

**Theorem 14.6** (Streinu and Whiteley, 2005)  *If $G$ is a rigid-foldable single-vertex crease pattern, then $\mathcal{C}(G)$ is connected.*

The proof of this follows directly from Theorem 13.28, and these two theorems, together, should really be called the Streinu–Whiteley Theorem. To recap, in (Streinu and Whiteley, 2005) Ileana Streinu and Walter Whiteley prove that $\mathcal{C}(G)$ is connected for single-vertex rigid foldings by proving that every spherical polygon of perimeter $2\pi$ drawn on a unit sphere can be unfolded to a great circle. This means that given any two points $p_1, p_2 \in \mathcal{C}(G)$, where $G$ is a single-vertex, rigid-foldable crease pattern, we can convert them to spherical polygons $s(p_1)$ and $s(p_2)$, then unfold $s(p_1)$ to a great circle, then fold this great circle to $s(p_2)$. Thus $\mathcal{C}(G)$ is path-connected and thus connected.

The proof of the Streinu–Whiteley Theorem is rather lengthy, so we will not reproduce it here. The idea behind the proof is to apply the **Carpenter's Rule Theorem**: every closed planar linkage (simple polygons) can be unfolded to a convex polygon (Connelly et al., 2003). The strategy of Streinu and Whiteley's proof is to first project the rigid origami vertex onto a sphere, which must lie within a hemisphere. Then project the spherical polygon to a plane tangent to the sphere at the pole of the hemisphere. Then this planar polygon can be expanded by the Carpenter's Rule Theorem, and Streinu and Whiteley prove that such rigid planar expansion motions correspond to rigid expansion motions of the spherical polygon, and thus the origami vertex. This correspondence establishes their theorem. Of course, see (Streinu and Whiteley, 2005) for the full details.

The Streinu–Whiteley Theorem raises the question of whether or not all rigid origami configuration spaces are connected. It turns out that this is not the case.

**Example 14.7** (Locked, rigid triangular crease pattern)  A rigid-foldable origami crease pattern $G$ on a region $R$ for which $\mathcal{C}(G)$ is not connected would have the interesting property that it could be put into a rigid origami form $\sigma$, perhaps by cutting the faces of $G$ apart and reassembling them into the folded image $\sigma(R)$, and not have a rigid folding that unfolds $\sigma(R)$ to the unfolded state. Such an origami object would be **locked** in its folded state (or in some rigid folding neighborhood of it). There are many origami models from the origami art world that fit into this category. The classic square twist seen in Figure 6.12 is one example, but any artistic origami model where the folding maneuvers are used to hold the paper into a desired 3-dimensional shape would likely have a disconnected configuration space (if they are rigid-foldable).

However, even the classic square twist can be made to have a connected configuration space if we add creases so as to make more of the crease pattern's faces triangles (see (Hull and Urbanski, 2018) for a proof of this). Thus one wonders if there are any rigidly foldable crease patterns made entirely of triangle faces that nonetheless have a disconnected configuration space. Such crease patterns would have a "strong" locked rigidly folded state in that no addition of creases to the pattern can be made to make it rigid-foldable to the unfolded state.

In (Abel et al., 2015a) the authors produce such a crease pattern $G_\theta$, shown in Figure 14.7, that is a modified triangle twist (compare with Figure 6.20). This is not a flat-foldable crease pattern, and the angle $\theta$ can be changed to modify $\mathcal{C}(G_\theta)$. In Figure 14.7 we have chosen $\theta = 66.67°$, and the right-most image shows a rigidly folded configuration that, we claim, cannot be unfolded with a rigid folding to the unfolded state. In fact, we will prove the following.

**Theorem 14.8** (Abel et al., 2015a)  *For $G_\theta$, the crease pattern in Figure 14.7, the configuration space $\mathcal{C}(G_\theta)$ is disconnected for $\theta = 65°$.*

*Proof*  Rather than compute the whole configuration space, we will focus on the creases surrounding the equilateral triangle in the center, letting $\rho_1$, $\rho_2$, and $\rho_3$ be the folding angles for these creases. If the configuration space $\mathcal{C}'(G_\theta) \subset \mathbb{R}^3$ of valid folding angles $(\rho_1, \rho_2, \rho_3)$ is disconnected, then so is $\mathcal{C}(G_\theta)$ since any path-connected pair of points in $\mathcal{C}(G_\theta)$ will result in path-connected points in $\mathcal{C}'(G_\theta)$.

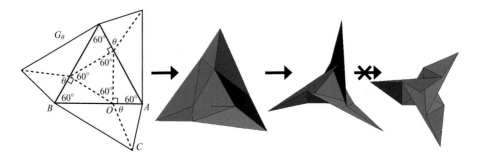

**Figure 14.7**  The crease pattern $G_\theta$ of a modified triangle twist with two folded states for which there is a rigid folding from the unfolded state and a third folded state for which there is not, making $\mathcal{C}(G_\theta)$ disconnected. Here $\theta = 66.67°$.

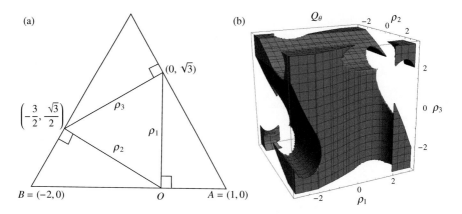

**Figure 14.8** (a) The coordination of the central part of $G_\theta$. (b) The configuration space $Q_\theta$ when $\theta = 65°$.

We choose a coordinate system that places $G_\theta$ in the $xy$-plane as shown in Figure 14.8(a) (suppressing their $z$-coordinates since they are all zero), and we assume that the central equilateral triangle is fixed in the rigid folding on $G_\theta$. Since the crease whose folding angle is $\rho_1$ lies on the $y$-axis, we have that the point $A$ will travel along a circle about $O$ in the $xz$-plane: $\overrightarrow{OA} = (\cos \rho_1, 0, \sin \rho_1)$. Then we can determine, using an appropriate affine transformation, that the location of the point $B$ is given by

$$\overrightarrow{OB} = \left( -\frac{1}{2}\cos\rho_2 - \frac{3}{2}, -\frac{\sqrt{3}}{2}\cos\rho_2 + \frac{\sqrt{3}}{2}, \sin\rho_2 \right).$$

Referring to the crease pattern in Figure 14.7 again, we can see that as the inner triangle of valley creases folds, whether or not $\triangle OAC$ and $\triangle OBC$ intersect can be determined by the angle between the vectors $\overrightarrow{OA}$ and $\overrightarrow{OB}$ as they fold. These triangles will lie in the same plane if and only if the angle between $\overrightarrow{OA}$ and $\overrightarrow{OB}$ becomes $180° - 2\theta$ when the paper is folded. This angle cannot be less that $180° - 2\theta$; it represents the smallest angle possible between $\overrightarrow{OA}$ and $\overrightarrow{OB}$ or else the paper would have to rip, say, along $OC$. A good way to check this is to use the dot product of $\overrightarrow{OA}$ and $\overrightarrow{OB}$, which will equal $2\cos\alpha$ where $\alpha$ is the angle between the vectors. (So $\alpha \geq 180° - 2\theta$.) Thus,

$$\overrightarrow{OA} \cdot \overrightarrow{OB} = -\frac{1}{2}\cos\rho_1\cos\rho_2 - \frac{3}{2}\cos\rho_1 + \sin\rho_1\sin\rho_2$$

$$= 2\cos\alpha \leq 2\cos(180° - 2\theta).$$

(14.1)

By symmetry, we obtain similar equations for the other pairs of the $\rho_i$:

$$-\frac{1}{2}\cos\rho_2\cos\rho_3 - \frac{3}{2}\cos\rho_2 + \sin\rho_2\sin\rho_3 \leq 2\cos(180° - 2\theta),$$

$$-\frac{1}{2}\cos\rho_3\cos\rho_1 - \frac{3}{2}\cos\rho_3 + \sin\rho_3\sin\rho_1 \leq 2\cos(180° - 2\theta).$$

The set $Q_\theta$ of triples $(\rho_1, \rho_2, \rho_3)$ that satisfy these three inequalities will be a superset of $C'(G_\theta)$; $Q_\theta$ only keeps track of one type of rigid foldability failure, when creases like $OA$ and $OB$ move too close together. There are other failures that can occur, like self-intersections. Still, if $Q_\theta$ is disconnected then so is $C'(G_\theta)$, and in Figure 14.8(b) we see that $Q_\theta$ appears to be disconnected when $\theta = 65°$. Indeed, the corner where $\rho_1$, $\rho_2$, and $\rho_3$ are all close to $\pi$ seems to be a disconnected component of $Q_{65°}$. We need to justify that this is, indeed, the case.

First, we can compute that the point $(2\pi/3, 2\pi/3, 2\pi/3)$ fails Inequality (14.1) when $\theta = 65°$, but that the point $(\pi, \pi, \pi)$ satisfies it. Also, if we set $\rho_1 = 2\pi/3$ then Inequality (14.1) will fail for $2\pi/3 \leq \rho_2 \leq \pi$. This means that all points in the square $s_x = \{(2\pi/3, y, z) : 2\pi/3 \leq y, z \leq \pi\}$ will fail Inequality (14.1) and thus not be in $Q_{65°}$. By symmetry, all points in the squares $s_y = \{(x, 2\pi/3, z) : 2\pi/3 \leq x, z \leq \pi\}$ and $s_z = \{(x, y, 2\pi/3) : 2\pi/3 \leq x, y \leq \pi\}$ will not be in $Q_{65°}$ either. Thus the point $(\pi, \pi, \pi) \in Q_{65°}$ is separated from the the component of $Q_{65°}$ that contains the origin by $s_x \cup s_y \cup s_z$, giving that $Q_{65°}$ is disconnected. One can check that points in $Q_{65°}$ near $(\pi, \pi, \pi)$ really are in $C'(G_{65°})$, giving us that $C(G_{65°})$ is disconnected. $\square$

Clearly there is some range of $\theta$ that makes $Q_\theta$, and thus $C(G_\theta)$, disconnected, but computing this range exactly is difficult. If we consider the line $\rho_1 = \rho_2 = \rho_3$, then this turns our dot product inequalities into

$$-\frac{3}{2}\left(\cos \rho_1 + \frac{1}{2}\right)^2 + \frac{11}{8} \leq 2\cos(180° - 2\theta).$$

The left-hand side of this inequality reaches its maximum of $11/8$ at $\rho_1 = 2\pi/3$, and the right-hand side will equal this for $\theta = (1/2) \arccos(-11/16) \approx 66.7163°$. Also, the point $(\pi, \pi, \pi)$ fails to satisfy the dot product inequalities for $\theta < 60°$, so we have that the line $\rho_1 = \rho_2 = \rho_3$ for $\rho_1 \in [-\pi, \pi]$ is not all in $Q_\theta$ for $60° \leq \theta \leq 66.7163°$. However, as can be seen in Figure 14.9, there are "bridges" in $Q_\theta$ along the boundary of the $[-\pi, \pi]^3$ cube that seem to keep $Q_\theta$ connected even after the isthmus line $\rho_1 = \rho_2 = \rho_3$ has become disconnected. This shows a disadvantage of using $Q_\theta$;

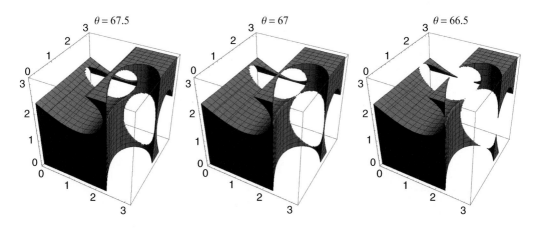

**Figure 14.9** Close-ups of the disconnecting isthmus of $Q_\theta$ near the critical value $\theta \approx 66.7163°$.

while $Q_\theta$ is relatively easy to compute, it does not capture all the ways that the crease pattern $G_\theta$ can fail rigid foldability. In fact, the "bridges" seen in Figure 14.9 are not present in the actual configuration space $C'(G_\theta)$. See (Abel et al., 2015a) for more details.

Rigid origami models that have disconnected configuration spaces have been of interest for origami applications in physics and engineering because often this results in the origami model having an interesting mechanical behavior. For instance, actual folded models of the locked triangle twist from Example 14.7 have the nice property of "snapping" into their locked state when folded. This happens because such physical models are typically made from a flexible material, like paper, card stock, or thin plastic. That is, the material is not perfectly rigid, and thus if enough force is applied to the material, it can be made to fold through the disconnected regions of the configuration space by bending some of the faces of the crease pattern. Physicists call this behavior **bistability** of origami models, where there are two different folded states that the model can rest in, and energy is required to fold the paper from one stable state to the other. In fact, (Silverberg et al., 2015) describes the bistability of square twists as well as the locked triangle twist of Example 14.7.

Another interesting example of an origami model whose disconnected configuration space leads to unusual mechanical behavior of the folded model is the Miura-ori ring, shown in Figure 14.10(a). This idea of wrapping a Miura-ori pattern into a ring has been explored (Nojima, 2002, 2003), but the ring we describe here, designed by Bin Liu in (Liu et al., 2018), uses a different geometry, outlined in Figure 14.10(b). A circle $C$ with radius $r_{in}$ is placed at the center of a polygon. Then, starting at one corner of the polygon, consider the line through this corner and tangent to the left side of $C$, drawing a segment from the corner of length $l_1$ along this line. At the end of this

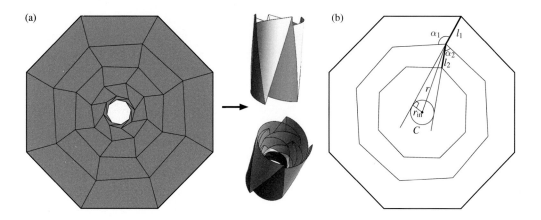

**Figure 14.10** (a) A Liu–Miura-ori octagonal ring. The folded images were produced using Lang's **Tessellatica** code (Lang, 2015). (b) The method of construction.

segment, do the same thing but this time make the line be tangent to the right side of $C$ and draw a segment of length $l_2$. Continuing this process a finite number of times, and performing it on all corners of the original polygon, creates congruent zig-zags whose vertices define a sequence of concentric polygons, and we cut out the smallest, central polygon to make the paper form an annulus. The concentric polygons together with the zig-zags form the crease pattern. An example starting with a regular octagon is shown in Figure 14.10(a). Note that the lengths $l_1, l_2, \ldots$ are design parameters and do not effect the inherent geometry of the model (within reason – they must be chosen so that the sequence of polygons have decreasing radii).

---

**Diversion 14.2**    Prove that, as constructed, every vertex of the Liu–Miura ring crease pattern will be flat-foldable. (That is, prove that in Figure 14.10(b) we have $\alpha_1 + \alpha_2 = \pi$.)

---

An interesting feature of this crease pattern is that it by no means folds rigidly. In fact, it **folds flat onto a cylinder** rather than a flat plane. This forces all of the faces of the crease pattern to bend in $\mathbb{R}^3$ as the paper collapses to a cylindrical shape. (Think of the unfolded crease pattern being in the $xy$-plane, centered at the origin, and then we a place cylinder $Cy$ with radius $r_{in}$ and axis the $z$-axis inside the annular hole, so that $Cy$ and $C$ coincide. Then the crease pattern will fold "flat" into $Cy$.) When physically folding such a model, the material resists being folded into a cylindrical shape. When force is applied to fold the structure, each polygonal ring of the crease pattern "snaps" into a cylindrical shape, with the outer rings having a more pronounced snap than the inner rings. The actual material used and the choices of the lengths $l_1, l_2, \ldots$ also affect the amount of snapping that happens.

In order to examine this non-rigid-foldable crease pattern with a rigid origami model, the authors of (Liu et al., 2018) add creases to triangulate each face, making each vertex a degree-6 vertex with four creases equal to the crease pattern creases and two "face bending" creases. Details of this and the disconnected configuration space that results from this model can be found in the "supplementary information" article associated with (Liu et al., 2018).

## 14.3    Self-Foldability

As mentioned in the previous section, many applications of origami intend to use folded structures to fold or unfold on their own. For example, they might deploy solar arrays in outer space (Miura, 1989), unfold mobile shelters (De Temmerman et al., 2007), or open up a tiny folded polyhedron to release cancer-killing medicine to a tumor site in a person's body (Na et al., 2015; Udomprasert and Kangsamaksin, 2017). Such applications involve designing a material or structure that will fold the creases using **actuators**. These actuators might be springs at the crease lines, or little motors, or a special material that swells and de-swells in response to heat or water. We say

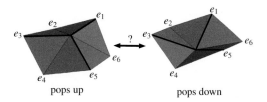

**Figure 14.11** A rigid vertex fold that can pop up or pop down with the same MV assignment.

that crease patterns **self-fold** when they fold or unfold using actuators at the creases, as opposed to being folded with the aid of human hands.

Designing self-folding materials can be very challenging, and one of the biggest challenges is making sure that the actuators drive the creases to fold the material into the target shape. For example, even if the crease actuators are programmed to force a certain MV assignment, that won't guarantee a unique folded shape, even for a single-vertex crease pattern! An example is shown in Figure 14.11, where a degree-6 vertex is rigidly folded so that creases $e_1$, $e_3$, and $e_5$ are mountains while the rest are valleys. This vertex when folded can pop up or pop down (as seen in Definition 13.33) under this MV assignment. If we desire the vertex to pop up, how do we program the actuators to do this reliably?

The first people to develop an answer to this question were Tomohiro Tachi and the author in (Tachi and Hull, 2017) using tools from vector calculus (see (Marsden and Tromba, 2011) for a reference). The idea is to consider the desired folding of a rigid origami as a curve $\rho(t)$ (parameterized with respect to arc-length and piecewise $C^1$) in the configuration space from an initial folded state $\rho(0)$ (often the unfolded state) to a target state $\rho(T)$. We call such a curve a **well-behaved rigid folding**. The piecewise $C^1$ property implies that there are at most two tangent vectors,

$$\vec{v}_+(t) = \lim_{s \to t+} \frac{d\rho(s)}{ds} \quad \text{and} \quad \vec{v}_-(t) = \lim_{s \to t-} \frac{d\rho(s)}{ds},$$

at every point $\rho(t)$. Note that since $\rho(t)$ is parameterized by arc-length, $\|\vec{v}_\pm(t)\| = 1$ for all $t$.

We then think of the forces created by the actuators on the creases as defining a vector field $\vec{f}$ on the parameter space of the crease pattern, and we want to determine if this vector field will "push," or drive the rigid origami along the curve $\rho(t)$ in the configuration space. To make this more formal, we will introduce some definitions.

Let our crease pattern $G$ have $n$ creases.

**Definition 14.9** A **driving force** is a continuous vector field $\vec{f}(x) = (f_1(x), \dots, f_n(x))$ in the parameter space, so $x \in \mathbb{R}^n$. A driving force is **conservative** if it is the negative gradient of some $C^1$ real-valued function $U(x)$ (i.e., $\vec{f}(x) = -\nabla U(x)$), and $U(x)$ is called the **potential function**.

A **rotational spring** driving force (and potential function) is a conservative driving force whose potential function is additively separable in each folding angle coordinate, so $U(x) = U_1(x_1) + \cdots + U_n(x_n)$.

Rotational spring driving forces are desirable because they are easier for engineers to implement in a physical model – a driving force represents a set of rotational moments applied to the crease hinges, and so if the potential function is separable into functions for each coordinate, then each crease can have its own actuating function as opposed to having the crease actuators changing relative to one another. In particular, if each coordinate potential function $U_i(x)$ were linear, then each crease could be actuated by a simple spring.

For a driving force $\vec{f}$, it will be useful to define the dot product function $d_{\vec{f}}(\vec{x}) = \vec{x} \cdot \vec{f}(\rho(t))$.

**Definition 14.10**    The **constrained forces** along a well-behaved rigid folding $\rho(t)$ over a driving force $\vec{f}$ are the **forward force** $f_+(t) = d_{\vec{f}}(\vec{v}_+(t)) = \vec{v}_+(t) \cdot \vec{f}(\rho(t))$ and the **backward force** $f_-(t) = d_{\vec{f}}(\vec{v}_-(t)) = \vec{v}_-(t) \cdot \vec{f}(\rho(t))$.

These constrained forces measure the work done by the driving force $\vec{f}$ to push the rigid folding along the curve $\rho(t)$. In order for a self-folding origami mechanism to rigidly fold along the $\rho(t)$ path, we want the forward force to be the maximum work possible so that the driving force won't push the folding along some other, more natural path. This motivates the next definition.

**Definition 14.11**    A well-behaved rigid folding $\rho(t)$ from starting rigid fold $\rho(0)$ to a target state $\rho(T)$ is **self-foldable** by a driving force $\vec{f}(\rho(t))$ if at every point $\rho(t)$ for $t \in [0, T]$ the forward force $f_+(t)$ is positive and the vector $v_+(t)$ gives the maximum value of $d_{\vec{f}}(\vec{x})$ over all unit tangent vectors in $\vec{x} \in S_{\rho(t)}(\mathcal{C}(G))$.

We call a well-behaved rigid folding $\rho(t)$ **uniquely self-foldable** by a driving force $\vec{f}$ if it is the only well-behaved rigid folding that is self-foldable by $\vec{f}$.

Here $S_x(\mathcal{C}(G)) = \{\vec{v} \in T_x(\mathcal{C}(G)) : \|\vec{v}\| = 1\}$, the subset of the tangent space of $\mathcal{C}(G)$ at $x$ consisting only of unit vectors (this is the same notation used in Section 10.2). The vectors in $S_x(\mathcal{C}(G))$ form a region or set of points on the unit sphere, which we also call the **valid tangents** at $x$. See Figure 14.12 for a visual of these definitions.

We will illustrate these definitions with an example and at the same time prove that if $G$ is any single-vertex, flat-foldable crease pattern and we wish to fold it to a target rigid configuration $A \in \mathcal{C}(G)$, then there exist a well-behaved rigid folding $\rho(t)$ and a rotational spring driving force $\vec{f}$ with $\rho(T) = A$ that make $\rho(t)$ uniquely self-foldable by $\vec{f}$.

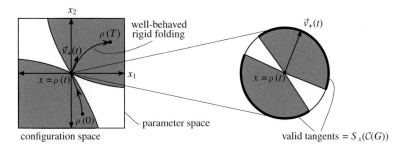

**Figure 14.12** An illustration of a well-behaved rigid folding $\rho(t)$ in a configuration space, its forward tangent vector $\vec{v}_+(t)$ at a point $x = \rho(t)$, and the set of valid tangents $S_x(\mathcal{C}(G))$.

**Example 14.12** (Self-folding degree-4 flat-foldable vertices)   Let $G$ be a single-vertex crease pattern with creases $e_1, \ldots, e_4$ that is flat-foldable, thus the sector angles between the creases can be taken to be $\pi - \alpha, \beta, \alpha,$ and $\pi - \beta$ as in Figure 13.9. Recall from Theorem 13.22 that if $\rho_i$ is the folding angle at crease $e_i$, then the configuration space contains two curves that intersect at the origin: mode 1 that satisfies $(\rho_1, \rho_2, \rho_3, \rho_4) = (\rho_1, \rho_2, -\rho_1, \rho_2)$ and mode 2 satisfying $(\rho_1, \rho_2, \rho_3, \rho_4) = (\rho_1, -\rho_2, \rho_1, \rho_2)$. Since $\mathcal{C}(G)$ consists only of these two curves, finding a rotational spring driving force that pushes the folding along one of the curves is not too difficult. However, the origin is a singular point, and thus any rigid folding path that either starts at or passes through the origin could inadvertently follow the wrong mode curve (see the configuration space in Figure 14.6(a)). Nonetheless, self-folding is possible through the singular point in this example.

**Theorem 14.13** (Tachi and Hull, 2017)   *For any flat-foldable degree-4 vertex crease pattern $G$ with arbitrary starting and target configurations $\rho_S$ and $\rho_T$, there exists a rotational spring driving force that makes the vertex uniquely self-foldable from $\rho_S$ to $\rho_T$.*

*Proof*   Assume that $\rho_T$ lies on the mode 1 configuration space curve; the mode 2 case follows similarly. Then

$$\rho_T = (\tau_1, \tau_2, \tau_3, \tau_4) = (\tau_1, \tau_2, -\tau_1, \tau_2).$$

Let us divide the mode 1 curve into two pieces separated by the origin, the 1+ curve that contains $\rho_T$ and the 1− curve.

**Claim:** If there exists a rotational spring potential function $U(x)$ satisfying the conditions

(1) $U$ monotonically decreases along mode 1 toward the target state, and
(2) $U$ monotonically decreases along mode 2 toward the unfolded state,

then the curve $\rho(t)$ starting at any rigid fold configuration of $G$ and ending at $\rho_T$ along the shortest path connecting them in $\mathcal{C}(G)$ is uniquely self-foldable by the driving force $\vec{f}(x) = -\nabla U(x)$.

*Proof of claim:* Assuming the claim's conditions, if we start from a point $\rho_S$ on the 1+ curve, then a shortest rigid folding curve $\rho(t)$ from $\rho_S$ to $\rho_T$ always has positive and maximum forward force $f_+(t) > 0$ because of condition (1). Therefore, this path is uniquely self-foldable by $-\nabla U$.

If $\rho_S$ is on curve 1−, then the shortest path to the origin, $\vec{0}$, is uniquely self-foldable by $-\nabla U$ by, again, condition (1). If $\rho_S$ is on the mode 2 curve, then the shortest path on this curve from $\rho_S$ to $\vec{0}$ is uniquely self-foldable by $-\nabla U$ by condition (2) (since the origin is the unfolded state). Once we arrive at the origin, traveling along curve 1+ is uniquely self-foldable under $-\nabla U$ again by condition (1). Traveling along any of the other three branches of $\mathcal{C}(G)$ at the origin increases $U$ by conditions (1) and (2) and thus will not give us the maximum forward force $\vec{f}_+(0)$ where $\vec{f} = -\nabla U$. Therefore, if we start anywhere on the mode 2 curve or the 1− curve, the path from $\rho_S$ to $\vec{0}$ and then through the 1+ curve to $\rho_T$, which is the shortest path in $\mathcal{C}(G)$ from $\rho_S$ to $\rho_T$, is uniquely self-foldable by $-\nabla U$, proving the claim.

We now provide a rotational spring potential function $U(x)$ that satisfies conditions (1) and (2). We do this in the most simple way possible by subtracting the target position from the current position in the configuration space:

$$U(x) = \frac{1}{2}\|x - \rho_T\|^2 = \sum_{i=1}^{4} \frac{1}{2}(x_i - \tau_i)^2. \tag{14.2}$$

To see that this satisfies condition (1), we note what it looks like on the mode 1 curve:

$$U(\rho_1, \rho_2, -\rho_1, \rho_2) = (\rho_1 - \tau_1)^2 + (\rho_2 - \tau_2)^2.$$

This is a concave-up parabolic bowl in $(\rho_1, \rho_2)$ and thus has a global minimum at its only critical point, the target state.

For condition (2), along the mode 2 curve we have

$$U(\rho_1, -\rho_2, \rho_1, \rho_2) = \frac{1}{2}\left((\rho_1 - \tau_1)^2 + (-\rho_2 - \tau_2)^2 + (\rho_1 + \tau_1)^2 + (\rho_2 - \tau_2)^2\right)$$
$$= \rho_1^2 + \tau_1^2 + \rho_2^2 + \tau_2^2.$$

This is also a concave-up parabolic bowl in $(\rho_1, \rho_2)$ whose minimum is at its only critical point, $\vec{0}$, proving condition (2). $\qquad\square$

The driving force generated by the potential function in Equation (14.2) can be easily made in a physical example by using, at each crease, rotational springs whose rest angles are set to the angles of the target state. Such a physical model would need faces made from very stiff material, but it would have the fun property that if one forced it into a mode 2 configuration and then let go of the model, it would snap first to the unfolded state and then to the target state (assumed to be in mode 1). Also note that this driving force is not unique; there are other driving forces and associated potential functions that satisfy conditions (1) and (2) and thus make flat-foldable vertices self-foldable.

---

We now present a series of lemmas that will help us determine if a given rigid-foldable crease pattern is self-foldable. We begin by examining the phenomenon seen in the Example 14.12, where following a desired path through the origin required making sure that the driving force didn't inadvertently turn down wrong paths at the singular point. This could easily happen; if there are many different paths at the origin, then there might not exist a driving force that uniquely self-folds to the target state. However, there is a straightforward way of checking if the unfolded state (the origin) will prevent unique self-foldability.

For a valid tangent $\vec{v} \in S_x(\mathcal{C}(G))$, let $W(\vec{v})$ denote the component of $S_x(\mathcal{C}(G))$ that contains $\vec{v}$.

**Lemma 14.14** (Perpendicular constraints)   *Suppose that a well-behaved rigid folding $\rho(t)$ starting from the unfolded state $\rho(0) = \vec{0}$ with tangent vector $\vec{v}_+(0)$ is uniquely self-foldable by a driving force $\vec{f}$. Then $\vec{f}(0)$ is orthogonal to every tangent vector in $S_{\vec{0}}(\mathcal{C}(G))$ that is not in $W(\vec{v}_+(0))$ or $W(\vec{v}_-(0))$.*

*Proof*   Note that the components of $S_{\vec{0}}(\mathcal{C}(G))$ on the unit sphere are closed sets, since $\mathcal{C}(G)$ is closed. Let $\vec{u} \in S_{\vec{0}}(\mathcal{C}(G))$ be a valid tangent that is not in either $W(\vec{v}_+(0))$ or $W(\vec{v}_-(0))$, and assume, for the sake of contradiction, that $\vec{u}$ is not orthogonal to $\vec{f}(0)$. By MV symmetry, we have $-\vec{u} \in S_{\vec{0}}\mathcal{C}(G))$ as well, and thus there are two folding paths with tangent vectors at the origin of $\pm\vec{u}$, one of which must make a positive dot product with $\vec{f}(0)$; assume it is $\vec{u}$. Then since $W(\vec{u})$ is closed, there is a vector $\vec{u}_{max} \in W(\vec{u})$ that maximizes $\vec{u}_{max} \cdot \vec{f}(0)$. Since $\vec{u}_{max}$ cannot be $\vec{v}_+(0)$ or $\vec{v}_-(0)$, we have found another path from the origin, started by $\vec{u}_{max}$, that is self-foldable by $\vec{f}$, contradicting the uniqueness of $\rho(t)$'s self-foldability.                □

It may seem like the necessary condition of self-foldability from Lemma 14.14 is prohibitively strong, but since configuration spaces typically exist in a high-dimensional parameter space, there could be enough "room" at the origin to have a driving force vector be orthogonal to all undesired configuration space paths. For example, the degree-4 flat-foldable vertex configuration space shown in Figure 14.6(a) looks like it would not satisfy this orthogonality condition, but that picture is misleading; the configuration space actually exists in $\mathbb{R}^4$, in which it is possible to find the required driving force, as we saw in the proof of Theorem 14.13.

Lemma14.14 implies a relationship between the space of valid tangents at the origin and the first-order (infinitesimal) approximation space at the origin.

**Lemma 14.15** (Infinitesimal dimension constraints)   *Let $G$ be a rigid-foldable crease pattern and $\rho(t)$ be a well-behaved rigid folding that is uniquely self-foldable by $\vec{f}$ from the unfolded state $\rho(0) = \vec{0}$ with tangent vector $\vec{v} = \rho'(0)$. Then the dimension $m$ of the solution space of the first-order infinitesimal constraints given by Equation (13.4)* *(this is the same as the tangent space $T_{\vec{0}}(\mathcal{C}(G))$) is strictly greater than the dimension n of the linear space S spanned by the vectors in $S_{\vec{0}}(\mathcal{C}(G))$ that are not in $W(\vec{v})$ or $W(-\vec{v})$.*

*Proof* The first-order approximation of a rigid folding at the origin defines a linear space, defined by Equation (13.4), that is tangent to $\mathcal{C}(G)$ at the origin (i.e., $T_{\vec{0}}(\mathcal{C}(G))$). Therefore $m \geq n$. Suppose that $m = n$. Then the first-order approximation space is exactly the linear space S defined in the statement of the lemma. In particular, it means that $\vec{v} \in S$. If $\vec{v}_i$ ($i = 1, \ldots n$) is a basis for S, then $\vec{v}$ can be written as a linear combination of the $\vec{v}_i$ vectors. However, we are given that $\rho(t)$ is uniquely self-foldable by the driving force $\vec{f}$, and so by Lemma 14.14 we have that $\vec{v}_i \cdot \vec{f}(0) = 0$ for all $i = 1, \ldots, n$, and therefore $\vec{v} \cdot \vec{f}(0) = 0$ as well. This contradicts the fact that $\rho(t)$ is uniquely self-foldable by $\vec{f}$ from the origin, and we conclude that $m > n$ must be true.     □

The previous lemma is useful because there are sometimes very simple ways to determine the dimension of the first-order, infinitesimal approximation space of a rigid folding. We state an example of this as another lemma.

**Lemma 14.16** (First-order dimension, quadrilateral grid)     *Let G be a crease pattern whose interior vertices form an $a \times b$ quadrilateral grid (not necessarily a tessellation) such that G has a rigid folding from the unfolded state. Then the space of the first-order infinitesimal approximations of the rigid folding at the origin has dimension $a + b$.*

*Proof* As per Equation (13.4), each interior vertex of G has first-order approximation constraints at the origin given by

$$\rho_1'(0)l_1 + \rho_2'(0)l_2 + \rho_3'(0)l_3 + \rho_4'(0)l_4 = \vec{0},$$

where the $l_i$ are the crease lines at the vertex. Let us suppose that $l_1$ is the "north" and $l_2$ the "west" edge at the vertex. Then for arbitrarily given values for $\rho_1'(0)$ and $\rho_2'(0)$, there will be a unique solution to $\rho_3'(0)$ and $\rho_4'(0)$ satisfying

$$\rho_3'(0)l_3 + \rho_4'(0)l_4 = -(\rho_1'(0)l_1 + \rho_2'(0)l_2). \tag{14.3}$$

Now consider the whole quadrilateral grid crease pattern. If we arbitrarily choose values for $\rho_i'(0)$ for the b crease lines on the top (north side) of G and the a crease lines on the left (west) side of G, then this will determine $\rho_i'(0)$ for the remaining edges of these top- and left-side vertices of G. This leaves an $(a - 1) \times (b - 1)$ section of G on which we may repeat this process inductively to determine the first-order approximation for the whole crease pattern.     □

**Diversion 14.3**     Fill in the detail of the proof of Lemma 14.16 arguing that Equation (14.3) will always have a unique solution for $\rho_3'(0)$ and $\rho_4'(0)$.

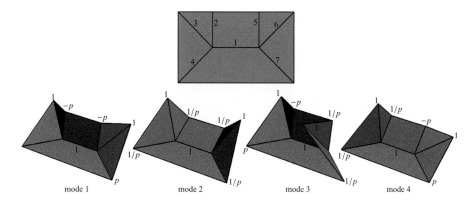

**Figure 14.13** A 2-vertex rigid- and flat-foldable crease pattern with four folding modes.

**Example 14.17** (2-vertex flat-foldable crease pattern)   We will now use the previous lemmas to prove that a certain crease pattern is **not** uniquely self-foldable by any driving force. Consider the 2-vertex crease pattern in Figure 14.13. Each vertex is flat-foldable, and so each vertex can be folded along the single-vertex degree-4 mode 1 or mode 2 configuration space from Example 14.12.

**Theorem 14.18**    *The crease pattern shown in Figure 14.13 is not uniquely self-foldable for any driving force.*

*Proof*   Since the vertices in this crease pattern are degree-4 and flat-foldable, we can use the tangent of half-angle parameterization of the folding angles so that the relative folding multipliers (i.e., folding speeds) of the creases can be easily calculated. We number the creases 1–7 as in Figure 14.13 and set the folding speed of crease 1 to be 1. Since the two vertices in this crease pattern have identical sector angles, the folding speeds of creases 3 and 6 will be $\pm 1$ and of creases 2, 4, 5, and 7 will be $\pm p$ or $1/p$ in accordance with the various folding modes, where $p = \cos((45° + 90°)/2)/\cos((45° - 90°)/2)$. The possible combinations of folding angles for the four folding modes of this crease pattern are shown in Figure 14.13.

If we write these folding speeds as a matrix,

$$\begin{pmatrix} 1 & -p & 1 & p & -p & 1 & p \\ 1 & 1/p & -1 & 1/p & 1/p & -1 & 1/p \\ 1 & 1/p & -1 & 1/p & -p & 1 & p \\ 1 & -p & 1 & p & 1/p & -1 & 1/p \end{pmatrix},$$

then each row represents an infinitesimal rigid folding mode in the 7-dimensional configuration space of this crease pattern. This matrix has rank 3. Also, if we delete any row from this matrix, the resulting $3 \times 7$ matrix still has rank 3, as can be checked via one's computer algebra system of choice.

Now, suppose that one of these folding modes is uniquely self-foldable through the origin by a driving force $\vec{f}$. Then the other modes form a linearly independent set of vectors of a 3-dimensional space, which is the space $S$ of Lemma 14.15. By Lemma 14.16, the dimension of the first-order infinitesimal approximations at the origin is $1 + 2 = 3$, and so by the contrapositive of Lemma 14.15 we have that this crease pattern is not uniquely self-foldable by any driving force. □

---

**Diversion 14.4**   Generalize Theorem 14.18 to prove that no crease pattern made from two degree-4, flat-foldable vertices will be uniquely self-foldable by any driving force.

---

Lemma 14.15 can be made into a necessary and sufficient condition in the case of rigid foldings that have only one degree of freedom (except at singular points).

**Theorem 14.19** (Hull and Tachi, 2018)   *Let G be a rigid-foldable crease pattern whose configuration space is the union of 1-manifolds, and let $\rho(t)$ be a rigid folding from the unfolded state, $\rho(0) = \vec{0}$. Then a driving force $\vec{f}$ exists to make $\rho(t)$ uniquely self-foldable if and only if $\rho'(0)$ is not contained in the linear space $S$ spanned by the valid tangents in $S_{\vec{0}}(C(G))$ that are not in $W(\rho'(0))$ or $W(-\rho'(0))$.*

*Proof*   Sufficiency follows from Lemma 14.15. For the other direction, we need to find a driving force $\vec{f}(t)$ for $\rho(t)$ such that $\vec{f}(0)$ is orthogonal to every vector in $S$. Let $\vec{f}(t)$ be a vector field such that $\vec{f}(0) = \rho'(0) - \rho'(0)^{\|}$, where $\rho'(0)^{\|}$ is the orthogonal projection of $\rho'(0)$ onto $S$. Then $\vec{f}(0)$ is orthogonal to the linear space $S$ and thus has zero dot product with all vectors in $S$. Also, $\vec{f}(0) \cdot \rho'(0) = \rho'(0) \cdot \rho'(0) - \rho'(0)^{\|} \cdot \rho'(0) > 0$ since $\rho'(0) \cdot \rho'(0)$ is the largest dot product that $\rho'(0)$ can have. Therefore, at the origin the forward force $f_+(0) = \vec{f}(0) \cdot \vec{v}$ will achieve its maximum when $\vec{v} = \rho'(0)$ because, since the configuration space is the union of 1-manifolds (i.e., the rigid folding has one degree of freedom), the component of $S_{\vec{0}}(C(G))$ that contains $\rho'(0)$ is just the vector $\rho'(0)$ itself, and the vectors in $S$ have a dot product of zero with $\vec{f}(0)$. Thus, we can make $\vec{f}(t)$ be the constant vector field $\vec{f}(0)$, and then at least in a neighborhood of the origin we will have that $\rho(t)$ is uniquely self-foldable by $\vec{f}$ (whereupon $\vec{f}$ can be extended to all of $C(G)$ in numerous ways). □

---

**Example 14.20** (The Miura-ori is not uniquely self-foldable)   It turns out that under our model of self-foldability, the standard rigid folding of the Miura-ori is not uniquely self-foldable from the unfolded state. To see why, let $G$ be the $3 \times 3$ Miura-ori crease pattern shown in Figure 14.14(a), where the folding multipliers of the creases for the standard folding of the Miura-ori are depicted on the creases. (Here $p = \cos\theta$ according to Theorem 13.22.) These folding multipliers can be used to represent the

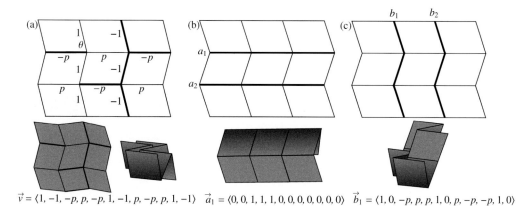

$\vec{v} = \langle 1, -1, -p, p, -p, 1, -1, p, -p, p, 1, -1 \rangle$   $\vec{a}_1 = \langle 0, 0, 1, 1, 1, 0, 0, 0, 0, 0, 0, 0 \rangle$   $\vec{b}_1 = \langle 1, 0, -p, p, p, 1, 0, p, -p, -p, 1, 0 \rangle$

**Figure 14.14** (a) A $3 \times 3$ Miura-ori standard folding with creases labeled by their folding multipliers (speeds) and written as a vector $\vec{v}$. (b) Highlighting the creases $a_1, a_2$ and showing the folding connfiguration $\vec{a}_1$. (c) Highlighting the creases $b_1, b_2$ and showing the configuration $\vec{b}_1$.

configuration space coordinates, and thus the vector $\vec{v}$ shown in Figure 14.14(a) is the tangent vector of the desired folding mode $\rho(t)$ from the origin in order to rigidly fold the Miura-ori in the standard way.

According to Lemma 14.16, the dimension of the linear space tangent to $\mathcal{C}(G)$ at the origin is $2 + 2 = 4$. A basis for this tangent space is

$$\vec{a}_1 = \langle 0, 0, 1, 1, 1, 0, 0, 0, 0, 0, 0, 0 \rangle,$$
$$\vec{a}_2 = \langle 0, 0, 0, 0, 0, 0, 0, 1, 1, 1, 0, 0 \rangle,$$
$$\vec{b}_1 = \langle 1, 0, -p, p, p, 1, 0, p, -p, -p, 1, 0 \rangle, \text{ and}$$
$$\vec{b}_2 = \langle 0, 1, -p, -p, p, 0, 1, p, p, -p, 0, 1 \rangle.$$

Each of these vectors is a valid rigid folding mode of the Miura-ori crease pattern from the origin (they just don't use all the creases). However, note that our desired vector away from the origin $\vec{v}$ can be written as

$$\vec{v} = -p\vec{a}_1 + p\vec{a}_2 + \vec{b}_1 - \vec{b}_2.$$

This means that $\vec{v}$ lies in the space spanned by the valid tangents of $S_{\vec{0}}(\mathcal{C}(G))$ excluding $W(\vec{v})$ and $W(-\vec{v})$. Thus, by Theorem 14.19 this crease pattern is not uniquely self-foldable in the direction of $\vec{v}$.

The previous two examples make the unique self-foldability of multiple-vertex crease patterns look like a rare property. But the problem with these examples is that their configuration spaces have too many branches at the origin. The next example describes a case where this does not happen.

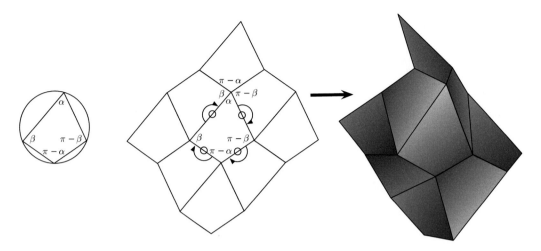

**Figure 14.15** A cyclic quadrilateral and the rotationally symmetric monohedral tiling it generates, which turns out to be rigid-foldable.

**Example 14.21** (Self-foldable monohedral tessellation)   A **monohedral tiling** is a tiling of the plane where all the tiles are congruent to a **prototile**. A very interesting exercise is to investigate when such tilings made from a quadrilateral prototile are rigid-foldable and characterize their configuration spaces. (See (Dieleman, 2018) for a very thorough study on this.) If we require that the prototile be a **cyclic quadrilateral** (can be inscribed in a circle) and make a **rotationally symmetric** tiling from it (where a 180° rotation about the midpoint of each side of a tile is a symmetry of the whole tessellation), then we will have that each vertex of the tessellation will satisfy Kawasaki's Theorem and thus be locally flat-foldable (see Figure 14.15).

Such origami tessellations (made from a cyclic quadrilateral in a rotationally symmetric pattern) are sometimes called **Huffman grids** (e.g., (Lang, 2018, p. 93; Evans et al., 2015c)) because David Huffman was the first to analyze them in full generality. In fact, in (Huffman, 1976) he argued that any crease pattern made from such a tiling has a rigid folding. His argument was simple: Each vertex has a rigid folding (being a flat-foldable vertex), and if we fold all the vertices using the same parameterization (e.g., make them all mode 1 as per Theorem 13.22), then the fold angles along a crease will be equal relative to its two vertices by the rotational symmetry of the crease pattern about the midpoint each crease. For example, if a crease $c = \{v_1, v_2\}$ is given folding angle $\rho_3$ in mode 1 by vertex $v_1$ using the notation and folding angle formulas of Theorem 13.22, then vertex $v_2$ will also use the $\rho_3$, mode 1 formula for crease $c$ as well, simply by the fact that the tiling is rotationally symmetric.

In addition, these origami tessellations self-fold very nicely.

**Theorem 14.22** (Hull and Tachi, 2018)   *Let G be a crease pattern that is a bounded section of a Huffman grid tessellation, and let $\rho(t)$ be a well-behaved rigid folding of G from the unfolded state $\rho(0) = \vec{0}$. Then there is a driving force $\vec{f}$ that makes $\rho(t)$ uniquely self-foldable.*

The proof of Theorem 14.22 is given in (Hull and Tachi, 2018). Rather than reproduce the proof here, we will outline the proof with a pair of diversions.

---

**Diversion 14.5**   Use Theorem 13.26 (that a vertex has a rigid folding from the flat state if and only it contains a bird's foot) to prove that in any rigid folding of a Huffman grid, the vertices must either all be in mode 1 or all be in mode 2.

---

**Diversion 14.6**   Finish the proof of Theorem 14.22 by using the previous diversion to count the number of folding modes of a Huffman grid at the origin and then applying Theorem 14.19.

---

**Example 14.23** (Symmetric degree-6 vertex self-folding)   As a final example for this section, we present a uniquely self-foldable vertex whose driving force offers rather counterintuitive instructions for programming the actuators at the creases.

Let $G$ be the degree-6 vertex where all of the sector angles between consecutive creases is $60°$. A degree-6 rigid-foldable vertex will have three degrees of freedom (by Theorem 13.20). Having more than one degree of freedom makes it difficult to control a rigid folding to follow a desired folding path $\rho(t)$. Therefore, let us assume that our degree-6 vertex $G$ folds **symmetrically** so that the creases alternate MVMVMV around the vertex and that the fold angles of the six creases will alternate in the form $(\rho_1, \rho_2, \rho_1, \rho_2, \rho_1, \rho_2)$ (where, say, the creases with $\rho_1 > 0$ are valleys and the creases with $\rho_2 < 0$ are mountains, or vice versa). Pictures of such symmetrically folding degree-6 vertices can be seen in Figure 14.11.

A rather remarkable feature of this symmetrically folding vertex is that the fold angles $\rho_1$ and $\rho_2$ have a nice relationship when re-parameterized. Unlike the degree-4 case where parameterizing with the tangent of the half-angle was required (as in Theorem 13.22), here we need the **tangent of the quarter-angles**:

$$\tan \frac{\rho_2}{4} = -p \tan \frac{\rho_1}{4} \text{ for mode 1} \quad \text{and} \quad \tan \frac{\rho_1}{4} = -p \tan \frac{\rho_2}{4} \text{ for mode 2,}$$

where $p = \cos((60° + 90°)/2)/\cos((60° - 90°)/2) = 2 - \sqrt{3}$, and modes 1 and 2 are pop up/down symmetric, as shown in Figure 14.11. (See (Tachi and Hull, 2017; Hull and Tachi, 2017) for proofs of this.) These two mode 1 and mode 2 curves are shown in Figure 14.16(a).

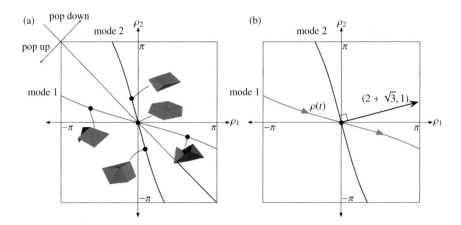

**Figure 14.16** (a) The configuration space (or rather, a $(\rho_1, \rho_2)$ slice) of the symmetrically folded degree-6 vertex, with a few points illustrated. (b) The driving force needed at the origin for a self-folding curve $\rho(t)$ to pass through the origin.

This configuration space is 6-dimensional (since there are six creases), and there are only two folding modes. Therefore, it should be possible to find a driving force $\vec{f}$ that will, for example, uniquely self-fold the curve $\rho(t)$ that follows mode 1 through the origin, and thus is perpendicular to the mode 2 curve at the origin. To do this, we compute the vector tangent to the mode 2 curve at the origin, using the tangent of the quarter-angle parameterization for simplicity:

$$\vec{v}_1 = \langle 1, 2 - \sqrt{3}, 1, 2 - \sqrt{3}, 1, 2 - \sqrt{3} \rangle.$$

An orthogonal vector (without any trivial coordinates) to this is

$$\vec{v}_2 = \langle 1, 2 + \sqrt{3}, 1, 2 + \sqrt{3}, 1, 2 + \sqrt{3} \rangle,$$

which was computed by taking the negative reciprocal of the slope of $\vec{v}_1$ in the $(\rho_1, \rho_2)$-plane. Therefore, we can take the driving force $\vec{f}(t)$ to be the constant vector $\langle 1, 2 + \sqrt{3} \rangle$, which has positive dot product with the mode 1 curve and is orthogonal to the mode 2 curve at the origin. This makes the crease pattern, under this symmetric rigid folding, uniquely self-foldable.

But notice that this driving force $\vec{f} = \langle 1, 2 + \sqrt{3} \rangle$ requires that we place actuators on the creases that **try to make all the creases be valleys**! The stronger rotational moment, $2 + \sqrt{3} \approx 3.732$, will dominate and make its creases be valleys, which will force the weaker forces of 1 to be mountains. The self-folding model indicates that having all positively oriented actuators is what is required to avoid the mode 2 curve as we fold through the unfolded state.

The author and Tomohiro Tachi created a physical model of such a degree-6 vertex, with spring hinges at the creases with strengths in accordance to our driving force $\vec{f}$. Indeed, it works, and rather surprisingly, springs loaded with this driving force served to uniquely self-fold to our target state globally, that is, even if we bend the crease pattern into a non-symmetric configuration. Verification as to why this should be so can be found in (Tachi and Hull, 2017), but it is an open question whether or not the unique self-foldability of a constrained crease pattern (like our symmetrically folding vertex) can have implications on self-foldability in the whole configuration space.

The utility of self-folding can be seen in the interest that it has generated in the physics and engineering literature on origami mechanics. The model of self-folding presented here is a very abstract model. It assumes perfect rigidity of the faces of the crease pattern and zero-thickness of the folded material. Real materials will have some bending of faces occur, which changes the configuration space and thus could alter any self-folding analysis. Because of this, many researchers have used an energy landscape model to explore self-foldability, for example (Kang et al., 2019; Evans et al., 2015a; Chen and Santangelo, 2018; Stern et al., 2017; Pinson et al., 2017). And several have confirmed using a variety of models what our examples hint at: that the number of branches in the configuration space of rigid foldings grows exponentially as the number of vertices in the crease pattern grows (Stern et al., 2017; Pinson et al., 2017; Chen and Santangelo, 2018).

Other approaches for aiding actuators to move the crease pattern along the correct folding path have been studied with promising results. Stern et al. (2017) placed actuators on select creases that then biased the folding path to configuration space branches where those creases' folding angles are larger than those of their neighbors at a vertex (thus leveraging a type of mechanical advantage in the crease pattern). Chen and Santangelo (2018) and Kang et al. (2019) looked at whether biasing vertices to pop up or down helps actuators follow a desired path (it does). There continues to be a lot of interest in self-folding, as well as much that we have yet to understand.

## 14.4  Open Problems

As stated in (Akitaya et al., 2019), there are still unanswered questions when it comes to rigid origami complexity.

**Open Problem 14.1**  Is exact rigid foldability, using all the creases in the crease pattern, NP-hard?

**Open Problem 14.2**  In (Akitaya et al., 2019) it is shown that the rigid foldability problem (in both variations, allowing unfolded creases and requiring all creases be folded) for flat-foldable crease patterns with all vertices of degree 4 is in NP. Is the rigid foldability of general crease patterns also in NP?

Like the previous chapter, these subjects of configuration spaces of rigid foldings and self folding are active areas of research. Many open questions may be asked, and we list only a few here.

**Open Problem 14.3**   Building off of Theorem 14.6, do all two-vertex, rigid-foldable crease patterns have connected configuration spaces?

**Open Problem 14.4**   Can a version of the locked, rigid-foldable triangle twist of Example 14.7 be tessellated into a larger pattern? The idea would be to make a crease pattern, all of whose faces are triangles, where each tiled piece of the crease pattern has a rigidly folded locked state (i.e., a disconnected configuration space). A possible practical application of such a crease pattern would be for the roof of a building – it could be easily manufactured in the flat, unfolded state, then (somehow) manipulated into to the rigid, locked state, from which it will be stable and not unfold again. All the better if the locked state forms a dome!

**Open Problem 14.5**   Can the configuration spaces of general, degree-5 rigid-foldable single-vertex crease patterns be characterized, beyond the statement of Theorem 14.4? Stated another way, are there reasonable equations that can be made for the folding angle relationships in the degree-5 single-vertex case? (Example 14.3 showed only a special case.)

**Open Problem 14.6**   Peter Dieleman's PhD thesis (Dieleman, 2018) characterizes the number of branches in the configuration space of degree-4, quadrilateral origami tessellations. Ivan Izmestiev (2016) classifies the configuration spaces of Kokotsakis polyhedra with quadrilateral bases, which offers another description of degree-4, quadrilateral origami tessellations. Can this prior work be leveraged to create a complete description of the self-foldability of quadrilateral origami tessellations (beyond, say, what was done in (Hull and Tachi, 2018))?

**Open Problem 14.7**   For rigid-foldable single-vertex crease patterns with more than one degree of freedom, under what conditions will a uniquely self-foldable 1-DOF restriction to the configuration space result in unique self-foldability in the whole configuration space (as we saw in Example 14.23)? And can this phenomenon be placed on more rigorous ground?

# References

CIM. 1997 (June). An interview [Stewart Alexander Robertson]. *Bulletin of the CIM (Centro Internacioinal de Mathematica)*, 4–6.

Abel, Zachary, Hull, Thomas C., and Tachi, Tomohiro. 2015a. Locked rigid origami with multiple degrees of freedom. Pages 131–138 of: Miura, Koryo, Kawasaki, Toshikazu, Tachi, Tomohiro, Uehara, Ryuhei, Wang-Iverson, Patsy, and Lang, Robert J. (eds), *Origami⁶: Proceedings of the Sixth International Meeting on Origami Science, Mathematics, and Education. I. Mathematics*. Providence, RI: American Mathematical Society.

Abel, Zachary, Connelly, Robert, Demaine, Erik D., Demaine, Martin L., Hull, Thomas C., Lubiw, Anna, and Tachi, Tomohiro. 2015b. Rigid flattening of polyhedra with slits. Pages 109–117 of: Miura, Koryo, Kawasaki, Toshikazu, Tachi, Tomohiro, Uehara, Ryuhei, Wang-Iverson, Patsy, and Lang, Robert J. (eds), *Origami⁶: Proceedings of the Sixth International Meeting on Origami Science, Mathematics, and Education. I. Mathematics*. Providence, RI: American Mathematical Society.

Abel, Zachary, Cantarella, Jason, Demaine, Erik D., Eppstein, David, Hull, Thomas C., Ku, Jason S., Lang, Robert J., and Tachi, Tomohiro. 2016. Rigid origami vertices: Conditions and forcing sets. *Journal of Computational Geometry*, **7**(1), 171–184.

Adams, Colin. 1994. *The Knot Book: An Elementary Introduction to the Mathematical Theory of Knots*. New York: W. H. Freeman.

Akitaya, Hugo, Cheung, Kenny, Demaine, Erik D., Horiyama, Takashi, Hull, Thomas C., Ku, Jason S., and Tachi, Tomohiro. 2020. Rigid foldability is NP-hard. *Journal of Computational Geometry*, **11**(1), 93–124.

Akitaya, Hugo A., Cheung, Kenneth C., Demaine, Erik D., Horiyama, Takashi, Hull, Thomas C., Ku, Jason S., Tachi, Tomohiro, and Uehara, Ryuhei. 2016. Box pleating is hard. Pages 167–179 of: Akiyama, Jin, Ito, Hiro, Sakai, Toshinori, and Uno, Yushi (eds), *Discrete and Computational Geometry and Graphs*. Cham: Springer International Publishing.

Albertson, Michael O., Alpert, Hannah, belcastro, sarah-marie, and Haas, Ruth. 2010. Grünbaum colorings of toroidal triangulations. *Journal of Graph Theory*, **63**(1), 68–81.

Alperin, Roger C. 2000. A mathematical theory of origami constructions and numbers. *New York Journal of Mathematics*, **6**, 119–133.

Alperin, Roger C. 2002. Mathematical origami: Another view of Alhazen's optical problem. Pages 83–93 of: Hull, Thomas C. (ed), *Origami³: Third International Meeting of Origami Science, Mathematics, and Education*. Natick, MA: A K Peters.

Alperin, Roger C. 2018. 3D folding axioms. Pages 347–358 of: Lang, Robert J., Bolitho, Mark, and You, Zhong (eds), *Origami⁷: The Proceedings from the 7th International Meeting on Origami in Science, Mathematics, and Education, Volume Two: Mathematics*. St. Albans, UK: Tarquin.

Alperin, Roger C., and Lang, Robert J. 2009. One-, two-, and multi-fold origami axioms. Pages 371–393 of: Lang, Robert J. (ed), *Origami⁴: Fourth International Conference on Origami in Science, Mathematics, and Education*. Natick, MA: A K Peters.

Andrilli, Stephen, and Hecker, David. 2003. *Elementary Linear Algebra, 5th ed.* Boston: Elsevier Academic Press.

De Temmerman, Niels, Mollaert, Marijke, Van Mele, Tom, and De Laet, Lars. 2007. Design and analysis of a foldable mobile shelter system. *International Journal of Space Structures*, **22**(3), 161–168.

Arkin, Esther M., Held, Martin, Mitchell, Joseph S. B., and Skiena, Steven S. 1996. Hamiltonian triangulations for fast rendering. *The Visual Computer*, **12**(9), 429–444.

Arnold, Vladimir I. 2005. *Arnold's Problems*. Berlin: Springer.

Assis, Michael. 2018. Exactly solvable flat-foldable quadrilateral origami tilings. *Phys. Rev. E*, **98**(Sep), 032112.

Auckly, David, and Cleveland, John. 1995. Totally real origami and impossible paper folding. *The American Mathematical Monthly*, **102**(3), 215–226.

Azevédo Breda, Ana M. R. 1992. A class of tilings of $S^2$. *Geom. Dedicata*, **44**(3), 241–253.

Balkcom, Devin J. 2002. *Robotic Origami Folding*. PhD thesis, Carnegie Mellon University.

Balkcom, Devin J., and Mason, Matthew T. 2008. Robotic origami folding. *The International Journal of Robotics Research*, **27**(5), 613–627.

Barr, Stephen. 1982. *Mathematical Brain Benders*. New York: Dover.

Barreto, Paulo T. 1997. Lines meeting on a surface: The "Mars" paperfolding. Pages 343–359 of: Miura, Koryo (ed), *Origami Science & Art: Proceedings of the Second International Meeting of Origami Science and Scientific Origami*. Otsu, Japan: Seian University of Art and Design.

Baxter, R. J. 1970. Colorings of a hexagonal lattice. *Journal of Mathematical Physics*, **11**(3), 784–789.

belcastro, sarah-marie, and Hull, Thomas C. 2002a. A mathematical model for non-flat origami. Pages 39–51 of: Hull, Thomas C. (ed), *Origami³: Third International Meeting of Origami, Science, Mathematics, and Education*. Natick, MA: A K Peters.

belcastro, sarah-marie, and Hull, Thomas C. 2002b. Modelling the folding of paper into three dimensions using affine transformations. *Linear Algebra and Its Applications*, **348**(1), 273–282.

Beloch, Margherita P. 1936. Sul metodo del ripiegamento della carta per la risoluzione dei problemi geometrici. *Periodico di Mathematiche, Ser. IV*, **16**, 104–108.

Bern, Marshall, and Hayes, Barry. 1996. The complexity of flat origami. Pages 175–183 of: *Proceedings of the 7th Annual ACM-SIAM Symposium on Discrete Algorithms*. Philadelphia: SIAM.

Bern, Marshall, Demaine, Erik D., Eppstein, David, and Hayes, Barry. 2002. A disk-packing algorithm for an origami magic trick. Pages 17–28 of: Hull, Thomas C. (ed), *Origami³: Third International Meeting of Origami, Science, Mathematics, and Education*. Natick, MA: A K Peters.

Bowers, John C., and Streinu, Ileana. 2015. Lang's universal molecule algorithm. *Annals of Mathematics and Artificial Intelligence*, **74**(3), 371–400.

Bowers, John C., and Streinu, Ileana. 2016. Geodesic universal molecules. *Mathematics in Computer Science*, **10**(1), 115–141.

Bricard, Raoul. 1897. Mémoire sur la théorie de l'octaèdre articulé. *Journal de Mathématiques Pures et Appliquées*, **3**, 113–150.

Bruckheimer, Maxim, and Hershkowitz, Rina. 1977. Constructing the parabola without calculus. *Mathematics Teacher*, **70**(8), 658–662.

Buhler, Joe, Butler, Steve, de Launey, Warwick, and Graham, Ron. 2012. Origami rings. *J. Aust. Math. Soc.*, **92**(3), 299–311.

Bumcroft, R. J. 1969. *Modern Projective Geometry*. New York: Holt, Rinehart, and Winston, Inc.

Burnside, W. 1904. On groups of order $p^a q^b$. *Proceedings of the London Mathematical Society*, **2**, 388–392.

Chang, Eric, and Hull, Thomas C. 2011. The flat vertex fold sequences. Pages 599–607 of: Wang-Iverson, Patsy, Lang, Robert J., and Yim, Mark (eds), *Origami$^5$: Fifth International Conference on Origami in Science, Mathematics, and Education*. Natick, MA: A K Peters.

Chen, Bryan Gin-ge, and Santangelo, Christian D. 2018. Branches of triangulated origami near the unfolded state. *Phys. Rev. X*, **8**(Feb), 011034.

Chiang, C. H. 2000. *Kinematics of Spherical Mechanisms, 2nd ed*. Malabar, FL: Krieger Publishing Company.

Chiu, Alvin, Hoganson, William, Hull, Thomas C., and Wu, Sylvia. 2019. *Counting locally flat-foldable origami configurations via 3-coloring graphs*. Preprint, https://arxiv.org/abs/1910.01278.

Connelly, Robert, Demaine, Erik D., and Rote, Günter. 2003. Blowing up polygonal linkages. *Discrete & Computational Geometry*, **30**(2), 205–239.

Courant, R., and Robbins, H. 1941. *What Is Mathematics?* New York: Oxford University Press.

Cox, David. 2004. *Galois Theory*. Hoboken, NJ: John Wiley & Sons.

Cox, David, and Shurman, Jerry. 2005. Geometry and number theory on clovers. *The American Mathematical Monthly*, **112**(8), 682–704.

Cox, David, Little, John, and O'Shea, Donal. 2005. *Ideals, Varieties, and Algorithms: An Introduction to Computational Algebraic Geometry and Commutative Algebra, 2nd ed*. New York: Springer.

Coxeter, H. S. M. 1973. *Regular Polytopes, 3rd ed*. New York: Dover.

Cromwell, Peter R. 2004. *Knots and Links*. Cambridge, UK: Cambridge University Press.

Dacorogna, Bernard, Marcellini, Paolo, and Paolini, Emanuele. 2008. Lipschitz-continuous local isometric immersions: Rigid maps and origami. *Journal de Mathématiques Pures et Appliquées*, **90**(1), 66–81.

Dacorogna, Bernard, and Marcellini, Paolo. 1999. *Implicit Partial Differential Equations*. Progress in Nonlinear Differential Equations and Their Applications, Vol. 37. Boston: Birkhäuser,.

Dayoub, Iris M., and Lott, Johnny W. 1977. What can be done with a Mira? *Mathematics Teacher*, **70**(5), 394–399.

De las Peñas, Ma. Louise Antonette N., Taganap, Eduard C., and Rapanut, Teofina A. 2015. Color symmetry approach to the construction of crystallographic flat origami. Pages 11–20 of: Miura, Koryo, Kawasaki, Toshikazu, Tachi, Tomohiro, Uehara, Ryuhei, Wang-Iverson, Patsy, and Lang, Robert J. (eds), *Origami$^6$: Proceedings of the Sixth International Meeting on Origami Science, Mathematics, and Education I. Mathematics*. Providence, RI: American Mathematical Society.

Demaine, Erik D., and O'Rourke, Joseph. 2007. *Geometric Folding Algorithms: Linkages, Origami, Polyhedra*. Cambridge, UK: Cambridge University Press.

Demaine, Erik D., Demaine, Martin L., and Lubiw, Anna. 1999 (January 17–19). Folding and one straight cut suffice. Pages 891–892 of: *Proceedings of the 10th Annual ACM-SIAM Symposium on Discrete Algorithms (SODA'99)*. Philadelphia: SIAM.

Demaine, Erik D., Demaine, Martin L., and Mitchell, Joseph S. B. 2000. Folding flat silhouettes and wrapping polyhedral packages: New results in computational origami. *Computational Geometry: Theory and Applications*, **16**(1), 3–21. Special issue of selected papers from the 3rd CGC Workshop on Computational Geometry, 1998.

Demaine, Erik D., Devadoss, Satyan L., Mitchell, Joseph S. B., and O'Rourke, Joseph. 2004 (August 9–11). Continuous foldability of polygonal paper. Pages 64–67 of: *Proceedings of the 16th Canadian Conference on Computational Geometry (CCCG 2004)*. Montreal: Concordia University.

Demaine, Erik D., Demaine, Martin L., Huffman, David A., Hull, Thomas C., Koschitz, Duks, and Tachi, Tomohiro. 2016 (September 26–30). Zero-Area Reciprocal Diagram of Origami. Pages 1–10 of: Kawaguchi, K., Ohsaki, M., and Takeuchi, T. (eds), *Proceedings of the IASS Annual Symposium 2016*. Madrid: IASS.

Denavit, J., and Hartenberg, R. S. 1955. A kinematic notation for lower-pair mechanisms based on matrices. *Journal of Applied Mechanics*, **22**(2), 215–221.

Dickson, L. E. 1904. Problem 219. *The American Mathematical Monthly*, **11**(4), 93.

Dieleman, Peter. 2018. *Origami Metamaterials: Design, Symmetries, and Combinatorics*. PhD thesis, Cleiden University.

Edwards, B. Carter, and Shurman, Jerry. 2001. Folding quartic roots. *Mathematics Magazine*, **74**(1), 19–25.

Emert, John, Meeks, Kay, and Nelson, Roger. 1994. Reflections on a Mira. *The American Mathematical Monthly*, **101**(6), 544–549.

Euler, Leonard. 1768. *Institutionum Calculi Integralis Volumen Primum (Instruction on Integral Calculus, First Volume, in Latin)*. Euler Archive–All Works. E342.

Evans, Arthur A., Silverberg, Jesse L., and Santangelo, Christian D. 2015a. Lattice mechanics of origami tessellations. *Phys. Rev. E*, **92**(Jul), 013205.

Evans, Thomas A., Lang, Robert J., Magleby, Spencer P., and Howell, Larry L. 2015b. Rigidity foldable origami twists. Pages 119–130 of: Miura, Koryo, Kawasaki, Toshikazu, Tachi, Tomohiro, Uehara, Ryuhei, Wang-Iverson, Patsy, and Lang, Robert J. (eds), *Origami⁶: Proceedings of the Sixth International Meeting on Origami Science, Mathematics, and Education. I. Mathematics*. Providence, RI: American Mathematical Society.

Evans, Thomas A., Lang, Robert J., Magleby, Spencer P., and Howell, Larry L. 2015c. Rigidly foldable origami gadgets and tessellations. *Royal Society Open Science*, **2**(9), 150067.

Fehlen, J. 1975. Paper folds and proofs. *Mathematics Teacher*, **68**(7), 608–611.

Francesco, Philippe Di, and Guitter, Emmanuel. 1994. Entropy of folding of the triangular lattice. *Europhysics Letters (EPL)*, **26**(6), 455–460.

Francesco, Philippe Di. 2000. Folding and coloring problems in mathematics and physics. *Bulletin of the American Mathematical Society*, **37**, 251–307.

Frigerio, Emma, and Huzita, Humiaki. 1989. A possible example of system expansion in origami geometry. Pages 53–69 of: Huzita, Humiaki (ed), *Proceedings of the First International Meeting of Origami Science and Technology*. Padova, Italy: Dipartimento di Fisica dell'Università di Padova.

Fuchs, Dmitri, and Tabachnikov, Sergei. 1999. More on paperfolding. *The American Mathematical Monthly*, **106**(1), 27–35.

Fujimoto, Shuzo. 2011. おりがみ あじさい折り// *Origami ajisai ori (Origami Hydrangea Folding, in Japanese)*. Tokyo: Seibundo Shinkosha.

Fujimoto, Shuzo, and Nishikawa, Masami. 1982. 創造する折り紙遊びへの招待/ *Sōzō suru origami asobi e no shōtai (Invitation to Creative Playing with Origami, in Japanese)*. Tokyo: Asahi Culture Center.

Fukagawa, Hidetoshi, and Pedoe, Dan. 1989. *Japanese Temple Geometry Problems*. Winnipeg, Canada: Charles Babbage Research Centre.

Fuse, Tomoko. 1990. *Unit Origami: Multidimensional Transformations*. New York: Japan Publications.

Gallivan, Britney C. 2002. *How to Fold Paper in Half Twelve Times: An "Impossible Challenge" Solved and Explained*. Pomona, CA: The Historical Society of Pomona Valley.

Geretschläger, Robert. 1997a. Euclidean constructions and the geometry of origami. *Mathematics Magazine*, **68**(5), 357–371.

Geretschläger, Robert. 1997b. Folding the regular heptagon. *Crux Mathematicorum*, **23**(2), 81–88.

Geretschläger, Robert. 1998. *Solving quartic equations in origami*. Preprint.

Geretschläger, Robert. 2002. *Geometric Constructions in Origami* (in Japanese). Tokyo: Morikita Publishing.

Geretschläger, Robert. 2008. *Geometric Origami*. Shipley, UK: Arbelos.

Ginepro, Jessica, and Hull, Thomas C. 2014. Counting Miura-ori foldings. *Journal of Integer Sequences*, **17**(10), Article 14.10.8.

Gjerde, Eric. 2008. *Origami Tessellations: Awe-Inspiring Geometric Designs*. Wellesley, MA: A K Peters.

Gleason, Andrew M. 1988. Angle trisection, the heptagon, and the triskaidecagon. *The American Mathematical Monthly*, **95**(3), 185–194.

Graver, Jack, Servatius, Brigitte, and Servatius, Herman. 1993. *Combinatorial Rigidity*. Graduate Studies in Mathematics, Vol. 2. Providence, RI: American Mathematical Society.

Grünbaum, Branko. 2003. *Convex Polytopes, 2nd ed.* Graduate Texts in Mathematics, Vol. 221. New York: Springer.

Haga, Kazuo. 2002. Fold paper and enjoy math: Origamics. Pages 307–328 of: Hull, Thomas C. (ed), *Origami³: Third International Meeting of Origami Science, Mathematics, and Education*. Natick, MA: A K Peters.

Hatori, Koshiro. 2003. *Origami Construction*. Available at http://origami.ousaan.com/library/conste.html.

Heerwart, E. 1920. *Course of Paper Folding: One of Froebel's Occupations for Children (English translation)*. London: Charles and Dibble.

Hilbert, David, and Cohn-Vossen, Stephan. 1956. *Geometry and the Imagination*. New York: Chelsea Publishing Co.

Hirsch, M. W. 1976. *Differential Topology*. Graduate Texts in Mathematics, Vol. 33. Berlin: Springer.

Houdini, Harry. 1922. *Paper Magic*. New York: E. P. Dutton & Co.

Huffman, David A. 1976. Curvature and creases: A primer on paper. *IEEE Transactions on Computers*, **C-25**(10), 1010–1019.

Huffman, David A. 1978. Surface curvature and applications of the dual representation. Pages 213–222 of: Hanson, A. R., and Riseman, E. M. (eds), *Computer Vision Systems*. New York: Academic Press.

Hull, Thomas C. 1994. On the mathematics of flat origamis. *Congressus Numerantium*, **100**, 215–224.

Hull, Thomas C. 1996. A note on "impossible" paperfolding. *The American Mathematical Monthly*, **103**(3), 242–243.

Hull, Thomas C. 2002. The combinatorics of flat folds: A survey. Pages 29–38 of: Hull, Thomas C. (ed), *Origami³: Third International Meeting of Origami Science, Mathematics, and Education*. Natick, MA: A K Peters.

Hull, Thomas C. 2003. Counting mountain-valley assignments for flat folds. *Ars Combinatoria*, **67**, 175–188.

Hull, Thomas C. 2005. Origametry part 6: Basic origami operations. *Origami Tanteidan Magazine*, 14–15.

Hull, Thomas C. 2009. Configuration spaces for flat vertex folds. Pages 361–370 of: Lang, Robert J. (ed), *Origami⁴: Fourth International Conference on Origami in Science, Mathematics, and Education*. Natick, MA: A K Peters.

Hull, Thomas C. 2011. Solving cubics with creases: The work of Beloch and Lill. *The American Mathematical Monthly*, **118**(4), 307–315.

Hull, Thomas C. 2012. *Project Origami: Activities for Exploring Mathematics, 2nd ed.* Boca Raton, FL: CRC Press/A K Peters.

Hull, Thomas C., and Tachi, Tomohiro. 2017. Double-line rigid origami. 11th Asian Forum on Graphic Science, The University of Tokyo Komaba Campus, Japan, August 6–10, 2017.

Hull, Thomas C., and Tachi, Tomohiro. 2018. Self-foldability of monohedral quadrilateral origami tessellations. Pages 521–532 of: Lang, Robert J., Bolitho, Mark, and You, Zhong (eds), *Origami⁷: The Proceedings from the 7th International Meeting on Origami in Science, Mathematics, and Education, Volume Two: Mathematics*. St. Albans, UK: Tarquin.

Hull, Thomas C., and Urbanski, Michael. 2018. Rigid foldability of the augmented square twist. Pages 533–543 of: Lang, Robert J., Bolitho, Mark, and You, Zhong (eds), *Origami⁷: The Proceedings from the 7th International Meeting on Origami in Science, Mathematics, and Education, Volume Two: Mathematics*. St. Albans, UK: Tarquin.

Husimi, Koji, and Husimi, Mitsue. 1979. *Origami no kikagaku (Geometry of Origami, in Japanese)*. Tokyo: Nihon-hyoron-sha.

Husimi, Koji. 1980. Origami no kagaku (The science of origami, in Japanese). *Saiensu (the Japanese edition of Scientific American)*, Oct., 8. appendix in separate volume.

Huzita, Humiaki, and Scimemi, Benedetto. 1989. The algebra of paper-folding (origami). Pages 205–222 of: Huzita, Humiaki (ed), *Proceedings of the First International Meeting of Origami Science and Technology*. Padova, Italy: Dipartimento di Fisica dell'Università di Padova.

Izmestiev, Ivan. 2016. Classification of flexible Kokotsakis polyhedra with quadrangular base. *International Mathematics Research Notices*, **2017**(3), 715–808.

Jensen, Tommy R., and Toft, Bjarne. 1995. *Graph Coloring Problems*. New York: John Wiley & Sons.

Johnson, D. A. 1957. *Paper Folding for the Mathematics Class*. Washington, DC: National Council of Teachers of Mathematics.

Justin, Jacques. 1984. Angle trisection. *British Origami*, **107**, 14–15.

Justin, Jacques. 1986a. Mathematics of origami, part 9. *British Origami*, **119**, 28–30.

Justin, Jacques. 1986b. Résolution par le pliage de l'équation du troisième degré et applications géométriques. *L'Ouvert: Journal of the APMEP of Alsace and the IREM of Strasbourg*, March, 9–19.

Justin, Jacques. 1989. Aspects mathematiques du pliage de papier. Pages 263–277 of: Huzita, Humiaki (ed), *Proceedings of the First International Meeting of Origami Science and Technology*. Padova, Italy: Dipartimento di Fisica dell'Università di Padova.

Justin, Jacques. 1997. Toward a mathematical theory of origami. Pages 15–29 of: Miura, Koryo (ed), *Origami Science & Art: Proceedings of the Second International Meeting of Origami Science and Scientific Origami*. Otsu, Japan: Seian University of Art and Design.

Justin, Jacques. 2012. *Folding of stamps*. Preprint.

Kang, Ji-Hwan, Kim, Hyunki, Santangelo, Christian D., and Hayward, Ryan C. 2019. Enabling robust self-folding origami by pre-biasing vertex duckling direction. *Advanced Materials*, **31**(39), 0193006.

Kantor, Yacov, and Jarić, M. V. 1990. Triangular lattice foldings—A transfer matrix study. *Europhysics Letters (EPL)*, **11**(2), 157–161.

Kantor, Yacov, Kardar, Mehran, and Nelson, David R. 1986. Statistical mechanics of tethered surfaces. *Phys. Rev. Lett.*, **57**(Aug), 791–794.

Kapovich, Michael, and Millson, John J. 1997. Hodge theory and the art of paper folding. *Publ. Res. Inst. Math. Sci.*, **33**(1), 1–31.

Kasahara, Kunhiko. 1988. *Origami Omnibus*. New York: Japan Publications.

Kasahara, Kunhiko, and Maekawa, Jun. 1983. *Viva! Origami*. Tokyo: Sanrio.

Kasahara, Kunhiko, and Takahama, Toshi. 1987. *Origami for the Connoisseur*. New York: Japan Publications.

Kawahata, Fumiaki. 1993. A challenge of constraints: Single square, no cuts (in Japanese). *ORU*, **1**(2), 100–104.

Kawahata, Fumiaki. 1997. The technique to fold free angles of formative art "origami". Pages 63–72 of: Miura, Koryo (ed), *Origami Science & Art: Proceedings of the Second International Meeting of Origami Science and Scientific Origami*. Otsu, Japan: Seian University of Art and Design.

Kawasaki, Toshikazu, and Yoshida, Masaaki. 1988. Crystallographic flat origamis. *Memoirs of the Faculty of Science, Kyushu University, Ser. A*, **42**(2), 153–157.

Kawasaki, Toshikazu. 1987. 立体的結晶折り紙について (On solid crystallographic origamis, in Japanese). *Sasebo National College of Technology Research Report*, **24**, 101–109.

Kawasaki, Toshikazu. 1989a. On high dimensional flat origamis. Pages 131–141 of: Huzita, Humiaki (ed), *Proceedings of the First International Meeting of Origami Science and Technology*. Padova, Italy: Dipartimento di Fisica dell'Università di Padova.

Kawasaki, Toshikazu. 1989b. On relation between mountain-creases and valley-creases of a flat origami. Pages 229–237 of: Huzita, Humiaki (ed), *Proceedings of the First International Meeting of Origami Science and Technology*. Padova, Italy: Dipartimento di Fisica dell'Università di Padova.

Kawasaki, Toshikazu. 1997. $R(\gamma) = 1$. Pages 31–40 of: Miura, Koryo (ed), *Origami Science & Art: Proceedings of the Second International Meeting of Origami Science and Scientific Origami*. Otsu, Japan: Seian University of Art and Design.

Kepler, Johannes. 1604. *Astronomiae pars Optica*. Frankfurt: Claudium Marnium & Haeredes Ioannis Aubrii.

Klein, Felix. 1897. *Famous Problems of Elementary Geometry*. Boston: Ginn & Co.

Koehler, J. 1968. Folding a strip of stamps. *Journal of Combinatorial Theory*, **5**, 135–152.

König, Joachim, and Nedrenco, Dmitri. 2016. Septic equations are solvable by 2-fold origami. *Forum Geometricorum*, **16**, 193–205.

Kritschgau, Jürgen, and Salerno, Adriana. 2017. Origami constructions of rings of integers of imaginary quadratic fields. *Integers*, **17**, Paper No. A34.

Lang, Robert J. 1988. Four Problems III. *British Origami*, **132**, 7–11.

Lang, Robert J. 1996. A Computational Algorithm for Origami Design. Pages 98–105 of: *Proceedings of the Twelfth Annual Symposium on Computational Geometry*. SCG '96. New York: ACM.

Lang, Robert J. 2003. *Origami and geometric constructions*. Preprint.

Lang, Robert J. 2004. *Angle quintisection*. Preprint.

Lang, Robert J. 2011. *Origami Design Secrets: Mathematical Methods for an Ancient Art, 2nd ed*. Boca Raton, FL: A K Peters/CRC Press.

Lang, Robert J. 2015. *Tessellatica*. Mathematica notebook. Available at https://langorigami .com/article/tessellatica.

Lang, Robert J. 2018. *Twists, Tilings, and Tessellations: Mathematical Methods for Geometric Origami*. Boca Raton, FL: A K Peters/CRC Press.

Lang, Robert J., and Montroll, John. 1991. *Origami Sea Life*. New York: Dover.

Lawrence, Jim, and Spingarn, Jonathan E. 1989. An intrinsic characterization of foldings of Euclidean space. *Annales de l'institut Henri Poincaré (C) Analyse non linéaire*, **S6**, 365–383.

Lawrence, Jim, and Spingarn, Jonathan E. 1987. On fixed points of non-expansive piecewise isometric mappings. *Proc. London Math. Soc.*, **55**(3), 605–624.

Lieb, Elliott H. 1967. Residual entropy of square ice. *Physical Review*, **162**(Oct), 162–172.

Liebschner, J. 1992. How does paper-folding fit into Froebel's educational theory? Pages 16–18 of: Smith, J. (ed), *Proceedings of the First International Conference on Origami in Education and Therapy*. London: British Origami Society.

Lill, Eduard. 1867. Résolution graphique des equations numériques d'un degré quelconque à une inconnue. *Nouvelles Annales de Mathématiques, Series 2*, **6**, 359–362.

Liu, Bin, Silverberg, Jesse L., Evans, Arthur A., Santangelo, Christian D., Lang, Robert J., Hull, Thomas C., and Cohen, Itai. 2018. Topological kinematics of origami metamaterials. *Nature Physics*, **14**(8), 811–815.

Lockwood, E. H. 1967. *A Book of Curves*. Cambridge, UK: Cambridge University Press.

Lotka, A. J. 1907. Construction of conic sections by paper-folding. *School Science and Mathematics*, **7**(7), 595–597.

Lucas, E. 1891. *Thèorie des Nombres*. Paris: Gauthier-Villars.

Lucero, Jorge C. 2018. Folding a 3D Euclidean space. Pages 331–346 of: Lang, Robert J., Bolitho, Mark, and You, Zhong (eds), *Origami$^7$: The Proceedings from the 7th International Meeting on Origami in Science, Mathematics, and Education, Volume Two: Mathematics*. St. Albans, UK: Tarquin.

Lunnon, W. F. 1968. A map-folding problem. *Mathematics of Computation*, **22**(101), 193–199.

Lunnon, W. F. 1971. Multi-dimensional map folding. *The Computer Journal*, **14**(1), 75–80.

Maekawa, Jun. 1997. Similarity in origami. Pages 109–118 of: Miura, Koryo (ed), *Origami Science & Art: Proceedings of the Second International Meeting of Origami Science and Scientific Origami*. Otsu, Japan: Seian University of Art and Design.

Marsden, Jerrold E., and Tromba, Anthony. 2011. *Vector Calculus, 6th ed*. New York: W. H. Freeman.

Martin, George E. 1979. Duplicating the cube with a Mira. *Mathematics Teacher*, **72**, 204–208.

Martin, George E. 1985. Paper-folding. *New York State Mathematics Teachers Journal*, **32**, 135–140.

Martin, George E. 1998. *Geometric Constructions*. New York: Springer.

Meguro, Toshiyuki. 1991–1992. Practical methods of origami design (in Japanese). *Origami Tanteidan Newsletter*, **2**, 7–14.

Meguro, Toshiyuki. 1994. Stag beetle and circle area molecule (in Japanese). *ORU*, **2**(1), 92–95.

Messer, Peter. 1984. *Summary of all irreducible cases of simultaneous superimpositions of elements in a folding plane that will produce a finite number of crease lines*. Preprint.

Messer, Peter. 1986. Problem 1054. *Crux Mathematicorum*, **12**(10), 284–285.

Milgram, R. James, and Trinkle, Jeff. 2004. The geometry of configuration spaces for closed chains in two and three dimensions. *Homology, Homotopy, and Applications*, **6**(1), 237–267.

Milnor, John W. 1965. *Topology from the Differentiable Viewpoint*. Charlottesville, VA: University Press of Virginia.

Miura, Koryo. 1989. A note on intrinsic geometry of origami. Pages 239–249 of: Huzita, Humiaki (ed), *Proceedings of the First International Meeting of Origami Science and Technology*. Padova, Italy: Dipartimento di Fisica dell'Università di Padova.

Montroll, John. 1979. *Origami for the Enthusiast*. New York: Dover.

Montroll, John. 1993. *Origami Inside-Out*. Maryland: Antroll Publishing Co.

Morassi, Roberto. 1989. The elusive pentagon. Pages 27–37 of: Huzita, Humiaki (ed), *Proceedings of the First International Meeting of Origami Science and Technology*. Padova, Italy: Dipartimento di Fisica dell'Università di Padova.

Morgan, Tom. 2012 (July). *Map Folding*. Master's thesis, Massachusetts Institute of Technology.

Morley, F. V. 1924. Discussions: A note on knots. *The American Mathematical Monthly*, **31**(5), 237–239.

Murata, Saburo. 1966a. The theory of paper sculpture, I (in Japanese). *Bulletin of Junior College of Art*, **4**, 29–37.

Murata, Saburo. 1966b. The theory of paper sculpture, II (in Japanese). *Bulletin of Junior College of Art*, **5**, 61–66.

Murray, Richard M. 1994. *A Mathematical Introduction to Robotic Manipulation*. Boca Raton, FL: CRC Press.

Na, Jun-Hee, Evans, Arthur A., Bae, Jinhye, Chiappelli, Maria C., Santangelo, Christian D., Lang, Robert J., Hull, Thomas C., and Hayward, Ryan C. 2015. Programming reversibly self-folding origami with micropatterned photo-crosslinkable polymer trilayers. *Advanced Materials*, **27**(1), 79–85.

Nedrenco, Dmitri. 2019. *On origami rings*. Preprint, https://arxiv.org/abs/1502.07995.

Nelson, D. R., and Peliti, L. 1987. Fluctuations in membranes with crystalline and hexatic order. *Journal de Physique*, **48**, 1085–1092.

Nishimura, Yasuzo. 2013. Solving quintic equations by two-fold origami. *Forum Mathematicum*, **27**(3), 1379–1387.

Nojima, Taketoshi. 2002. Modelling of folding patterns in flat membranes and cylinders by origami. *JSME International Journal Series C Mechanical Systems, Machine Elements and Manufacturing*, **45**(1), 364–370.

Nojima, Taketoshi. 2003. Modelling of compact folding/wrapping of flat circular membranes. *JSME International Journal Series C Mechanical Systems, Machine Elements and Manufacturing*, **46**(4), 1547–1553.

Noma, Masamichi. 1992. ええもんめーっけ// Eemon mēkke (I found a good one, in Japanese). *Origami Tanteidan Newsletter*, **3**(14), 2.

Palmer, Chris K. 1997. Extruding and tesselating polygons from a plane. Pages 323–331 of: Miura, Koryo (ed), *Origami Science & Art: Proceedings of the Second International*

*Meeting of Origami Science and Scientific Origami.* Otsu, Japan: Seian University of Art and Design.

Pinson, Matthew B., Stern, Menachem, Carruthers Ferrero, Alexandra, Witten, Thomas A., Chen, Elizabeth, and Murugan, Arvind. 2017. Self-folding origami at any energy scale. *Nature Communications*, **8**(1), 15477.

Propp, Jim. 2001. The many faces of alternating-sign matrices. Pages 43–58 of: Cori, R., Mazoyer, J., Morvan, M., and Mosseri, R. (eds), *Discrete Models: Combinatorics, Computation, and Geometry*. DMTCS Proceedings, Vol. AA, Maison Inform. Math. Discrèt.

Riaz, M. 1962. Geometric solutions of algebraic equations. *The American Mathematical Monthly*, **69**(7), 654–658.

Robertson, Stewart A. 1977–1978. Isometric folding of Riemannian manifolds. *Proceedings of the Royal Society of Edinburgh*, **79**(3–4), 275–284.

Robertson, Stewart A., and El-Kholy, E. 1986. Topological foldings. *Comm. Fac. Sci. Univ. Ankara Ser. $A_1$ Math. Statist.*, **35**(1–2), 101–107.

Row, T. S. 1901. *Geometric Exercises in Paper Folding*. Chicago: The Open Court Publishing Co.

Rupp, C. A. 1924. On a transformation by paper folding. *The American Mathematical Monthly*, **31**(9), 432–435.

Scher, Daniel P. 1996. Folded paper, dynamic geometry, and proof: A three-tier approach to the conics. *Mathematics Teacher*, **89**(3), 188–193.

Schief, Wolfgang K., Bobenko, Alexander I., and Hoffmann, Tim. 2008. On the integrability of infinitesimal and finite deformations of polyhedral surfaces. Pages 67–93 of: Schief, Wolfgang K., Bobenko, Alexander I., and Hoffmann, Tim (eds), *Discrete Differential Geometry*, Oberwolfach Seminars, Vol. 38. Basel: Birkhäuser Basel.

Scimemi, Benedetto. 1989. Draw of a regular heptagon by the folding. Pages 71–77 of: Huzita, Humiaki (ed), *Proceedings of the First International Meeting of Origami Science and Technology*. Padova, Italy: Dipartimento di Fisica dell'Università di Padova.

Sen, K. C. 1721. *Wakoku Chiyekurabe (Mathematical Contests)*. in Japanese.

Silverberg, Jesse L., Na, Jun-Hee, Evans, Arthur A., Liu, Bin, Hull, Thomas C., Santangelo, Christian D., Lang, Robert J., Hayward, Ryan C., and Cohen, Itai. 2015. Origami structures with a critical transition to bistability arising from hidden degrees of freedom. *Nature Materials*, **14**(4), 389–393.

Smith, Scott. 2003. Paper folding and conic sections. *Mathematics Teacher*, **96**(3), 202–207.

Soedel, Werner, and Foley, Vernard. 1979. Ancient catapults. *Scientific American*, **240**(3), 150–161.

Stern, Menachem, Pinson, Matthew B., and Murugan, Arvind. 2017. The complexity of folding self-folding origami. *Phys. Rev. X*, **7**(Dec), 041070.

Streinu, Ileana, and Whiteley, Walter. 2005. Single-vertex origami and spherical expansive motions. Pages 161–173 of: Akiyama, Jin, Kano, Mikio, and Tan, Xuehou (eds), *Discrete and Computational Geometry*. Berlin: Springer.

Tabachnikov, Sergei. 2007. Book review of "Arnold's Problems". *The Mathematical Intelligencer*, **29**(1), 49–52.

Tachi, Tomohiro. 2010. Origamizing polyhedral surfaces. *IEEE Transactions on Visualization and Computer Graphics*, **16**(2), 298–311.

Tachi, Tomohiro. 2010a. Freeform rigid-foldable structure using bidirectionally flat-foldable planar quadrilateral mesh. Pages 87–102 of: Ceccato, Cristiano, Hesselgren, Lars, Pauly,

Mark, Pottmann, Helmut, and Wallner, Johannes (eds), *Advances in Architectural Geometry 2010*. Vienna: Springer Vienna.

Tachi, Tomohiro. 2010b. Geometric considerations for the design of rigid origami structures. Pages 458–460 of: *Proceedings of the International Association for Shell and Spatial Structures (IASS) Symposium 2010*, Vol. 12. Madrid, IASS.

Tachi, Tomohiro. 2012. Design of infinitesimally and finitely flexible origami based on reciprocal figures. *Journal of Geometry and Graphics*, **16**(2), 223–234.

Tachi, Tomohiro. 2020. *Rigid Origami*. Unpublished manuscript.

Tachi, Tomohiro, and Demaine, Erik D. 2011. Degenerative coordinates in 22.5° grid system. Pages 489–497 of: Wang-Iverson, Patsy, Lang, Robert J., and Yim, Mark (eds), *Origami$^5$: Fifth International Conference on Origami in Science, Mathematics, and Education*. Natick, MA: A K Peters.

Tachi, Tomohiro, and Hull, Thomas C. 2017. Self-foldability of rigid origami. *Journal of Mechanisms and Robotics*, **9**(2), 021008.

Thurston, William P. 1997. *Three-Dimensional Geometry and Topology, Vol. 1*. Princeton, NJ: Princeton University Press.

Tolman, Kyler A., Lang, Robert J., Magleby, Spencer P., and Howell, Larry L. 2017. *Split-vertex technique for thickness-accommodation in origami-based mechanisms*. Paper No. V05BT08A054 of: Schmiedeler, James, and Voglewede, Philip (eds), *International Design Engineering Technical Conference and Computers and Information in Engineering*, Vol. 5B: 41st Mechanisms and Robotics Conference. New York: ASME Press.

Touchard, Jacques. 1950. Contributions a l'étude du problème de timbres poste. *Canadian Journal of Mathematics*, **2**, 385–398.

Udomprasert, Anuttara, and Kangsamaksin, Thaned. 2017. DNA origami applications in cancer therapy. *Cancer Science*, **108**(8), 1535–1543.

Vacca, Giovanni. 1930. Della piegatura della carta applicata alla geometria. *Periodico di Mathematiche, Ser. 4*, **10**, 43–50.

Videla, Carlos R. 1997. On points constructible from conics. *The Mathematical Intelligencer*, **19**(2), 53–57.

Watanabe, Naohiko, and Kawaguchi, Kenichi. 2009. The method for judging rigid foldability. Pages 165–174 of: Lang, Robert J. (ed), *Origami$^4$: Fourth International Conference on Origami in Science, Mathematics, and Education*. Natick, MA: A K Peters.

Weisstein, E. W. *Cubic Formula*. Available at http://mathworld.wolfram.com/CubicFormula.html.

Whitney, Hassler. 1937. On regular closed curves in the plane. *Compositio Mathematica*, **4**, 276–284.

Wilcox, H. K. W. 1873. National standards and emblems. *Harper's New Monthly Magazine*, **47**, 171–181.

Yaschenko, Ivan. 1998. Make your dollar bigger now!!! *The Mathematical Intelligencer*, **20**(2), 38–40.

Yates, R. C. 1943. Folding the conics. *The American Mathematical Monthly*, **50**(4), 228–230.

# Index

active path, 167
actuators, 290, 304, 316, 317
Alperin, Roger, 43, 64–66
angle defect, 242
angle quintisection, 60–62
Angle Sum Theorem, 203
Arnold, Vladimir, 172
Assis, Michael, 158

backward force (of self-folding driving force), 306
Balkcom, Devin, 276
belcastro, sarah-marie, 255
Beloch's fold, 45, 47, 51, 55, 56, 63
Beloch, Margherita, 10, 29, 31, 39, 41, 42, 44, 46, 63, 66
Bern, Marshall, 127
Big-Little-Big Lemma, 91, 108, 272
bird's foot, 281, 315
bistability, 303
boundary creases, 122
boundary strata, 206
boundary vertices, 75
box-pleated crease pattern, 57, 132–135
Bricard, Raoul, 273, 289
bun-shi, 168

c-creases, 122, 206
Carpenter's Rule Theorem, 299
Cauchy's rigidity theorem, 289
circle decomposition, 166
circle packing, 168, 175
circle-river packing, 170
computational complexity, 127–135, 152, 290–295
cone, 81
cone angle, 81, 239
cone fold, 88–90, 298
configuration space
   of flat vertex folds, 96–103
   of rigid folding, 261, 290, 296–304, 312, 316
conservative driving force, 305
constant in first (or $k$th) order, 257
constrained forces (of self-folding driving force), 306
convex polyhedral facet, 219
crease pattern, 75

of isometric folding, 192
cross (mountain/valley), 281
cross product operator, 262
crossing diagonals method, 14
curved creases, 69–71
cyclic quadrilateral, 314

Dacorogna, Bernard, 217, 222
deformation parameter, 257
deformed surface, 257
degenerate (spherical linkage), 298
degree of freedom (DOF), 24, 271
Denavit, J., 255
Descartes' Theorem, 246
Dieleman, Peter, 318
Dirichlet problems, 217, 223–226
discrete surface, 257
disk packing, 168, 175
down vertex, 113, 253
driving force, 305
dual discrete surface, 257
dual graph, 257

Edwards, B. Carter, 46
entropy, 156
Euler, Leonhard, 275

$\mathcal{F}(\mathcal{M}, \mathcal{N})$, 193
flat cone fold, 89
flat origami (definition), 77
flat paper, 81
flat vertex fold, 75–106
   definition, 81
   folded disc, 84, 87, 89, 90
   local, 83
   necessary and sufficient conditions, 85–88
   on cone, 81
   pointy, 84, 87, 89, 90
flat-foldable (definition), 77
   isometric folding, 208
folded disc, see flat vertex fold
folding angle, 76, 231
folding knots, 59–60
folding manifolds, 191–216
folding map, 77, 79, 89, 90, 104, 105, 112, 238

folding multiplier, 275
folding *n*ths, 13–16
folding thirds, 13, 14, 58
folding toward, 112, 181
forward force (of self-folding driving force), 306
frieze group, 189
Froebel, Fedrich, 10, 46
$f_x$, 197

Gallivan, Britany, 159–161
Gauss map, 239–251, 265, 266, 272
Gaussian curvature, 240
generalized Kawasaki's Theorem, 111, 214
generalized Maekawa's Theorem, 112–114, 208
generic (plane curve), 115
generic (vertex), 92, 102, 103
geometric dual graph, 257
Geretschläger, Robert, 46
germ (of manifold), 297
Grünbaum coloring, 155
graph colorings, 137–145
gutter wire, 291

Haga's Theorem, 14, 15
Haga, Kazuo, 14
Hartenberg, R. S., 255
Hatori, Koshiro, 43
Hayes, Barry, 127
homogeneous coordinates, 56, 236, 238
homomorphism, origami, 182
Huffman grids, 314–315
Huffman, David, 105, 239, 247–249, 273, 279, 289, 314
Husimi, Kôdi, 105
Huzita, Humiaki, 47
hyperbolic paper, 81

infinitesimal isometric deformation, 257
infinitesimal rigid foldability, 262, 309, 310
inner angle, 203
interior crease, 113
interior vertices, 75
isometric folding, 192
Izmestiev, Ivan, 289, 318

Jacobian matrix, 217
Justin's Theorem, 114–118
Justin, Jacques, 76, 79, 81, 85, 90, 94, 104, 106, 149–151, 191

König, Joachim, 69
Kapovich, Michael, 298
Kawaguchi, Kenichi, 256
Kawasaki's Theorem, 85–88, 90, 97–99, 101–106, 114
    generalized for isometric foldings, 214

generalized for multiple-vertex crease patterns, 111
Kawasaki, Toshikazu, 104, 106, 181, 190, 191, 216, 255
Klein, Felix, 46
Koehler, J., 146
Kokotsakis polyhedra, 289
*k*th-order rigid-foldable, 262

Lang, Robert J., 14, 16, 61, 62, 64–66, 169–172, 272
Lawrence, Jim, 191, 215
Lill's method, 39, 66–69
Lill, Eduard, 38–40
Lipschitz continuous, 218
Liu–Miura ring, 303–304
local flat vertex fold, 83
locally finite polyhedral set, 219
locked crease pattern, 300
Lunnon, W. F., 146–149

Maekawa's Theorem, 81–84, 88, 90–93, 105, 106, 114
    generalized for isometric foldings, 208
    generalized for multiple-vertex crease patterns, 112–114
    generalized for rigid origami vertices, 251–254
Maekawa, Jun, 106
major creases, 247, 272
manifolds, 191–216
Marcellini, Paolo, 217, 222
Margoulis napkin problem, 172–174
Meguro, Toshiyuki, 80
Messer, Peter, 42, 44
Millson, John J., 298
minor creases, 247, 272
Miura, Koryo, 139, 239
Miura-ori, 139–145, 152, 189, 239, 312–313
modes (of rigid folding vertex), 272
monohedral tiling, 314
Montroll, John, 14
Morgan, Tom, 152
mountain crease, 76
mountain-valley (MV) assignment, 79
    of isometric folding, 206
multifold, 58–69
Murata, Saburo, 105
MV Dimension Principle, 208

Nedrenco, Dmitri, 69
Nishimura, Yasuzo, 69
Noma's method, 15, 16
Noma, Masamichi, 15
non-crossing conditions, 123
    for isometric folding, 207
nonexpansive map, 192
Not-All-Equal 3-Satisfiability (NAE 3-SAT), 127

NP-hard, 127–135, 290–295
number of layers, 83

one-in-three satisfiability problem (1-in-3 SAT),
   291
origami (definition), 76
origami homomorphism, 182
origami line graph, 108, 137, 157
origami tessellations, 137–139, 186–189
orthogonal matrix, 218

Paolini, Emanuele, 217, 222
parameter space, 261
phantom fold, 108
piecewise $C^1$, 218
pointy, *see* flat vertex fold
polyhedral set, 219
   locally finite, 219
polyhedral vertex, 240
pop up/down (vertex), 287
potential function (of conservative driving force),
   305
prototile, 314

Rabbit-Ear Theorem, 168
radial function, 194
reciprocal diagram, 264
reciprocal-parallel dual discrete surfaces, 258
Recovery Theorem, 204, 222
rigid foldability with optional creases problem, 290
rigid folding, 256–289
   definition of, 261
   well-behaved, 305
rigid map, 218
rigid origami, 231–255
rigid-foldable crease pattern, 231
Robertson, Stewart, 80, 85, 106, 191, 215
Rodrigues' rotation matrix formula, 262
rotational spring driving force, 306
rotationally symmetric tiling, 314
Row, T. Sundara, 10, 29, 35, 46
rumpled ruble problem, 172–174

s-creases, 122, 206
s-faces, 121
s-net, *see* superposition pattern
sangaku, 46
Scimemi, Benedetto, 47
self-foldable, 306
self-folding, 290, 304–317
Shurman, Jerry, 46
signed area, 240
single-vertex fold, 81
singular point, 297
solid angle vertex fold, 251
speed (of rigid folding), 275
Spingarn, Jonathan, 191, 215
splitter gadget, 292

square twist, 79, 119, 137–139, 275, 288, 300
square twist tessellation, 137–139, 187
staircase splitter gadget, 292
staircase wire, 291
stamp-folding problem, 135, 145–152
standard splitter gadget, 292
statistical mechanics, 145
strata, 197
Streinu, Ileana, 283
Streinu–Whiteley Theorem, 283, 299
superposition order, 123
   of isometric folding, 207
superposition pattern, 120
   of isometric folding, 206
supporting plane, 219
symmetry group, 181–190

Tachi, Tomohiro, 256, 273, 305, 316
Taco-Tortilla Conditions, 123
   for isometric folding, 207
tangent of half-angle, 273–275, 315
tangent of quarter-angle, 315
tangent space, 193, 306
tessellations, 137–139, 186–189
tethered membrane folding, 152–157
Thurston, William, 235
trace (of Gauss map), 240
transfer matrix, 144
TreeMaker, 169–172
triangle twist, 128, 139, 275, 300
triangle twist tessellation, 138
tripod (mountain/valley), 281
turning number, 115, 298

uniaxial base, 169
uniquely self-foldable, 306
universal molecule, 170–172
unspecified cross, 287
up vertex, 113, 253

Vacca, Giovanni, 46
valid (MV assignment), 79
   for isometric folding, 208
valid tangents, 306
valley crease, 76
virtual vertices, 236

wallpaper group, 188
Watanabe, Naohiko, 256
Weierstrass substitution, 275
well-behaved rigid folding, 305
Whiteley, Walter, 283
widening splitter gadget, 292

Yaschenko, Ivan, 172
Yoshida, Masaaki, 181

zero-area reciprocal diagram, 266

Printed in the United States
by Baker & Taylor Publisher Services